MW00979332

Wireless Sensor Network Designs

Wireless Sensor Network Designs

Anna Hać
University of Hawaii at Manoa, Honolulu, USA

John Wiley & Sons, Ltd

Telephone (+44) 1243 779777

Email (for orders and customer service enquiries): cs-books@wiley.co.uk
Visit our Home Page on www.wileyeurope.com or www.wiley.com

Other Wiley Editorial Offices

John Wiley & Sons Inc., 111 River Street, Hoboken, NJ 07030, USA

Jossey-Bass, 989 Market Street, San Francisco, CA 94103-1741, USA

Wiley-VCH Verlag GmbH, Boschstr. 12, D-69469 Weinheim, Germany

John Wiley & Sons Australia Ltd, 33 Park Road, Milton, Queensland 4064, Australia

John Wiley & Sons (Asia) Pte Ltd, 2 Clementi Loop #02-01, Jin Xing Distripark, Singapore 129809

John Wiley & Sons Canada Ltd, 22 Worcester Road, Etobicoke, Ontario, Canada M9W 1L1

Wiley also publishes its books in a variety of electronic formats. Some content that appears in print may not be available in electronic books.

Library of Congress Cataloging-in-Publication Data

Hać, Anna.
Wireless sensor network designs / Anna Hac.
p. cm.
Includes bibliographical references and index.
ISBN 0-470-86736-1
1. Sensor networks. 2. Wireless LANs. I. Title.

TK7872.D48.H33 2003
621.382'1 – dc22

2003057612

British Library Cataloguing in Publication Data

A catalogue record for this book is available from the British Library

ISBN 0-470-86736-1

Typeset in 11/13pt Palatino by Laserwords Private Limited, Chennai, India
Printed and bound in Great Britain by TJ International, Padstow, Cornwall
This book is printed on acid-free paper responsibly manufactured from sustainable forestry in which at least two trees are planted for each one used for paper production.

Contents

Preface ix

About the Author xv

1 Networked Embedded Systems **1**
1.1. Introduction 1
1.2. Object-Oriented Design 3
1.3. Design Integration 4
1.4. Design Optimization 6
1.5. Co-design and Reconfiguration 9
1.6. Java-Driven Co-design and Prototyping 12
1.6.1. Java-Based Co-design 13
1.6.2. Run-Time Management 15
1.6.3. Embedded Systems Platform 17
1.7. Hardware and Software Prototyping 20
1.8. Multiple Application Support 23
1.8.1. FPGA-Based System Architecture 25
1.9. Summary 27
Problems 28
Learning Objectives 28
Practice Problems 29
Practice Problem Solutions 29

2 Smart Sensor Networks **31**
2.1. Introduction 31
2.2. Vibration Sensors 32
2.3. Smart Sensor Application to Condition Based Maintenance 34
2.4. Smart Transducer Networking 42
2.5. Controller Area Network 46
2.6. Summary 58

Problems 60
Learning Objectives 60
Practice Problems 60
Practice Problem Solutions 60

3 Power-Aware Wireless Sensor Networks 63
3.1. Introduction 63
3.2. Distributed Power-Aware Microsensor Networks 65
3.3. Dynamic Voltage Scaling Techniques 71
3.4. Operating System for Energy Scalable Wireless Sensor Networks 75
3.5. Dynamic Power Management in Wireless Sensor Networks 79
3.6. Energy-Efficient Communication 81
3.7. Power Awareness of VLSI Systems 85
3.8. Summary 95
Problems 97
Learning Objectives 97
Practice Problems 97
Practice Problem Solutions 98

4 Routing in Wireless Sensor Networks 101
4.1. Introduction 101
4.2. Energy-Aware Routing for Sensor Networks 102
4.3. Altruists or Friendly Neighbors in the Pico Radio Sensor Network 109
4.3.1. Energy-Aware Routing 111
4.3.2. Altruists or Friendly Neighbors 114
4.3.3. Analysis of Energy Aware and Altruists Routing Schemes 116
4.4. Aggregate Queries in Sensor Networks 120
4.4.1. Aggregation Techniques 125
4.4.2. Grouping 133
4.5. Summary 135
Problems 136
Learning Objectives 136
Practice Problems 137
Practice Problem Solutions 137

5 Distributed Sensor Networks 141
5.1. Introduction 141
5.2. Bluetooth in the Distributed Sensor Network 142
5.2.1. Bluetooth Components and Devices 144
5.2.2. Bluetooth Communication and Networking 146
5.2.3. Different Technologies 151
5.3. Mobile Networking for Smart-Dust 154
5.3.1. Smart-Dust Technology 154
5.3.2. Communication and Networking 159
5.4. Summary 162
Problems 163
Learning Objectives 163

Practice Problems 163
Practice Problem Solutions 163

6 Clustering Techniques in Wireless Sensor Networks 165
6.1. Introduction 165
6.2. Topology Discovery and Clusters in Sensor Networks 166
6.2.1. Topology Discovery Algorithm 169
6.2.2. Clusters in Sensor Networks 171
6.2.3. Applications of Topology Discovery 177
6.3. Adaptive Clustering with Deterministic Cluster-Head Selection 181
6.4. Sensor Clusters' Performance 185
6.4.1. Distributed Sensor Processing 187
6.5. Power-Aware Functions in Wireless Sensor Networks 192
6.5.1. Power Aware Software 196
6.6. Efficient Flooding with Passive Clustering 198
6.6.1. Passive Clustering 203
6.7. Summary 207
Problems 208
Learning Objectives 208
Practice Problems 209
Practice Problem Solutions 209

7 Security Protocols for Wireless Sensor Networks 213
7.1. Introduction 213
7.2. Security Protocols in Sensor Networks 214
7.2.1. Sensor Network Security Requirements 216
7.2.2. Authenticated Broadcast 219
7.2.3. Applications 223
7.3. Communication Security in Sensor Networks 225
7.4. Summary 230
Problems 230
Learning Objectives 230
Practice Problems 231
Practice Problem Solutions 231

8 Operating Systems for Embedded Applications 235
8.1. Introduction 235
8.2. The Inferno Operating System 236
8.3. The Pebble Component-Based Operating System 242
8.3.1. Protection Domains and Portals 246
8.3.2. Scheduling and Synchronization 250
8.3.3. Implementation 253
8.3.4. Embedded Applications 258
8.4. Embedded Operating System Energy Analysis 264
8.5. Summary 270
Problems 271
Learning Objectives 271

Practice Problems 272
Practice Problem Solutions 272

9 Network Support for Embedded Applications 275
9.1. Introduction 275
9.2. Bluetooth Architecture 277
9.3. Bluetooth Interoperability with the Internet and Quality of Service 283
9.4. Implementation Issues in Bluetooth-Based Wireless Sensor Networks 288
9.5. Low-Rate Wireless Personal Area Networks 297
9.6. Data-Centric Storage in Wireless Sensor Networks 306
9.7. Summary 314
Problems 315
Learning Objectives 315
Practice Problems 315
Practice Problem Solutions 316

10 Applications of Wireless Sensor Networks 323
10.1. Introduction 323
10.2. Application and Communication Support for Wireless Sensor Networks 325
10.3. Area Monitoring and Integrated Vehicle Health Management Applications 334
10.3.1. Development Platform 338
10.3.2. Applications 343
10.4. Building and Managing Aggregates in Wireless Sensor Networks 345
10.5. Habitat and Environmental Monitoring 349
10.5.1. Island Habitat Monitoring 350
10.5.2. Implementation 355
10.6. Summary 360
Problems 362
Learning Objectives 362
Practice Problems 362
Practice Problem Solutions 363

References 369

Index 385

Preface

The emergence of compact, low-power, wireless communication sensors and actuators in the technology supporting the ongoing miniaturization of processing and storage, allows for entirely new kinds of embedded system. These systems are distributed and deployed in environments where they may not have been designed into a particular control path, and are often very dynamic. Collections of devices can communicate to achieve a higher level of coordinated behavior.

Wireless sensor nodes deposited in various places provide light, temperature, and activity measurements. Wireless nodes attached to circuits or appliances sense the current or control the usage. Together they form a dynamic, multi-hop, routing network connecting each node to more powerful networks and processing resources.

Wireless sensor networks are application-specific, and therefore they have to involve both software and hardware. They also use protocols that relate to both the application and to the wireless network.

Wireless sensor networks are consumer devices supporting multimedia applications, for example personal digital assistants, network computers, and mobile communication devices. Emerging embedded systems run multiple applications, such as web browsers, and audio and video communication applications. These include capturing video data, processing audio streams, and browsing the World Wide Web (WWW). There is a wide range of data gathering applications, energy-agile applications, including remote climate monitoring, battlefield surveillance, and intra-machine monitoring. Example applications are microclimate control in buildings, environmental monitoring, home automation, distributed monitoring of factory plants or chemical processes, interactive museums, etc. An application of collective awareness is a credit card anti-theft mode. There is also a target tracking application, and applications ranging from medical monitoring and diagnosis to target detection, hazard detection, and automotive and industrial control. In short, there are applications in military (e.g. battlefields), commercial (e.g. distributed mobile computing, disaster discovery systems, etc.), and educational environments (e.g. conferences, conventions, etc.) alike.

This book introduces networked embedded systems, smart sensors, and wireless sensor networks. The focus of the book is on the architecture, applications, protocols, and distributed systems support for these networks.

Wireless sensor networks use new technology and standards. They involve small, energy-efficient devices, hardware/software co-design, and networking support. Wireless sensor networks are becoming an important part of everyday life, industrial and military applications. It is a rapidly growing area as new technologies are emerging, and new applications are being developed.

The characteristics of modern embedded systems are the capability to communicate over the networks and to adapt to different operating environments.

Designing an embedded system's digital hardware has become increasingly similar to software design. The wide spread use of hardware description languages and synthesis tools makes circuit design more abstract. A cosynthesis method and prototyping platform can be developed specifically for embedded devices, combining tightly integrated hardware and software components.

Users are demanding devices, appliances, and systems with better capabilities and higher levels of functionality. In these devices and systems, sensors are used to provide information about the measured parameters or to identify control states. These sensors are candidates for increased built-in intelligence. Microprocessors are used in smart sensors and devices, and a smart sensor can communicate measurements directly to an instrument or a system. The networking of transducers (sensors or actuators) in a system can provide flexibility, improve system performance, and make it easier to install, upgrade and maintain systems.

The sensor market is extremely diverse and sensors are used in most industries. Sensor manufacturers are seeking ways to add new technology in order to build low-cost, smart sensors that are easy to use and which meet the continuous demand for more sophisticated applications. Networking is becoming pervasive in various industrial settings, and decisions about the use of sensors, networks, and application software can all be made independently, based on application requirements.

The IEEE (Institute of Electrical and Electronics Engineers) 1451 smart transducer interface standards provide the common interface and enabling technology for the connectivity of transducers to microprocessors, control and field networks, and data acquisition and instrumentation systems. The standardized Transducer Electronic Data Sheet (TEDS) specified by IEEE 1451.2 allows for self-description of sensors. The interfaces provide a standardized mechanism to facilitate the plug and play of sensors to networks. The network-independent smart transducer object model, defined by IEEE 1451.1, allows sensor manufacturers to support multiple networks and protocols. This way, transducer-to-network interoperability can be supported. IEEE standards P1451.3 and P1451.4 will meet the needs of analog transducer users for high-speed applications. Transducer vendors and users, system integrators, as well as network providers can benefit from the IEEE 1451 interface standards. Networks of distributed microsensors are emerging as a solution for a wide range of data gathering applications. Perhaps the most substantial challenge faced by designers of small but long-lived microsensor nodes, is the need for significant reductions in energy consumption. A power-aware design methodology emphasizes the graceful scalability of energy consumption with factors such as available resources, event frequency, and desired output quality, at all levels of the system hierarchy. The architecture for a power-aware microsensor node highlights the collaboration between software that is capable of energy-quality tradeoffs and hardware with scalable energy consumption.

Power-aware methodology uses an embedded micro-operating system to reduce node energy consumption by exploiting both sleep state and active power management. Wireless distributed microsensor networks have gained importance in a wide spectrum of civil and military applications. Advances in MEMS (Micro Electro Mechanical Systems) technology, combined with low-power, low-cost, Digital Signal Processors (DSPs) and Radio Frequency (RF) circuits have resulted in feasible, inexpensive, wireless microsensor networks. A distributed, self-configuring network of adaptive sensors has significant benefits. They can be used for remote monitoring in inhospitable and toxic environments. A large class of benign environments also requires the deployment of a large number of sensors, such as intelligent patient monitoring, object tracking, and assembly line sensing. The massively distributed nature of these networks provides increased resolution and fault tolerance as compared with a single sensor node. Networking a large number of low-power mobile nodes involves routing, addressing and support for different classes of service at the network layer. Self-configuring wireless sensor networks consist of hundreds or thousands of small, cheap, battery-driven, spread-out nodes, bearing a wireless modem to accomplish a monitoring or control task jointly. Therefore, an important concern is the network lifetime: as nodes run out of power, the connectivity decreases and the network can finally be partitioned and become dysfunctional.

Deployment of large networks of sensors requires tools to collect and query data from these networks. Of particular interest are aggregates whose operations summarize current sensor values in part or all of an entire sensor network. Given a dense network of a thousand sensors querying for example, temperature, users want to know temperature patterns in relatively large regions encompassing tens of sensors, and individual sensor readings are of little value.

Networks of wireless sensors are the result of rapid convergence of three key technologies: digital circuitry, wireless communications, and MEMS. Advances in hardware technology and engineering design have led to reductions in size, power consumption, and cost. This has enabled compact, autonomous nodes, each containing one or more sensors, computation and communication capabilities, and a power supply. Ubiquitous computing is based on the idea that future computers merge with their environment until they become completely invisible to the user. Ubiquitous computing envisions everyday objects as being augmented with computation and communication capabilities. While such artifacts retain their original use and appearance, their augmentation can seamlessly enhance and extend their usage, thus opening up novel interaction patterns and applications. Distributed wireless microsensor networks are an important component of ubiquitous computing, and small dimensions are a design goal for microsensors. The energy supply of the sensors is a main constraint of the intended miniaturization process. It can be reduced only to a specific degree since energy density of conventional energy sources increases slowly. In addition to improvements in energy density, energy consumption can be reduced. This approach includes the use of energy-conserving hardware. Moreover, a higher lifetime of sensor networks can be accomplished through optimized applications, operating systems, and communication protocols. Particular modules of the sensor hardware can be turned off when they are not needed. Wireless distributed microsensor systems enable fault-tolerant monitoring and control of a variety of applications. Due to the large number of microsensor nodes that may be deployed, and the long system lifetimes required, replacing the battery is not an option. Sensor systems must utilize minimal energy while operating over a wide range of operating scenarios. These include power-aware computation and communication component technology, low-energy signaling and networking, system partitioning considering computation and communication trade-offs, and a power-aware software infrastructure. Routing and data dissemination in sensor networks requires a simple and scalable solution. The topology discovery algorithm for wireless sensor networks selects a set of distinguished nodes, and constructs a reachability map based on their information. The topology discovery algorithm logically organizes the network in the form of clusters and forms a tree of clusters rooted at the monitoring node. The topology discovery algorithm is completely distributed, uses only local information, and is highly scalable.

To achieve optimal performance in a wireless sensor network, it is important to consider the interactions among the algorithms operating at the different layers of the protocol stack. For sensor networks, one question is how the self-organization of the network into clusters affects the sensing performance. Thousands to millions of small sensors form self-organizing wireless networks, and providing security for these sensor networks is not easy since the sensors have limited processing power, storage, bandwidth, and energy. A set of Security Protocols for Sensor Networks (SPINS), explores the challenges for security in sensor networks. SPINS include: μTESLA (the micro version of the Timed, Efficient, Streaming, Loss-tolerant Authentication Protocol), providing authenticated streaming broadcast, and SNEP (Secure Network Encryption Protocol) providing data confidentiality, two-party data authentication, and data freshness, with low overhead. An authenticated routing protocol uses SPINS building blocks. Wireless networks, in general, are more vulnerable to security attacks than wired networks, due to the broadcast nature of the transmission medium. Furthermore, wireless sensor networks have an additional vulnerability because nodes are often placed in a hostile or dangerous environment, where they are not physically protected. The essence of ubiquitous computing is the creation of environments saturated with computing and communication in an unobtrusive way. WWRF (Wireless World Research Forum) and ISTAG (Information Society Technologies Advisory Group) envision a vast number of various intelligent devices, embedded in the environment, sensing, monitoring and actuating the physical world, communicating with each other and with humans. The main features of the IEEE 802.15.4 standard are network flexibility, low cost, and low power consumption. This standard is suitable for many applications in the home requiring low data rate communications in an *ad hoc* self-organizing network.

The IEEE 802.15.4 standard defines a low-rate wireless personal area network (LR-WPAN) which has ultra-low complexity, cost, and power, for low data rate wireless connectivity among inexpensive fixed, portable, and moving devices. The IEEE 802.15.4 standard defines the physical (PHY) layer and Media Access Control (MAC) layer specifications. In contrast to traditional communication networks, the single major resource constraint in sensor networks is power, due to the limited battery life of sensor devices. Data-centric methodologies can be used to solve this problem efficiently. In Data Centric Storage (DCS) data dissemination frameworks, all event data are stored by type at designated nodes in the network and can later be retrieved by distributed mobile access points in the network. Resilient Data-Centric Storage (R-DCS) is a method of achieving scalability and resilience by replicating data at strategic locations in the sensor network. Various wireless technologies, like simple RF, Bluetooth, UWB (ultrawideband) or infrared can be used for communication between sensors. Wireless sensor networks require low-power, low-cost devices that accommodate powerful processors, a sensing unit, wireless communication interface and power source, in a robust and tiny package. These devices have to work autonomously, to require no maintenance, and to be able to adapt to the environment. *Wireless Sensor Network Designs* focuses on the newest technology in wireless sensor networks, networked embedded systems, and their applications. A real applications-oriented approach to solving sensor network problems is presented. The book includes a broad range of topics from networked embedded systems and smart sensor networks, to power-aware wireless sensor networks, routing, clustering, security, and operating systems along with networks support. The book is organized into ten chapters, with the goal to explain the newest sensor technology, design issues, protocols, and solutions to wireless sensor network architectures.

As previously discussed, Chapter 1 describes networked embedded systems, their design, prototyping, and application support. Chapter 2 introduces smart sensor networks and their applications. Chapter 3 introduces power-aware wireless sensor networks. Routing in wireless sensor networks and the aggregation techniques are discussed in Chapter 4. Distributed sensor networks are presented in Chapter 5, and clustering techniques in wireless sensor networks are introduced in Chapter 6. Chapter 7 presents security protocols in sensor networks. Operating systems for embedded applications are discussed in Chapter 8. Chapter 9 presents network support for embedded applications. Applications of wireless sensor networks are studied in Chapter 10.

A. H.

ACKNOWLEDGEMENTS

The author wishes to thank John Wiley Publisher Birgit Gruber for her support and excellent handling of the preparation of this book. Many thanks to the entire staff of John Wiley, particularly Irene Cooper and Daniel Gill, for making publication of this book possible. Thanks also to the six reviewers for their thorough reviews.

About the Author

Anna Hać received MS and PhD degrees in computer science from the Department of Electronics, Warsaw University of Technology, Poland, in 1977 and 1982, respectively.

She is a professor in the Department of Electrical Engineering, University of Hawaii at Manoa, Honolulu. During her long and successful academic career she was a Visiting Scientist at the Imperial College, University of London, England, a postdoctoral fellow at the University of California at Berkeley, an assistant professor of Electrical Engineering and Computer Science at The Johns Hopkins University, a Member of Technical Staff at AT&T Bell Laboratories, and a senior summer faculty fellow at the Naval Research Laboratory.

Her research contributions include system and workload modeling, performance analysis, reliability, modeling process synchronization mechanisms for distributed systems, distributed file systems, distributed algorithms, congestion control in high-speed networks, reliable software architecture for switching systems, multimedia systems, wireless networks, and network protocols.

She has published more than 130 papers in archival journals and international conference proceedings, and she is the author of two textbooks *Multimedia Applications Support for Wireless ATM Networks* (2000), and *Mobile Telecommunications Protocols for Data Networks* (2003).

She is a member of the Editorial Board of the *IEEE Transactions on Multimedia*, and is on the Editorial Advisory Board of Wiley's *International Journal of Network Management*.

1

Networked Embedded Systems

1.1. INTRODUCTION

The characteristics of modern embedded systems are the capability to communicate over the networks and to adapt to different operating environments. Embedded systems can be found in consumer devices supporting multimedia applications, for example, personal digital assistants, network computers, and mobile communication devices. The low-cost, consumer-oriented, and fast time-to-market objectives dominate embedded system design. Hardware and software codesign is used to cope with growing design complexity.

Designing an embedded system's digital hardware has become increasingly similar to software design. The widespread use of hardware description languages and synthesis tools makes circuit design more abstract. Market pressures to reduce development time and effort encourage abstract specification as well, and promote the reuse of hardware and software components. Therefore, a specification language should provide a comfortable means for integrating reuse libraries.

Java is an object-oriented, versatile language of moderate complexity that can be used as the specification language in the design flow. Java has built-in primitives for handling multiple threads, and supports the concurrency and management of different control flows. New applications can be rapidly developed in Java.

Wireless Sensor Network Designs A. Hać
© 2003 John Wiley & Sons, Ltd ISBN: 0-470-86736-1

A cosynthesis method and prototyping platform can be developed specifically for embedded devices that combine tightly integrated hardware and software components.

Exploration and synthesis of different design alternatives and co-verification of specific implementations are the most demanding tasks in the design of embedded hardware and software systems. Networked embedded systems pose new challenges to existing design methodologies as novel requirements like adaptivity, run-time, and reconfigurability arise. A codesign environment based on Java object-oriented programming language supports specification, cosynthesis and prototype execution for dynamically reconfigurable hardware and software systems.

Networked embedded systems are equipped with communication capabilities and can be controlled over networks. JaCoP (Java driven Codesign and Prototyping environment) is a codesign environment based on Java that supports specification, cosynthesis and prototyping of networked embedded systems.

Emerging embedded systems run multiple applications such as web browsers, audio and video communication applications, and require network connectivity.

Networked embedded systems are divided into:

- multifunction systems that execute multiple applications concurrently, and
- multimode systems that offer the users a number of alternative modes of operation.

In multifunction systems, the embedded systems can execute multiple applications concurrently. These applications include capturing video data, processing audio streams, and browsing the WWW (World Wide Web). Embedded systems must often adapt to changing operating conditions. For example, multimedia applications adapt to the changing network rate by modifying video frame rate in response to network congestion. A trade-off between quality of service (QoS) and the network rate is applied. Audio applications use different compression techniques, depending on the network load and quality of service (QoS) feedback from the client applications.

In multimode systems, the embedded multimode systems experience a number of alternative modes of operation. For example, a mobile phone performing a single function can change the way it operates to accommodate different communication protocols supporting various functions and features. Flexible, multimode devices are mandatory for applications such as electronic banking and electronic commerce. Depending on the type of connection and the security level required, devices apply different encryption algorithms when transmitting data.

Another use of embedded multimode systems is influenced by the rapid evolution of web-based applications. The application change causes the requirements for devices such as set-top boxes to change within months. For certain application domains, designers can alleviate the problem of short product lifetime by designing hardware and software system components that users can configure or upgrade after production. However, many embedded devices do not support these tasks.

Remote administration of electronic products over the Internet has become an important feature: there are printers or copiers with embedded Web servers. By using reconfigurable hardware components, vendors can change hardware implemented functionality after installing networked devices at the customer site.

1.2. OBJECT-ORIENTED DESIGN

Embedded systems require hardware and software codesign tools and frameworks. However, designing configurable hardware and software systems creates a problem. A complete design environment for embedded systems should include dynamically reconfigurable hardware components. The object-oriented programming language, Java, is used for specification and initial profiling, and for the final implementation of the system's software components.

Several trends influence the embedded systems development and shape requirements for the optimal development tool. Designing an embedded system's digital hardware has become increasingly similar to software design. The widespread use of hardware description languages and synthesis tools makes circuit design more abstract. Market pressures to reduce development time and effort encourage abstract specification as well, and promote the reuse of hardware and software components. Therefore, a specification language should provide a comfortable means for integrating reuse libraries.

Object-oriented programming has proven to be an efficient paradigm for the design of complex software systems. Although 'object-oriented' may imply performance degradation, it comes with significant benefits: not only does it provide a better means for managing complexity and reusing existing modules, it also reduces problems and costs associated with code maintenance. These benefits far outweigh the performance degradation.

In embedded-system design, a major trend is increasingly to implement functionality in software. This allows for faster implementation, more flexibility, easier upgradability, and customization with additional features. Unlike hardware, software incurs no manufacturing costs, although the costs of software maintenance cause increasing concern.

The design of networked embedded systems requires support for a diverse feature set that includes Internet mobility, network programming, security, code reuse, multithreading, and synchronization. However, Java has not been designed for specifying systems with hard real-time constraints.

The need for communicating with embedded systems over the Internet pushes more designers towards Java, which has been adopted as the premier design platform for implementing set-top box applications. Set-top boxes create a rapidly growing embedded-systems market that promises to reach millions of homes.

1.3. DESIGN INTEGRATION

A cosynthesis method and prototyping platform can be developed specifically for embedded devices that combine tightly integrated hardware and software components.

A cosynthesis is used to make the most efficient assignment of tasks to software or hardware, and the cosynthesis method uses an initial Java specification of the desired functionality. Software profiling by the Java virtual machine identifies bottlenecks and computation-intensive tasks. Using a graphical visualization tool that displays each task's relative and absolute execution times, the designer quickly uncovers an application's most computationally demanding tasks.

The candidate tasks for hardware implementation are selected on the basis of profiling results and the reuse library of available hardware components. A high-level synthesis tool is used to transform Java methods into register transfer level VHDL (Very High-Speed Integrated Circuit Hardware Description Language). The tool generates an appropriate interface description for each hardware block.

The target architecture for synthesis is the prototyping and exploration platform shown in Figure 1.1. In this architecture software and hardware parts of the design are handled separately. A run-time environment implemented in software on a PC (Personal Computer) is used to prototype software. An additional configurable hardware extension, the dynamically reconfigurable field-programmable gate array (DPGA) board, handles hardware.

In the run-time environment, the Java virtual machine forms the core component of the software execution engine. A database stores information about the classes and methods used by the design. Software synthesis takes the form of generated byte code, compiler encoded, platform independent code. The

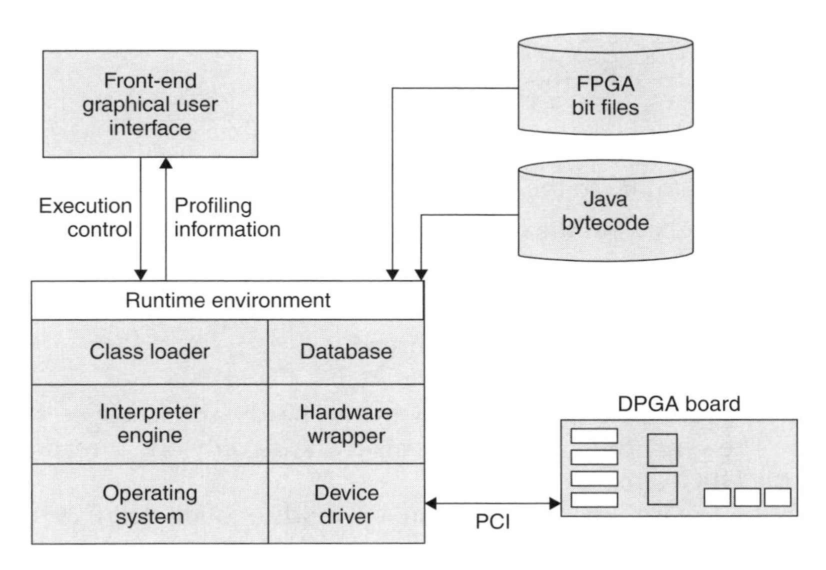

Figure 1.1 Design exploration platform for networked embedded systems.

virtual machine contains the necessary interface mechanisms (the hardware wrapper and device driver) to communicate with hardware modules.

This approach yields a smooth migration from a software implementation to a hardware and software system without modifying the Java source code. The components for hardware object handling to the Java virtual machine can be added and interfaced to an external hardware component. This design automates the configurable hardware device management.

A prototyping board connected to a PC via a PCI (PC Interface) bus is used to design the exploration platform. The board consists of a DPGA chip and a local memory. The DPGA offers short reconfiguration times and full access to implemented circuit internal registers, which allows mapping of multiple hardware objects onto a single chip. The parts of the chip can be reconfigured even when the other parts of the chip are operating, allowing execution of multiple methods on the DPGA board when running a system prototype.

In practice, the run-time environment in Figure 1.1 reads in a table with the desired function partitions routed to hardware or software implementations. The environment directly handles these tasks executed as software on the Java virtual machine. For hardware modules, it configures the DPGA (with information from the DPGA bit files) and manages its communication with the DPGA.

During execution of the application, the interpreter must activate the hardware call module whenever the control flow reaches the hardware method. Depending on the DPGA board's current state, a hardware call can trigger one or more of the following actions:

- complete or partial reconfiguration of the DPGA;
- transfer of input data to the board;
- transfer of data back to the calling thread;
- transmission of an enable signal that in turn starts the emulation of the hardware design, and
- transfer of data back to the calling thread.

These procedures are implemented in the hardware wrapper shown in Figure 1.1. The synchronization mechanism allows only one thread at a time to access the DPGA board.

The initial software specification cannot provide enough details to optimize the hardware–software boundary because there is not enough detail about the final implementation. During prototyping, the most interesting are different protocols and interface mechanisms.

1.4. DESIGN OPTIMIZATION

Embedded-system designers seek to maximize performance within the constraints of limited hardware resources. During optimization, the focus is on those parts of the design that could be alternatively implemented in software or hardware, and their corresponding interfaces. The prototyping environment gathers characteristic information about hardware and software modules, and stores it in a library of reusable modules. The most important parameters that are measured include:

- execution times of a module for both hardware and software implementations;
- required area for hardware implementation, and
- specific interface costs (in terms of additional execution time for data transfer, hardware area, or software code).

In systems that depend on both hardware and software, efficient interface design is crucial to achieve maximum performance. For this reason, the optimization process emphasizes an efficient design of interfaces while searching for the optimum mapping of modules to hardware and software.

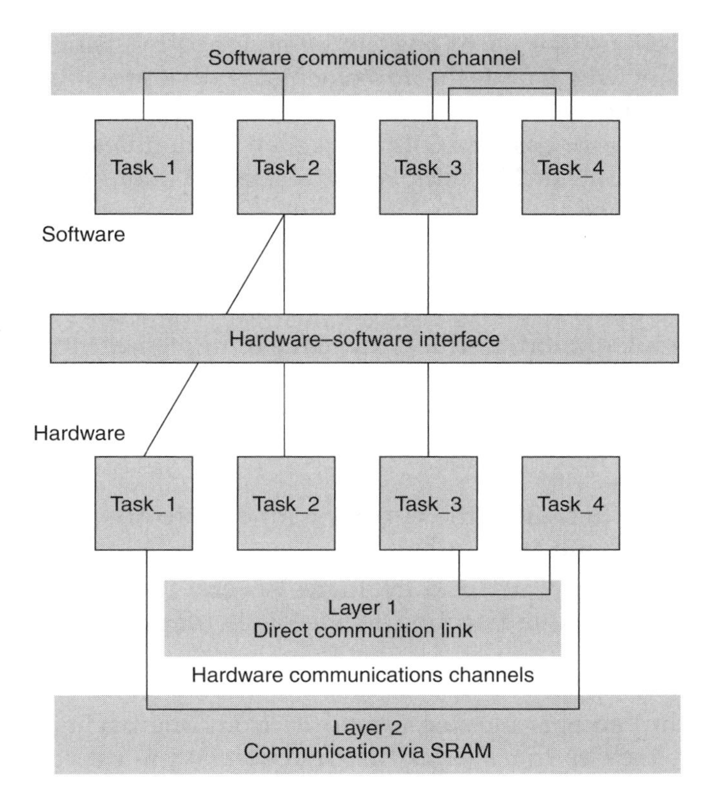

Figure 1.2 System communication model.

The basic structure and interface types are shown in Figure 1.2. The interface model contains a communication channel for software modules, an interface for data exchange between the hardware and software domain, and a two-layer communication channel for hardware entities. Layer 1 represents a direct communication link that is implemented on chip. Layer 2 represents communication via scratch-pad memory SRAM.

During the design optimization, the prototyping environment calculates the interface overhead for the current mapping of modules to the hardware and software partitions. If tightly coupled modules are mapped to hardware, the communication is implemented by using on-chip registers or through the shared memory, using static RAM (random access memory).

The prototyping environment optimizes a system design for minimum execution times within the constraints of limited DPGA chip area and communication bandwidth. To represent the quality of the implementation by a single numerical value, the cost function is used. The cost function is a weighted sum of the squares of the costs of hardware, software, and communication.

For small designs, the partitions can be assigned by using an exhaustive search algorithm, which finds the implementation that provides the minimum cost function. The complexity of this solution grows exponentially and becomes infeasible for systems with more than 25 modules. This partitioning belongs to the group of NP-complete problems. A more suitable, heuristic optimization method is simulated annealing.

The simulated-annealing method models the physical process of melting a material and then cooling it so that it crystallizes into a state of minimal energy. This method has been successfully applied to several problems in very large systems integration. Further, it is easy to implement for differing cost functions and usually delivers good results. The algorithm, a probabilistic search method that can climb out of a local minimum, consists of two nested loops:

- The outer loop decreases the current temperature according to a certain user-specified cooling schedule.
- The inner loop generates and evaluates several new partitions by shifting modules from one partition to the other (depending on the current temperature).

The algorithm accepts moving a module from one partition to the other if doing so decreases the overall cost function. With a certain probability that depends on the current temperature, it may also accept cost increases. This way, the algorithm can escape from a local minimum. The inner loop repeats until the algorithm detects a steady state for the current temperature. Depending on the temperature schedule and the stopping criterion, a trade-off between the computation time and the result quality is possible. Computation times can be decreased significantly by choosing an initial partition instead of a random partition before starting the optimization. This way the algorithm begins with a lower starting temperature, thus reducing the number of iterations.

For example, the optimization process for a reasonably sized system with about 100 modules takes less than one minute with a Pentium II processor. Because simulated annealing is a probabilistic method, two executions of the algorithm may produce slightly different results, but with such short run-times, the optimization process can be repeated several times, thereby increasing the likelihood of obtaining a near optimal result.

Java-based rapid prototyping of embedded systems offers several benefits. It allows system-level testing of novel designs and architectures using a flexible platform for development of both software and hardware components. Java-based prototyping also combines profiling results from the specification level and characteristic measures of the prototype implementation. Combining these data gives a more solid prediction of the final system's performance and cost, making possible a reliable optimization strategy.

1.5. CO-DESIGN AND RECONFIGURATION

Exploration and synthesis of different design alternatives and co-verification of specific implementations are the most demanding tasks in the design of embedded hardware and software systems. Networked embedded systems pose new challenges to existing design methodologies as novel requirements like adaptivity, run-time, and reconfigurability arise. A codesign environment based on Java object-oriented programming language supports specification, cosynthesis and prototype execution for dynamically reconfigurable hardware and software systems.

A design exploration and prototyping platform has been developed for embedded hardware and software systems with reconfiguration capabilities. The target architecture for such systems consists of a microprocessor running a Java virtual machine, and a hardware processor consisting of one or more FPGAs. An overview of the design flow for cosynthesis is illustrated in Figure 1.3. Starting from an initial Java specification, profiling data is gathered

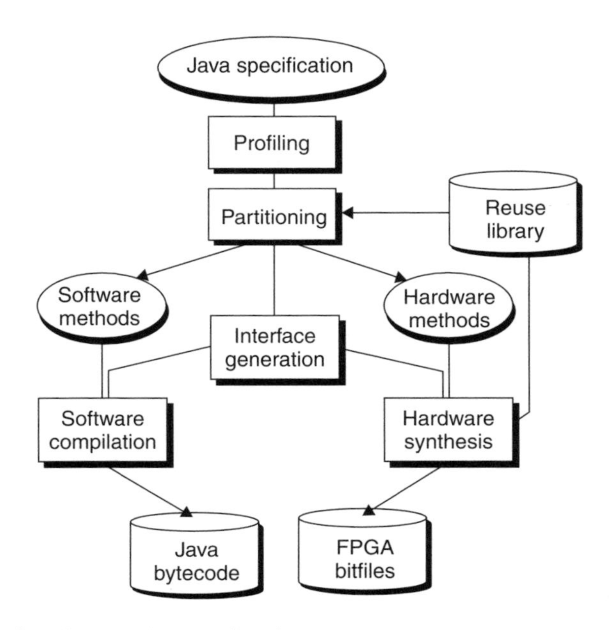

Figure 1.3 Specification and cosynthesis.

while executing the program with typical input data. This profiling data is then visualized to guide the designer in the partitioning process. Partitioning is done at the method level of granularity using a graphical user interface. Functions to be implemented in hardware are synthesized using high-level and logic synthesis tools. Previously designed hardware components are accessible through a database of parameterizable VHDL components. After the cosynthesis, Java byte code for all methods of the initial specification is stored in the pool of software methods. For all methods that are candidates for implementation in reconfigurable hardware, the FPGA configuration data as well as interface information is stored in the pool of hardware methods. The target hardware platform consists of dynamically reconfigurable FPGAs (DPGAs). These new FPGA architectures can be partially reconfigured at run time, i.e. a portion of the chip can be reprogrammed while other sections are operating without interruption. The target software platform for system prototyping is Linux PC.

A run-time manager controls the dynamic behavior of a reconfigurable system during execution. The run-time manager schedules methods for execution either as software on the Java virtual machine (JVM) of the host processor or as hardware on the reconfigurable DPGA hardware. The scheduling depends on the dynamic behavior of the application and on the current partitioning table chosen by the designer. In contrast to traditional prototyping systems, execution on this platform is a highly dynamic process. The execution flow of the hardware and software system is dominated by the software part. Software methods are executed on the JVM. Whenever the control flow reaches a hardware method, the time system determines whether the appropriate configuration file has already been downloaded. If not, then the manager chooses a DPGA and starts configuration. If there is already a DPGA configured with the desired functionality, or if only partial reconfiguration is necessary, the address and parameters of the communication channel to the target DPGA are loaded.

The core component of the run time environment is the Java virtual machine shown in Figure 1.1. It basically consists of a class loader for dynamically loading Java byte code and an execution engine for interpreting the byte codes on the host processor. Design framework can use the KAFFE JVM which comes with a complete source code. The execution framework integrates reconfigurable DPGA hardware, and several extensions to the class loader and the interpreter are necessary. The class loader is extended to read in the current hardware and software partitioning table and to handle hardware methods, i.e. methods which have to be executed on the DPGA board. These hardware methods and information about their corresponding interfaces, which are necessary to transfer data to and from the DPGA, are accessed through a database.

The execution engine needs to know whether a method will be interpreted as byte code or executed in hardware. Therefore, the class loader assigns a special flag to every hardware method. During execution of the application, the interpreter has to activate the hardware call module whenever flow of control reaches a hardware method. Depending on the current state of the DPGA board, several actions are triggered with every hardware call. If necessary, a new configuration data is downloaded to the chip, input data is transferred to the board, the hardware design is executed and the resulting data is transferred back to the calling thread. These procedures are implemented in the hardware wrapper (Figure 1.1). Furthermore, a strict synchronization mechanism is implemented. In this implementation, one thread at a time is allowed to access the DPGA board.

Extending the Java virtual machine for interaction with reconfigurable hardware resources allows implementation of a completely user-transparent mechanism for execution of a mixed hardware and software applications. In the prototyping phase, the user can explore design alternatives and different hardware and software mappings via a graphical user interface. The Java code remains unchanged, as the run-time system and the extended interpreter completely manage execution of hardware and software components. However, the drawback of this approach is that only Java virtual machines can be used where the source code is available. For different VMs or releases, extensions and customizations have to be made.

In an alternative implementation, the Java class and native methods are used to interface with the hardware part of the system. The main focus is on implementing all necessary functionality for hardware interfacing and reconfiguration in Java. Therefore, the platform specific API (Application Programming Interface) of the reconfigurable hardware board can be kept very small. In this case the board API basically consists of native functions to write a value to, and read a value from, a certain address of the board. These functions are implemented via the Java Native Interface (JNI). This means that all methods for managing the configuration process and execution are implemented in Java, and all communication to the hardware board is based on the native implementations of the read and write functions. For communicating with the external board via the PCI bus, a dedicated device driver is implemented as a kernel loadable module under Linux.

The benefits of this approach are clear. The Java VM does not have to be modified and the hardware interface is clearly defined within the Java language. This means that the designer has complete control over all methods for accessing and managing the reconfigurable hardware. The drawback is that the application code has to be modified. The DPGA interface class has to be included in the application source and the designer has to call the appropriate functions for using the DPGA. However, this method can be used with any virtual machine. Therefore, it is relatively simple to integrate and test different commercial implementations of the JVM.

1.6. JAVA-DRIVEN CO-DESIGN AND PROTOTYPING

In embedded systems an increasing share of functionality is implemented in software, and the flexibility or reconfigurability is added to the list of nonfunctional requirements. Networked embedded systems are equipped with communication capabilities and can be controlled over networks. JaCoP (Java driven Codesign and Prototyping environment) is a codesign environment based on Java that supports specification, cosynthesis and prototyping of networked embedded systems.

The rapidly growing market for web-enabled consumer electronic devices introduces a paradigm shift in embedded system design. Traditionally, embedded systems have been designed to perform a fixed set of previously specified functions within a well-known operating environment. The functionality of the embedded system remains unchanged during product lifetime. However, with shorter time-to-market windows and increasing product functionality, this design philosophy has exhibited its shortcomings. Hardware and software codesign tools are increasingly used to alleviate some of the problems in the design of complex heterogeneous systems.

The key feature of next-generation embedded devices is the capability to communicate over networks and to adapt to different operating environments. There is an emerging class of systems that concurrently execute multiple applications, such as processing audio streams, capturing video data and web browsing. These systems need to be adaptive to changing operating conditions. For instance, in multimedia applications the video frame rate has to be adjusted depending on the network congestion. Likewise, for audio streams different compression techniques are applied depending on the network load. Besides this class of multifunction system there are multimode systems, i.e. systems that know several alternative modes of operation, for example a mobile phone that is able to switch between different communication protocols or a transmitter that can toggle between different encryption standards.

This paradigm shift in both functional and nonfunctional requirements of embedded appliances not only holds for consumer devices. In industrial automation there is a growing demand for sensor and actuator devices that can be remotely controlled and maintained via the Internet.

Several system-level design languages and codesign frameworks have been proposed by researchers and are gaining acceptance in industry, but there is a lack of methods and tools for investigating issues that are raised when designing run-time reconfigurable hardware and software systems. A complete design environment for embedded systems includes dynamically reconfigurable hardware components. JaCoP (Java driven Codesign and Prototyping environment) is based on Java, which is used for specification and initial profiling as well as for the final implementation of system software.

1.6.1. Java-Based Co-design

Designing the digital hardware part of a system has become increasingly similar to software design. With widespread use of hardware description languages and synthesis tools, circuit design has moved to higher levels of abstraction. For managing complexity of future designs, abstract specification and reuse of previously developed hardware and software components is essential. Therefore, a specification language should provide means for integrating reuse libraries, that is, packages of previously developed components. Furthermore, object-oriented programming has proven to be a very efficient paradigm in the design of complex software systems. Object-oriented programming provides better means for managing complexity and for reusing existing modules. Object-oriented programming also reduces costs associated with code maintenance. In embedded system design, a major trend is increasingly to implement functionality in software. The reasons for this are faster implementation, better flexibility and easier upgradability. Consequently, the cost of software maintenance is an issue of growing importance. Java is useful as a specification language, and it is an object-oriented, versatile language of moderate complexity. Java has built-in support for handling multiple threads, and expressing concurrency and managing different flows of control is well supported.

The Java Beans specification provides a standard concept for reuse of software components. With respect to networked embedded system design, features like Internet mobility, network programming, security, and synchronization are of great importance. However, Java has not been designed for specification of real-time systems. Therefore, there have been proposed extensions (and restrictions) to the language for specifying such systems. The need to communicate with embedded systems over the Internet pushes more designers towards Java. Personal Java has been adopted as the premier design platform for implementing applications for set-top boxes.

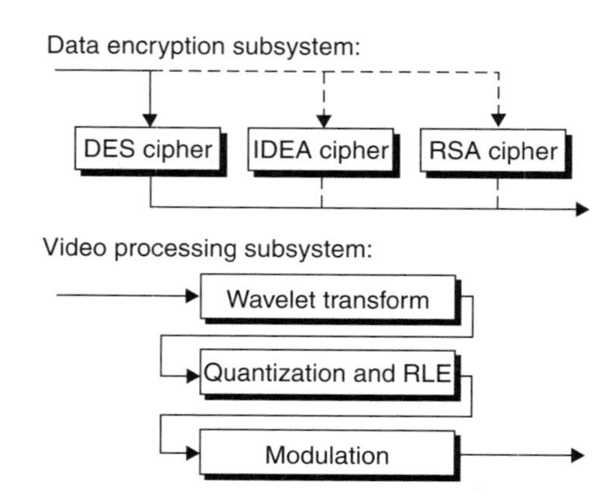

Figure 1.4 Reconfigurable system set-top box.

The framework for specification and design exploration of mixed hardware and software implementations is targeted for reconfigurable embedded systems. In Figure 1.4, two examples of application of run-time reconfigurations are given. The first task graph shows an encryption system which can switch between different operation modes, i.e. different encryption algorithms. Typically, only one application at a time is active and reconfiguration of the system is not time-critical. The second task graph represents a part of a video communication system. The task of processing a video stream is decomposed into a series of subtasks, which are executed on the same piece of silicon. The chip is reconfigured periodically. Rapidly reconfigurable hardware components are needed to meet soft deadlines.

The design flow is as follows: starting from an initial Java specification, profiling data is gathered while executing the program with typical input data. This profiling data is then analyzed and animated to guide the designer in the partitioning process. Partitioning is done at the method level of granularity using a graphical user interface. Functions that are implemented in hardware are synthesized using high-level and logic synthesis tools. Previously designed hardware components are integrated by using a database of parameterizable VHDL components. After cosynthesis, Java byte code for all methods of the initial specification is stored in the pool of software methods. For all methods that are candidates for implementation in reconfigurable hardware, the FPGA configuration data as well as interface information is stored in the pool of hardware methods. The target hardware platform consists of dynamically reconfigurable FPGAs (DPGAs). These new FPGA architectures may be partially reconfigured at run-time, i.e. a portion of the chip can be reprogrammed while other sections are operating without interruption. The target software platform for system prototyping is currently a Linux Pentium PC.

1.6.2. Run-Time Management

The interactions between the hardware and software parts of the system, as well as the reconfiguration process, are managed by the run-time environment (Figure 1.1). The run-time manager schedules methods for execution either as software on the Java virtual machine of the host processor or as hardware on the reconfigurable DPGA hardware. The scheduling depends on the dynamic behavior of the application and on the current partitioning table chosen by the designer. In contrast to traditional FPGA-based prototyping systems, execution on this platform is a highly dynamic process. The execution flow of the hardware and software system is dominated by the software part. Software methods are executed on the Java virtual machine. Whenever the control flow reaches a hardware method, the run-time system determines whether the appropriate configuration file has already been downloaded. If not, then the manager chooses an available DPGA and starts configuration. If there is already a DPGA configured with the desired functionality, or if only partial reconfiguration is necessary, the address and parameters of the communication channel to the target DPGA are loaded.

In virtual machine, a Java class and native methods are used for interfacing with the hardware part of the system. The main focus is to implement all necessary functionality for interfacing hardware and for reconfiguration in Java. Therefore, the platform specific API of the reconfigurable hardware board is kept very small. In this case, the board API basically consists of native functions to write a value to, and read a value from, a certain address on the board. These functions are implemented via the Java Native Interface (JNI). All methods for managing the reconfiguration process and execution are implemented in Java and all communication to the hardware board is based on the native implementations of the read and write functions. For communicating with the external board via the PCI bus, a dedicated Linux device driver is used.

The benefits of this approach are as follows. The Java VM does not have to be modified and the hardware interface is clearly defined within the Java language. A designer has complete control over all methods for accessing and managing the reconfigurable hardware. The drawback is that the applications source code has to be modified. The interface class DPGA_circuit has to be included in the application source, and the designer has to call the appropriate functions for using the DPGA. However, this methodology can be used with any virtual machine. Therefore, it is relatively simple to integrate and test different commercial implementations of the JVM. To support a better interface for the user of the JaCoP system, a special class (HW_base) for managing hardware designs and for controlling the reconfiguration process

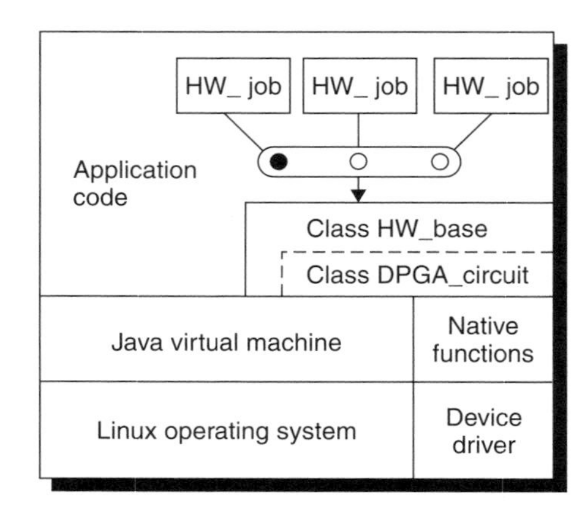

Figure 1.5 Thread access to hardware.

is used (Figure 1.5). This class encapsulates all functionality specific to the underlying hardware resource, and also hides details of the hardware and software interfaces from the designer. Communication to the DPGA board is done by method invocations of the board API (DPGA_circuit) class. To present the dynamic reconfiguration capabilities of the hardware, it is desirable to implement several individual hardware designs on the chip. Each of these designs is accessed by a corresponding Java thread. With the JaCoP native implementation, a mechanism is provided for multiple threads to make concurrent use of the external hardware resource. Therefore, the class HW_job is developed. An object of this type (Figure 1.5) represents a thread that makes use of hardware methods. These objects use a single instance of type HW_base when accessing the DPGA. To avoid racing conditions or invalid hardware configuration, the operations that access the DPGA are defined as critical sections. The methods for reconfiguration and for register reads and writes are synchronized. The currently active HW_job cannot be preempted by other threads while executing such critical sections. In network embedded system applications, threads are used in two different scenarios. A typical example is a system task that operates on certain blocks of data, consuming a significant amount of time. After processing this data, the thread terminates. Another example is a task that is periodically activated. To save the cost of repeated reconfigurations, the corresponding hardware design is kept on the FPGA. Whenever data is available for processing, the corresponding thread becomes active and returns to a wait state afterwards.

Reuse of hardware beans occurs when the software engineering techniques are increasingly used to lever the concurrent design of hardware and software of embedded systems based on a single system-level description. When designing a set of systems within a certain application domain (e.g. set-top boxes), reuse of standard components (modulators, video and audio processing tasks, ciphering algorithms, data compression, and error control) is especially attractive. To reduce the development cost and the time to market, the tool and methodology for reuse is integrated into JaCoP. This methodology is based on the Java Beans mechanism for reusing software components. Therefore, the concept is extended to allow for hardware components (hardware beans).

In the introspection and serialization, the beans are used for composing applications by using a visual builder tool. In this design environment, the different parameters of a component are exhibited (i.e. introspection) and interactively customized. After customization, the chosen configuration has to be saved (i.e. serialization) for use within the run-time environment. This mechanisms are also used for hardware designs. For example when composing a system in the JaCoP environment, the weights of a Finite Impulse Response (FIR)-filter bean can be chosen by the system designer without the need for an additional synthesis step.

Properties and events are used in the implemented functionality of a hardware and software basic component, which is reflected in the methods of a bean. Events are used for intercomponent communication during run-time. Properties are basically named attributes of a component, which can be accessed at design time and at run-time. According to the strict naming convention, property 'X' can be accessed by so-called 'get X' and 'set X' methods. With respect to hardware components, properties and events are both used for transferring data between components. For example, by setting a property of a decoder bean, a block of input data can be transferred to the decoder. After processing of the data block, an event is fired and the result can be obtained by retrieving the corresponding bean property. Multiple threads can access both hardware and software beans at run-time. Therefore, the aspects of synchronization need to be taken into account when developing a bean.

Reuse methodologies typically require additional effort during the design of a specific component. They are useful when implementing a variety of systems within a certain application domain. One of the benefits of Java is that it encourages design for reuse by providing the Java Beans mechanism and also by providing a built-in concept and a tool for hypertext documentation of Java classes.

1.6.3. Embedded Systems Platform

The target architecture of an embedded system platform consists of a standard microprocessor tightly interfaced with a dynamically reconfigurable FPGA.

They are connected to static random access memory (RAM). This is a prototype of a single chip solution of a reconfigurable system. Such systems are available, for example, the Siemens Tricore or the National Napa1000 Reconfigurable Processor. The Napa1000 is a single-chip implementation for signal processing which provides both fixed logic (a 32-bit RISC (Reduced Instruction Set Computer) core) and a reconfigurable logic part (a 50-k gate Adaptive Logic Processor). For developing the codesign methodology and the corresponding cosynthesis flow, a Linux PC is used as development platform and host processor and the XC6200DS board as a reconfigurable hardware resource. The main components of this board are a PCI interface and a reconfigurable processing unit XC6216. Furthermore, there are two banks of memory included, which can be accessed from both the DPGA and the PCI bus. The most prominent feature of this DPGA is its microprocessor interface. Direct write and read operations over an address and data bus are supported to every logic cell or register on the chip. This provides a mechanism for transferring data between hardware and software components of the design. The hardware wrapper can read from, and write to, every register of the hardware design at run-time.

This implementation is referred to as the JaCoP native interface. An alternative implementation for the run-time system is referred to as the JaCoP interpreter. Performance of the JaCoP interpreter is defined by the costs of DPGA reconfiguration and communication between hardware and software. A complete reconfiguration of the chip is too slow for dynamic applications with soft deadlines. Typical times for standard reconfiguration ranges between 30 ms and 400 ms depending on the size of the circuit. In order to reduce this overhead for reconfiguration, a mechanism for compression and improved transmission of the DPGA configuration files has been developed. Basically, all redundant address and data pairs are omitted and the necessary configuration data is transmitted in a binary format. By using this optimization, reconfiguration time is reduced to about 4 ms to 31 ms. Besides reconfiguration, the second important factor that introduces overhead is the hardware and software communication. An efficient implementation of the hardware and software interface is the most important factor for overall performance of the combined system. In this architecture communication is accomplished by writing and reading internal registers on the DPGA. For the XC6216, a single write or read operation can only access one specific column of logic cells. To minimize communication costs, a layout optimization is used so that individual registers are placed in individual columns whenever possible. Furthermore, for the used DPGA, a so-called map register is configured before accessing an individual register to mask out the corresponding rows of logic cells. Profiling shows that this is a very time consuming process, therefore a second optimization for the layout of the hardware designs is used. Whenever possible, all I/O (input/output) registers of a design have to be placed in corresponding rows. By using this layout constraint, configuration of the map register can be avoided before read or write accesses. Consequently, register access is improved drastically from about 70 microseconds down to 8 microseconds.

By using these optimizations, the total overhead for integrating external hardware can be attributed to the different components involved in the DPGA execution process. About 78 % of the total time for hardware integration occurs in the virtual machine and the device driver. About 22 % of time is used to execute the DPGA design and transfer data over the PCI bus. For this reason, speedup of a mixed hardware and software implementation can only be achieved by implementing a method of significant complexity in hardware. Moving simple operations like additions or multiplications to DPGA hardware cannot deliver speedup because of the overhead involved. More complex examples include an algorithm for error detection and correction. Hamming codes are typically used in conjunction with other codes for detection and correction of single bit faults. Both the Hamming coder and decoder are implemented on the DPGA. As this application includes more complex bit level operations, a significant speedup of the hardware and software implementation is experienced in comparison with the execution of the software prototype on the host CPU. The main advantage of the native interface is that it can be used with any Java VM, for example, an optimized JVM2, which is only available in binary form. A significant speedup can be achieved by integrating the reconfigurable hardware platform. If an application has computational complex components that are executed on the DPGA board, hardware accelerated execution is possible regardless of the necessary overhead for reconfiguration and communication. The results also show that the implementation using native functions has a small performance drawback when compared to modifying the virtual machine. On the other hand, this implementation is especially attractive when using commercial JVMs.

A codesign environment for Java-based design of reconfigurable networked embedded systems, JaCoP includes tools that aid in specification, hardware and software partitioning, profiling, cosynthesis and prototype execution. This approach provides means for implementation of hardware and software threads. Multiple threads can concurrently use the features of a reconfigurable DPGA architecture. Based on the Java Beans specification, a suitable methodology for design reuse is integrated. The design flow is implemented and tested on a PC connected to a reconfigurable hardware board.

1.7. HARDWARE AND SOFTWARE PROTOTYPING

A design flow of the prototyping environment is shown in Figure 1.6. This design flow includes a complete synthesis flow and accommodates fast prototyping. The hardware and software interface is generated by an interface generator. The software part is instrumented and byte code is generated for execution on the Java virtual machine. The hardware part is synthesized and mapped to the reconfigurable FPGA target platform. The interface is done in both software and hardware.

During the specification phase, only the software methods (left branch in Figure 1.6) are used for functional validation and profiling. Then the initial specification is partitioned into a part for execution on the host PC (software methods) and a part which is executed on the FPGA hardware platform (hardware methods). Partitioning can be based on Java methods, on loops, or on basic block level. During code generation, byte code is generated for all methods, whereas FPGA configuration files are generated for the individual hardware methods only. The run-time system (RTS) reads information from the partitioning step and decides whether, according to the partitioning method, to schedule it on the host PC or on the FPGA hardware. The RTS also manages dynamic reconfiguration of the hardware at run-time.

For each hardware method, an interface description is automatically generated. It consists of a RT-level VHDL frame for inclusion into the hardware building block, and a hardware method call for inclusion into the software code.

Code generation for the hardware part consists of the following steps:

- High-level synthesis (HLS) of the selected Java methods implemented in hardware;
- Register-transfer-level synthesis of the VHDL description generated by the HLS tool in the previous step and logic optimization of the resulting netlist;
- Layout synthesis and generation of the configuration data for the target FPGA hardware platform.

The exchange of data between the hardware and software part occurs through an interface. The interface generator uses information provided by the high-level synthesis tool and automatically generates a register-transfer level description of the VHDL frame with the correct size. The structure of this interface frame is shown in Figure 1.7. The frame consists of a set of registers for storing input and output data of the corresponding Java method. It also contains a small logic block for handling control signals that are necessary for interaction with the run-time system. This includes signals for starting and

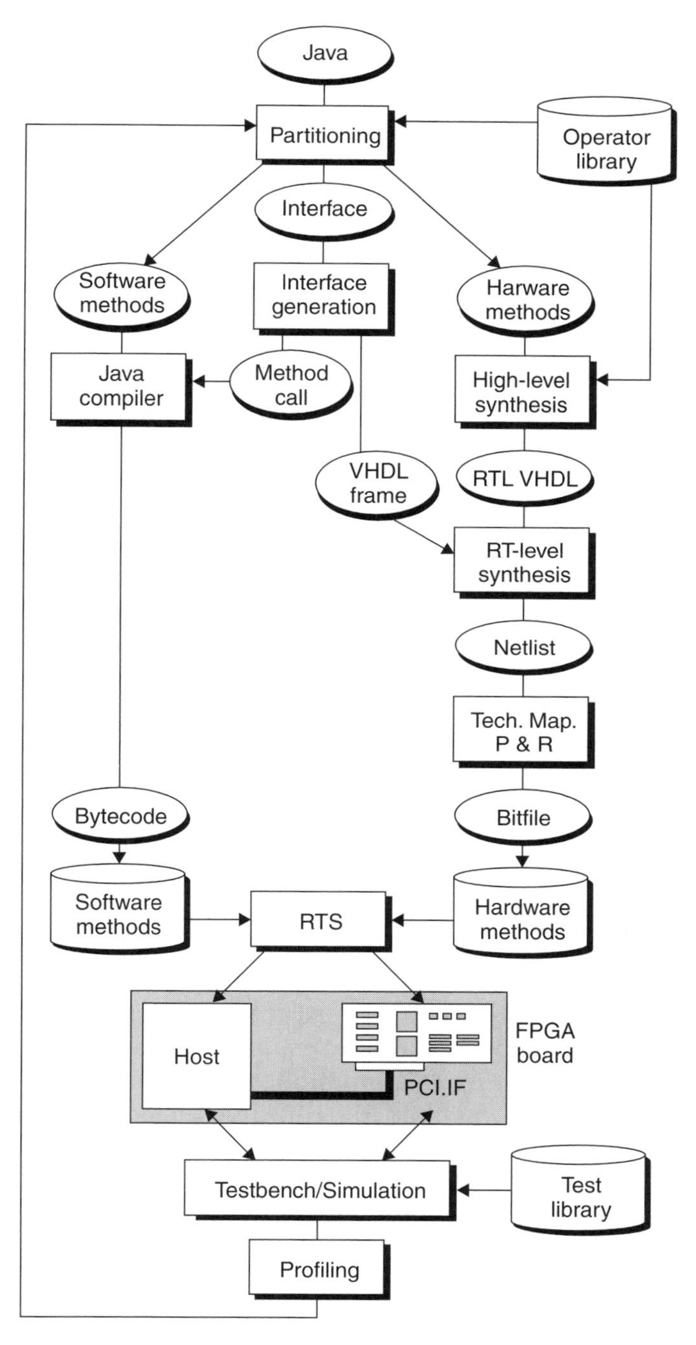

Figure 1.6 Prototyping environment.

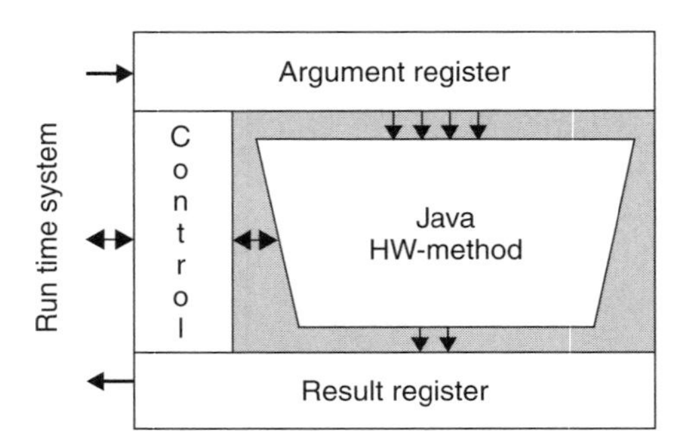

Figure 1.7 VHDL frame.

resetting the hardware process, and a ready signal that indicates when the computation in hardware is completed and the results can be read back.

The VHDL component for the Java hardware method generated during high-level synthesis is embedded into the VHDL frame produced by the interface generator. The complete design is synthesized for the FPGA platform.

The run-time system (RTS) is responsible for managing execution and interaction of hardware and software methods. RTS relies on a database that contains the set of methods executable either in hardware or software, and the methods for which both hardware and software implementations are available. The software executables are defined by Java classes that are stored in Java byte code format. The corresponding hardware blocks are emulated on the FPGA board and stored as configuration bit files.

During cosimulation, the run-time system schedules methods for execution according to the current partitioning table. Software methods are executed on the Java Virtual Machine (JVM) on the host workstation as shown in Figure 1.6. In a software-oriented approach, the execution flow of the combined system is dominated by the software part of the system.

The run-time system also manages execution of tasks on the emulation platform. The set of available FPGAs and communication channels is specified in a configuration file. During cosimulation, when control flow in one of the threads reaches a hardware method, the run-time system determines whether the corresponding configuration bit file has already been downloaded to an FPGA.

At the beginning of the cosimulation procedure, none of the FPGAs is configured with a bit stream. The RTS determines an available FPGA and triggers the download mechanism. When the bit file has been downloaded to an FPGA, the RTS determines the address of the FPGA containing the requested hardware method, and starts to transfer data which has to be processed by this method. After the emulated method has finished processing,

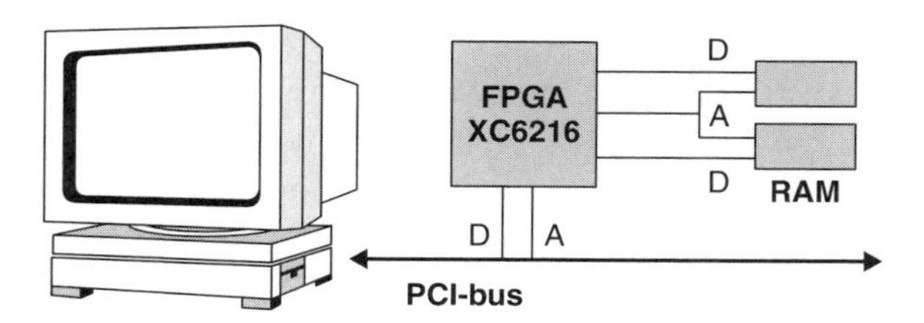

Figure 1.8 Experimental platform.

it sends an interrupt signal to the RTS and the results from the computation can be read from the result register of the FPGA design. During emulation of a hardware method on the FPGA, the calling thread is suspended and any other software thread can be executed on the Java virtual machine at the same time.

A hardware and software prototyping platform is shown in Figure 1.8. The workstation is used to enter the Java specification, to compile Java to byte code, and to execute the byte code for the software part of the system. Different synthesis tasks necessary to move dedicated Java methods to the FPGA emulation board are carried out on the host workstation. The RTS is implemented in software on the host. The hardware part is implemented by using a dedicated FPGA board connected to the workstation over the PCI bus.

1.8. MULTIPLE APPLICATION SUPPORT

In traditional embedded systems design, the key objective is to find an optimal architecture to perform a single, specific application. In this system architecture, ASICs (Application Specific Integrated Circuits) and processors are typical building blocks. The designer performs partitioning of the application onto ASICs and processors according to the performance metrics such as processing power, flexibility and power consumption. Emerging embedded systems in the area of information appliance (e.g. PDAs and IMT2000 (International Mobile Telecommunication) terminals) are different from traditional embedded systems in the following aspects:

- they run multiple applications such as web browsers, audio and video communication applications, etc., and
- they require network connectivity.

To design a multiple-application embedded system that requires considerable processing capability for each application, is a challenging task from the viewpoint of the traditional design approach, which gives, in general, an architecture that is good for only one application but is inferior or not suitable for other applications. To meet the cost and power consumption constraints, the designer may have to use costly redesign loops that often do not converge or produce over-design. Resolving this problem requires programmability of the system architecture so that an embedded system can be adapted to new applications.

Programmability is introduced to the architecture by embedding FPGAs into the system. FPGA is a viable option for the following reasons:

- FPGAs have processing capability and logic capacity close to those of ASICs.
- By reprogramming the FPGAs, the embedded system can adapt itself to a new application.
- Although available FPGAs are not very power-efficient, there is continuous effort to achieve low-power consumption through methods such as applying low supply voltage and power-down mode.

Embedded system management uses network connectivity to download new applications from remote servers. This connectivity can be served by using Java as a software platform. A programmable embedded system uses Java to download and execute new application codes without shutting down the entire system.

Java-based embedded system architecture includes FPGA coupled with a standard processor. In this architecture, Java code and FPGA bit-stream of new application is downloaded over the network. FPGA is programmed and dynamic reconfiguration of FPGA is supported. Communication between hardware (FPGA) and software (Java application code) uses a set of native APIs to access the underlying hardware from Java applications. Native APIs from the Java application are invoked by Java Native Interface (JNI). A partitioning granularity is a Java method which is implemented into hardware components (hardware methods).

Designing embedded system architecture for emerging information appliances, such as PDAs and IMT2000 terminals, requires multiple-application support and network connectivity. Programmable architecture is an efficient way of implementing the multiple-application support. FPGA-based embedded system architecture uses Java as a software platform. A set of native communication APIs is used for the communication between Java application and the FPGA, where from 80 to 90 % of the system's execution time is consumed for communication over the PCI bus.

The Y-chart approach to exploring the design space of multiple-application embedded systems uses a stream-based function (SBF) model of a set of applications that produces quantitative performance numbers for parameterized target architectures. A simulator performs interpreted or noninterpreted simulation while extracting the performance results.

Java is used in the embedded system design, for example:

- system specification for expressing real-time constraints, and
- software implementation and execution platform with enhanced real-time capability.

Programmability of an FPGA requires run-time reconfiguration method that dynamically changes the functionality of an FPGA during system execution. A run-time partial reconfiguration strategy is applied to FPGA so that programming and execution can be performed concurrently. Configuration overhead has significant impact on the overall system performance and needs to be reduced.

1.8.1. FPGA-Based System Architecture

An example of a hardware component running in the FPGA, which corresponds to a Java method, is shown in Figure 1.9. An input buffer and an output buffer are used to receive arguments and to store the results, respectively. A control signal buffer is used to manipulate and check (e.g. to give a

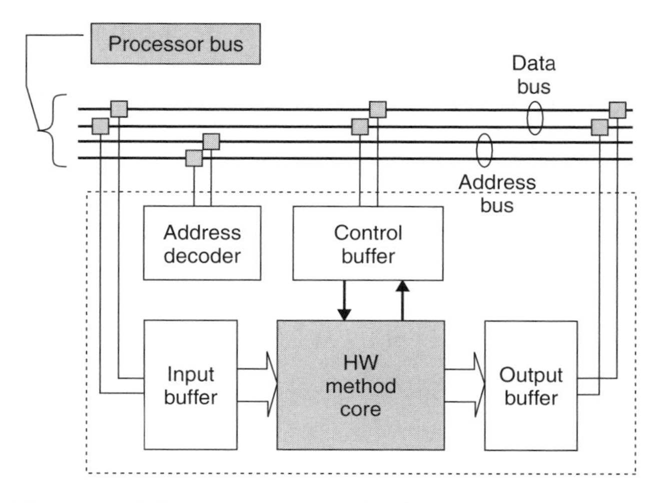

Figure 1.9 Architecture of the hardware method.

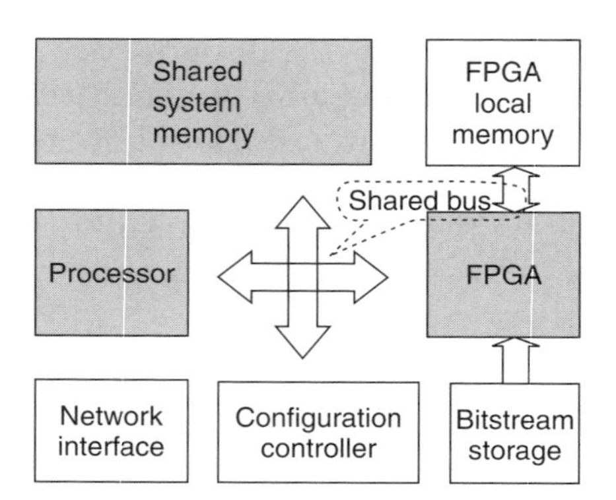

Figure 1.10 The hardware platform.

start signal and to check the done signal) the state of the hardware method. These buffers are mapped onto the processor address space and the processor can access them with memory read and write instructions.

The hardware platform of the architecture consists of a standard processor, an FPGA, and a system memory, which are connected through a shared processor bus as shown in Figure 1.10.

When the processor initiates a read or write operation to some hardware method, the corresponding hardware method responds to the bus activity with the help of its own address decoder. The processor initiates hardware and software communication and the FPGA responds as a slave in the bus activity. A configuration controller with bit-stream storage is included in the architecture to support dynamic reconfiguration of the FPGA. Since the configuration controller takes the responsibility of programming the FPGA with the initiation from the processor, the processor can perform other useful jobs in parallel with an FPGA programming job.

Java is used as a software platform as shown in Figure 1.11. An embedded OS provides services, for example thread service and other hardware management service, for Java run-time environment. Java applications in remote places can be downloaded into the embedded system through the system manager. The system manager implements three basic protocols as follows:

- application code (including FPGA bit stream) download protocol;
- system-maintenance-related protocol for remote management;
- authentication protocol to control access to the embedded system.

The system manager contains a custom class loader that is based on a socket connection. An application in a remote place can be downloaded and executed in the following procedures:

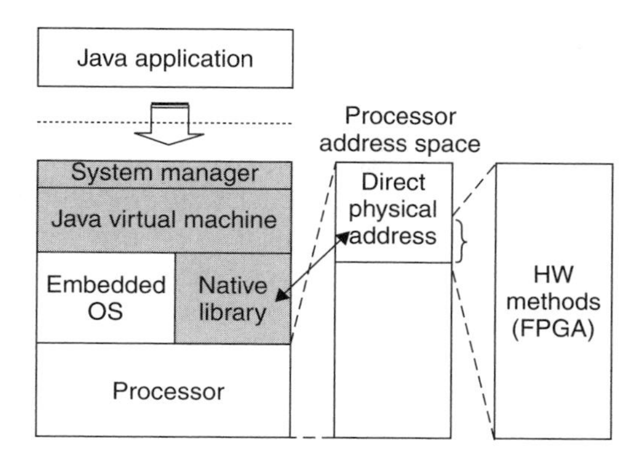

Figure 1.11 The software platform.

- get an authority to make the system enter management mode;
- download new applications code, and then perform initialization such as placing FPGA bit stream into bit-stream storage;
- execute new application.

For simplicity, it is assumed that buffers of hardware methods are mapped onto predefined and fixed physical address regions.

When Java is used as a software platform, the application cannot access the underlying hardware directly. The application running in the JVM on the processor cannot access the physical address region mapped on to buffers of hardware methods in the FPGA directly. This can be solved by:

- modifying the JVM to have direct access service to the physical address of the processor, or
- developing a native communication library so that an application can access it through Java Native Interface (JNI) with some overhead.

The first solution results in minimal overhead, however, modifying the JVM internals is not easy and may not follow the latest releases of JVM. Developing an extra-native communication library outside of JVM introduces a JNI layer.

1.9. SUMMARY

The key feature of next-generation embedded devices is the capability to communicate over networks and to adapt to different operating environments. There is an emerging class of system that concurrently executes multiple applications, such as processing audio streams, capturing video data and web browsing. Such systems need to be adaptive to changing operating conditions.

A specification language should provide means for integrating reuse libraries, that is, packages of previously developed components. Object-oriented programming has proven to be a very efficient paradigm in the design of complex software systems. Object-oriented programming provides better means for managing complexity and for reusing existing modules. Object-oriented programming also reduces costs associated with code maintenance. In embedded system design, a major trend is increasingly to implement functionality in software. The reasons for this are faster implementation, better flexibility and easier upgradability.

Programmability is introduced to the architecture by embedding FPGAs into the system. FPGAs have processing capability and logic capacity close to those of ASICs, and by reprogramming the FPGAs, the embedded system can adapt itself to a new application.

Programmable architecture is an efficient way of implementing the multiple-application support. FPGA-based embedded system architecture uses Java as a software platform. A set of native communication APIs is used for the communication between Java application and the FPGA.

PROBLEMS

Learning Objectives

After completing this chapter you should be able to:

- demonstrate understanding of the object-oriented design;
- discuss what is meant by design integration;
- explain what design optimization is;
- demonstrate understanding of codesign and reconfiguration;
- explain what Java driven codesign and prototyping is;
- discuss Java based codesign;
- explain what run-time management is;
- demonstrate understanding of an embedded systems platform;
- discuss the issues of hardware and software prototyping;
- explain what the multiple application support is;
- demonstrate understanding of FPGA-based system architecture.

Practice Problems

Problem 1.1: What is the difference between multifunction and multimode networked embedded systems?
Problem 1.2: Why is cosynthesis used for embedded devices?
Problem 1.3: What is the role of a run-time manager?
Problem 1.4: What is Java driven codesign and prototyping environment?
Problem 1.5: What functionality is added by JaCoP?
Problem 1.6: What steps are taken for code generation in hardware prototyping?
Problem 1.7: What are the basic protocols implemented by a system manager in the software platform?

Practice Problem Solutions

Problem 1.1:

Multifunction networked embedded systems can execute multiple applications concurrently. Multimode networked embedded systems offer the users a number of alternative modes of operation.

Problem 1.2:

A cosynthesis method and prototyping platform developed specifically for embedded devices combines tightly integrated hardware and software components. A cosynthesis is used to make the most efficient assignment of tasks to software or hardware.

Problem 1.3:

A run-time manager controls the dynamic behavior of a reconfigurable system during execution. The run-time manager schedules methods for execution, either as software on the Java virtual machine (JVM) of the host processor or as hardware on the reconfigurable DPGA hardware.

Problem 1.4:

JaCoP (Java driven codesign and prototyping environment) is a codesign environment based on Java which supports specification, cosynthesis and prototyping of networked embedded systems.

Problem 1.5:

A codesign environment for Java-based design of reconfigurable networked embedded systems, JaCoP includes tools that aid in specification, hardware and software partitioning, profiling, cosynthesis, and prototype execution.

Problem 1.6:

Code generation for the hardware prototyping includes high-level synthesis (HLS) of the selected Java methods implemented in hardware, register-transfer-level synthesis of the VHDL description generated by the HLS tool and logic optimization of the resulting netlist, and layout synthesis and generation of the configuration data for the target FPGA hardware platform.

Problem 1.7:

The system manager implements three basic protocols, namely the application code (including FPGA bit-stream) download protocol, the system maintenance-related protocol for remote management, and the authentication protocol to control access to the embedded system.

2

Smart Sensor Networks

2.1. INTRODUCTION

Users are demanding devices, appliances, and systems with better capabilities and higher levels of functionality. Sensors in these devices and systems are used to provide information about the measured parameters or to identify control states, and these sensors are candidates for increased built-in intelligence. Microprocessors are used in smart sensors and devices. A smart sensor can communicate measurements directly to an instrument or a system. The networking of transducers (sensors or actuators) in a system can provide flexibility, improve system performance, and make it easier to install, upgrade and maintain systems.

The sensor market is extremely diverse and sensors are used in most industries. Sensor manufacturers are seeking ways to add new technology for building low-cost, smart sensors that are easy to use and which meet the continuous demand for more sophisticated applications. Networking is becoming pervasive in various industrial settings. Decisions about the use of sensors, networks, and application software can all be made independently, based on the application requirements. In reality, however, all these function modules cannot be easily integrated due to the lack of a set of common interfaces.

A typical sensor or control network consists of network nodes comprising up to 256 units linked by multiwire cables. Each network node contains a microprocessor device, and a sensor or multiple sensors can be connected to each node through an electronic interface. Every network has its own

Wireless Sensor Network Designs A. Hać
 ISBN: 0-470-86736-1

custom-designed interface for sensors, and sensor manufacturers have to support various networks and protocols. The purpose of the IEEE (Institute of Electrical and Electronics Engineers) 1451 *Standards for Smart Transducer Interface for Sensors and Actuators*, is to define a set of common interfaces for connecting transducers to microprocessor-based systems, instruments, and field networks in a network-independent fashion.

The standardized Transducer Electronic Data Sheet (TEDS) specified by IEEE 1451.2 allows for self-description of sensors. The interfaces provide a standardized mechanism to facilitate the plug-and-play of sensors to networks. The network-independent smart transducer object model defined by IEEE 1451.1 allows sensor manufacturers to support multiple networks and protocols. This way, transducer-to-network interoperability can be supported. IEEE P1451.3 and P1451.4 standards will meet the needs of the analog transducer users for high-speed applications. Transducer vendors and users, system integrators, and network providers can benefit from the IEEE 1451 interface standards.

2.2. VIBRATION SENSORS

The ability of modern condition maintenance systems to provide smart interactive control over potentially dangerous or production sensitive machinery, assumes that the data received from connected sensors is correct for all possible fault conditions. No matter what measurement parameter is monitored, there is usually a multitude of different sensors available to choose from, and it would be rare to find just one model for each parameter that could accurately and reliably measure all required ranges. Selection of the correct sensor and correct installation is of paramount importance, especially when it comes to vibration monitoring. High frequency gear mesh measurements, for instance, would require a sensor with a suitably high frequency range, whereas at the other end of the scale, very low speed machinery or monitoring structural movements would require a low frequency accelerometer with high mechanical gain and good resolution.

Smart sensors communicate with their outside world by using the data capture and analysis or control system. Smart sensors use digital communication. There are many alternative paths along which to develop the potential benefits of an agreed protocol. The proposed IEEE 1451 standard has four levels, three of which focus purely on digital interfaces as shown in Figure 2.1, while the fourth level, known as P1451.4, defines an interface for mixed mode sensors with analog signals as well as digital information.

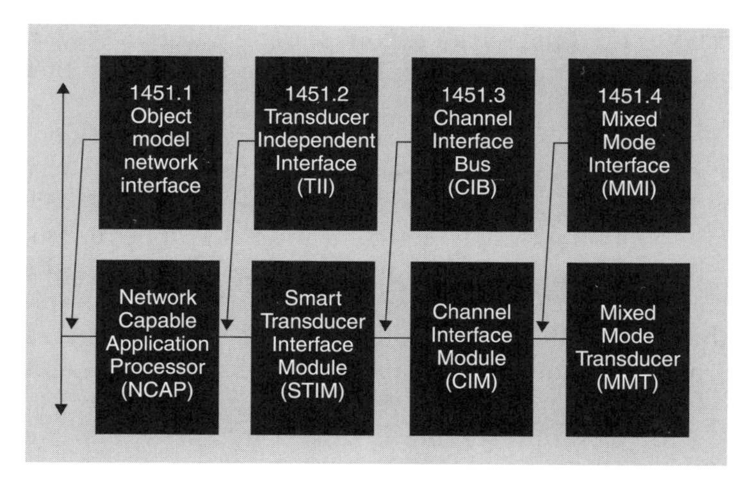

Figure 2.1 IEEE 1451 architecture.

The standard specifies data sets and formats that allow for each sensor to contain an electronic data sheet of information, such that it can be readily identified by the computer from a whole array of other sensors. This additional electronic package uses Transducer Electronic Data Sheet (TEDS). It has to be easy to use, support all different types of transducers, be flexible enough to meet individual needs, and remain compatible with level 1451.3 of the standard. TEDS data should include the following parameters:

- identification, e.g. model number;
- device, e.g. sensor type, sensitivity, and measurement units;
- calibration, e.g. date of last calibration and correction factors;
- application, e.g. channel ID (identifier) and measurement coordinates.

Apart from size restrictions or sensors with special outputs or for high temperature environments, TEDS can be put inside almost any vibration sensor during manufacture, and some data collection systems are available with the requisite TEDS interface. A TEDS sensor enables a system automatically to check on the status, exact position, and any other relevant detail put into the memory, during the normal data collection process. The TEDS function is the first step towards the truly intelligent vibration sensor. With additional integration of an ADC (analog-to-digital converters), and signal analysis such as FFT (Fast Fourier Transform) or frequency band monitoring within the sensor itself, important monitoring decisions can be made directly at the measurement location, and with local node or even individual sensor telemetry, cabling problems can also be eliminated.

No matter how smart the sensor becomes, there will always be the problem of correct sensor selection and deployment, in order to obtain the best information about potential failure.

Once the accelerometer has been correctly sourced, the next step is to mount and position it correctly in order to measure what is required. Accelerometers are designed to give an output in one axis only, so positioning can sometimes be essential for obtaining the best signal. Hand held probes are to be avoided if possible, due to the effect on frequency range and the positional errors that can occur with their use. Stud mounting on a properly prepared surface is always the best method, especially for high frequency measurements. Any effects on signal integrity should be understood and allowed for, or compensated for, in the measurement system.

The market for vibration sensors is driven by application and customer demand towards lower cost, and yet still be rugged, reliable and even intelligent transducers. Production machines with built-in vibration sensors are already available.

2.3. SMART SENSOR APPLICATION TO CONDITION BASED MAINTENANCE

IEEE 1451 is the proposed standard for interfacing sensors and actuators to digital microcontrollers, processors and networks. This standard reduces the complexities in establishing digital communication with transducers and actuators. IEEE 1451 defines the bus architecture, addressing protocols, wiring, calibration, and error correction, thus enabling a building-block approach to system design with plug-and-play modules. System integrators, instrument developers, engineers, and end users can plug IEEE 1451 compliant sensors and actuators together with measurement and communication modules to form a measurement system that allows transducers to interface directly with established networks and control systems.

Techniques for machinery fault prediction under development use multiple sensors with algorithms to extract useful information from the spectral properties of signals. Methods such as wavelet analysis, Hilbert transform analysis, adaptive neural networks, performance analysis, nonlinear characterization and multifunction data fusion with embedded sensors, are applied. In a plug-and-play architecture, sensors and actuators are linked together through a series of common interfaces to modules designed not only to process the signals, but to interface to existing communication networks. This approach eliminates full featured, more expensive components such as computers and stand alone instruments.

In the process control industry, sensors and transducers are connected directly to digital networks, over a common interface, and used in factory automation and closed loop control. The growth in slow speed sensors for measuring temperature, pressure, and position, contributed to the development of digital bus architectures. These systems have bandwidth limitations, proprietary hardware, and require design work to interface them with existing sensors.

Microprocessors, microcontrollers, ADCs (analog-to-digital converters) and their related electronics have become smaller, more powerful, and less expensive. There are advantages in including increased functionality into the transducer. The proposed industry standard interface for the connection of transducers and actuators to microcontrollers and to connect microcontrollers to networks is a logical extension of the General Purpose Interface Bus (GPIB or IEEE 488), except that this brings standardization to sensors instead of instruments. Figure 2.2 shows the building blocks for the implementation of a smart sensor interface.

An example application is a condition-monitoring system for milling machines in a large factory. A measurement system is needed for monitoring the health of bearings inside the milling machine and to detect tool wear or damage. Traditional methods of vibration measurement, using portable data collectors to monitor bearing health, have failed in this application, due to the varying operating conditions of the milling machine. Spindle movement and intermittent cutting operations affect the vibration signatures and can mask the vibration measurements of the bearings completely. The measurement

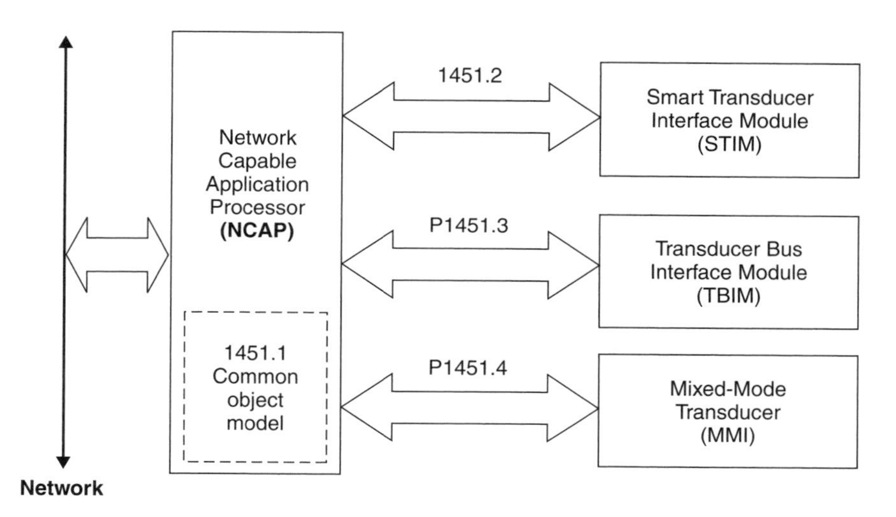

Figure 2.2 Functional block diagram of IEEE P1451.

system must take into account the various states of the machine and record vibration measurements at predetermined intervals.

An intelligent tool-condition-monitoring method is needed to detect tool wear or damage automatically instead of replacing the tool at regular intervals or discovering defects in material after operation. Direct sensing methods have been developed using multiple sensors to detect vibration, force, acoustic emission, temperature, and motor current. Tool wear is a very complex process and can be detected with sensor fusion, feature extraction, and pattern recognition.

A spectrum analyzer or data acquisition system with a dedicated personal computer can be adapted to work in this environment, with various inputs from the control system for timing. This was considered to be too expensive and cumbersome to be implemented across the factory, on every line every machine. A low-cost system needed to be developed that could accept inputs from various sensors, process the information, and notify operators of impending failures or problems. With IEEE 1451 compatible components such as vibration sensors and actuators, a smart transducer interface module and a communications module, a measurement system can be constructed to implement the functions needed and communicate throughout the factory's network. A solution is illustrated in Figure 2.3.

The IEEE 1451.1 standard defines the Network Capable Application Processor (NCAP). The NCAP is the smart sensor's window to the external control network that is connected to any transducer, or a group of transducers, with an appropriately configured NCAP. This building block of the IEEE 1451 standard typically consists of a processor with an embedded operating system and a sense of time. The processor has a communication stack for a network protocol. If the NCAP is used with the Ethernet, for example, it will have a TCP/IP protocol stack. Figure 2.4 illustrates an example of an NCAP. In this

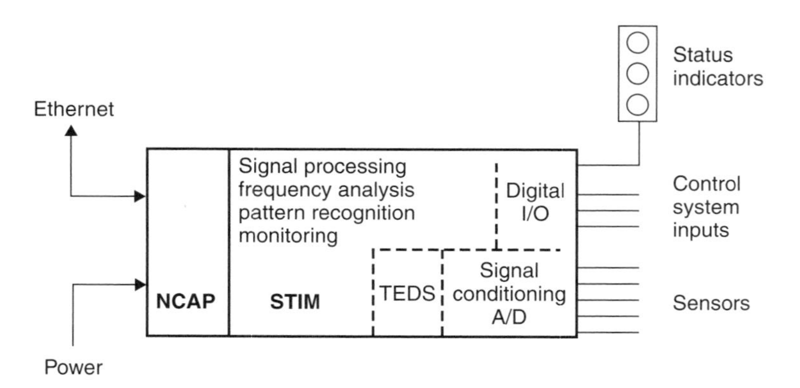

Figure 2.3 IEEE P1451 implementation of machine condition monitoring system.

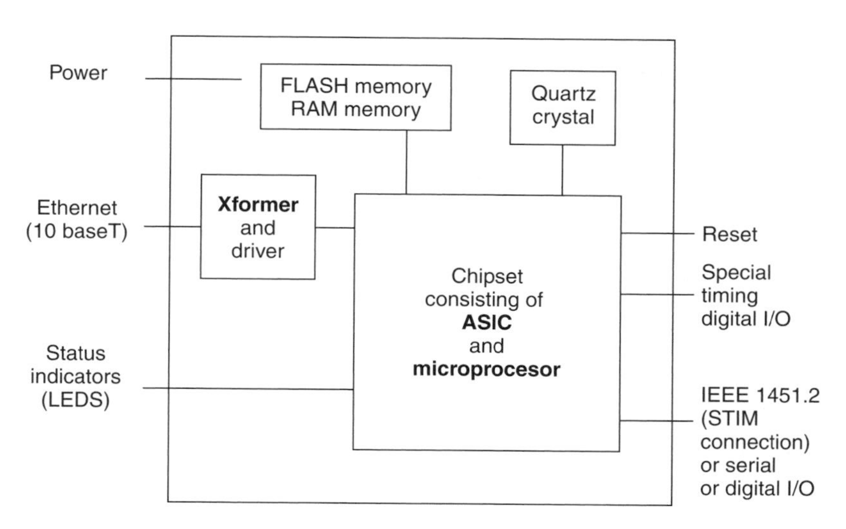

Figure 2.4 Network capable application processor (NCAP).

approach the design of the building blocks is done by the experts in this field who develop the modules to interface smart transducers with the existing networks. A system designer implementing a solution for a process chooses the module for a particular application and plugs it into the design. With additional code built into the NCAP, this module can be used as a micro web server with web pages providing information about the transducers connected to it.

The IEEE 1451.2 standard specifies Smart Transducer Interface Module (STIM). This is a digital interface and serial communication protocol that allows any transducer, or group of transducers, to receive and send digital data using a common interface. This common interface, called the Transducer Independent Interface (TII), is a 10-wire serial I/O bus that is similar to the IEEE 488 bus. The TII implements a serial data exchange with allowances for handshaking and interrupts. TII has defined power supply lines and permits hot-swapping of modules for plug and play capability. Any transducer can be adapted to the 1451.2 protocol with a Smart Transducer Interface Module (STIM). This building block of the 1451 standard is the measuring system. It can be as simple as a switch connected to a 4-bit processor, or as complex as a 255-channel device running an individual process. The STIM performs the tasks of signal conditioning, signal conversion and linearization. With added hardware it can perform functions such as spectrum analysis, fuzzy pattern recognition, adaptive noise canceling or a specific algorithm. The development of STIMs focuses on how to meet individual needs and special applications. Figure 2.5 illustrates one example of a STIM and how it interfaces with an NCAP through the TII.

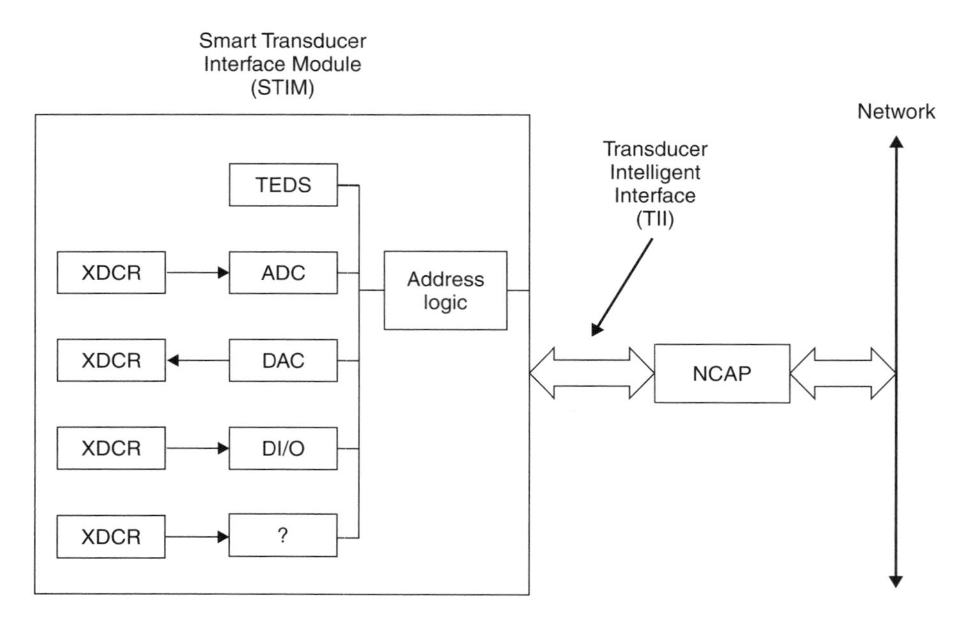

Figure 2.5 Overview of a STIM and how it associates through the TII to an NCAP.

Information about the STIM and the attached transducers is digitally stored in the format for which is integral to this standard Transducer Electronic Data Sheet (TEDS). This includes transducer identification, channel information, physical location, calibration, and correction data. TEDS provides a standardized set of mechanisms and information that can be used by applications to adapt automatically to device changes, thus supporting plug and play devices.

The IEEE P1451.3 standard defines Distributed Multidrop System (DMS), a digital interface for connecting multiple physically separated transducers, which allows for time synchronization of data. This transducer bus facilitates communications, data transfer, triggering, and synchronization.

A representation of IEEE P1451.3 with the functional blocks of the NCAP, transducer bus controller, and the transducer bus interface modules is shown in Figure 2.6. A single transmission line is used to supply power to the TBIMs and to provide communication to the bus controller. The NCAP contains the controller for the bus and the interface to the broader network. A TBIM supports different transducers and the bus may contain many TBIMs. This allows a distributed network of sensors and actuators to be connected through a common interface.

The IEEE P1451.4 standard defines Mixed-Mode Communication Protocol and Interface to bridge the gap between legacy systems and IEEE 1451

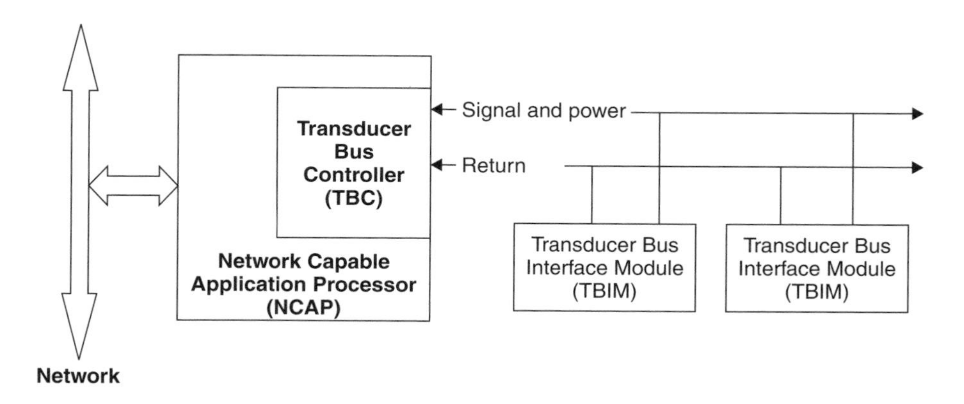

Figure 2.6 Physical representation of the IEEE P1451.3: multidrop system interface.

architectures. This standard allows analog transducers to communicate digital information, for the purposes of self-identification and configuration, over the same medium. A TEDS is defined for traditional analog sensors to store information such as model number, serial number, sensitivity and calibration parameters, inside the transducer. The term 'mixed-mode' refers to the operation of the transducer in either its traditional analog (sensing) mode or in its digital (communication) mode, during which transducer can be reconfigured, or its TEDS can be retrieved or updated. The transducer functions normally when the voltage supply is forward biased and will output its analog measurement signal. When the sensor is reverse biased, the traditional analog circuitry is disabled and the TEDS memory can be accessed. The circuit schematic is outlined in Figure 2.7 illustrating the reverse bias technique.

Although this example is specific to IEPE (Integrated Electronics, Piezo-Electric) devices, the preliminary standard generalizes the configuration of this mixed mode interface for a wide range of transducers. Some legacy transducers systems may require more than one line for operation, for instance, certain devices may require a constant voltage source, and a separate line for the transducer output signal. In this case, the analog power line is defined as the data line while in the digital mode. A similar reverse polarization scheme disables the analog circuitry, and activates the digital communication.

While in digital mode, the P1451.4 transducer can identify itself by transmitting the contents of its memory. This is the capability most commonly associated with P1451.4. However, part of this memory may contain information as to how the P1451.4 transducer may be configured. After receiving this information, a host to the P1451.4 bus (likely a 1451.2 STIM) can issue a

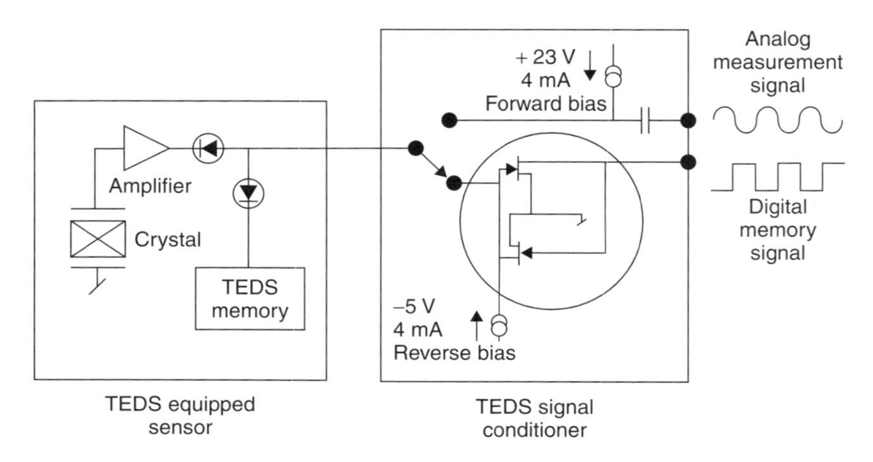

Figure 2.7 Circuit schematic for IEEE P1451.4.

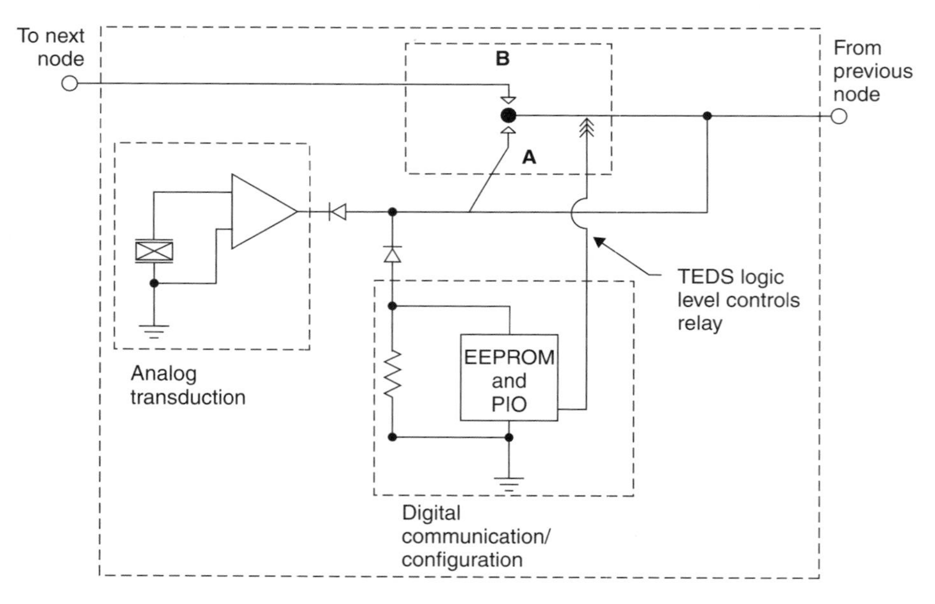

Figure 2.8 Self configuring P1451.4 transducer.

command to the transducer to configure itself into a number of different configurations. One immediate use of this capability is to implement a multidrop sensor bus. Figure 2.8 outlines a transducer with such capability.

The transducer in Figure 2.8 contains three distinct components, each of which is enclosed by a dashed boundary. The first, labeled 'analog transduction' represents an IEPE type sensor. The second, labeled 'Digital

Communication/Configuration' adds the mixed-mode capability promised by IEEE P1451.4. Together with the analog transduction component, it forms the transducer outlined in Figure 2.7. However, the digital hardware in this particular transducer has an extra pin, which is held at logical high or logical low upon command. This logic level, in turn, controls the position of switching hardware found in the third component of the transducer. The A position of this switching hardware directs the power/signal line to this particular node. The B position of the switching hardware directs the power/signal line to another transducer.

By arranging these self-configuring transducers appropriately as shown in Figure 2.9, we can construct a multidrop sensor bus of mixed mode transducers. The digital circuitry in Figure 2.8 is always connected to the bus. When the bus is pulled low, all nodes of the network are visible to the controller (this is likely to be a 1451.2 STIM). The protocol of this P1451.4 network allows each node shown in Figure 2.9 to have a unique identification. The network protocol also permits the master to poll the entire bus to identify each node uniquely. With this data, the master consecutively toggles each node to its B (or pass-through) position.

The master commands node 1 to toggle to its A position. The master releases the bus from negative bias. All digital circuitry is then disabled, and only the analog circuitry of node 1 is exposed to the constant current, positively polarized, line bias. The analog transduction section of node 1 ensues to bias and operate in its traditional manner, and high fidelity measurements possible with IEPE sensors can be taken by the master (STIM).

When the measurements phase for that particular node is complete, the master pulls the line low to disable the analog circuitry and wake up the digital circuitry of the entire bus. The master commands node 1 to toggle to its B position, then commands node 2 to its A position. Analog measurements can then be made from node 2. This process is repeated for each of the N nodes on the network.

The hardware interfaces and communication protocols, defined under IEEE 1451, will enable instrumentation manufacturers to design and produce solutions for machinery-condition-analysis systems at a significantly lower

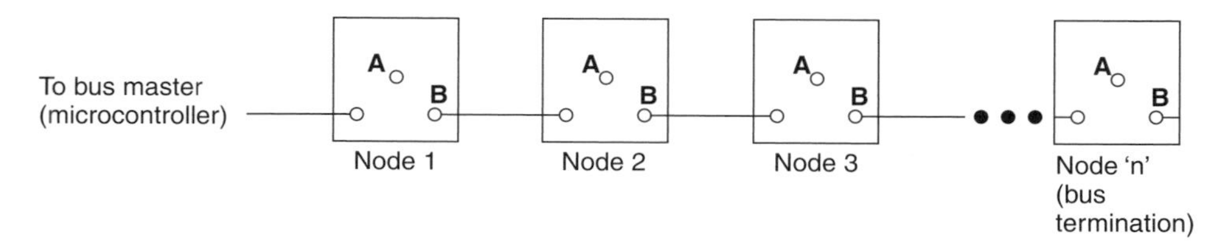

Figure 2.9 P1451.4 multidrop sensor bus.

cost than traditional methods. The proposed standard takes advantage of established networks so that sensors and transducers can be leveraged onto networks with familiar, inexpensive, off-the-shelf wiring and networking components. The IEEE standard's plug-and-play approach allows freedom of choice between transducers, field networks and interface modules. Standard Internet and intranet links allow access to distributed devices from any remote site, and enable customized and familiar IP (Internet Protocol) addressing.

2.4. SMART TRANSDUCER NETWORKING

IEEE 1451 defines hardware and software standardized methods for supporting smart sensor and network connectivity. The standard's specifications place no restrictions on the use of signal conditioning and processing schemes, analog-to-digital converters, microprocessors, network protocols, and network communication media. IEEE 1451 reduces industry's effort to develop and migrate towards networked smart transducers. This standard provides the means to achieve transducer-to-network interchange ability and transducer-to-network interoperability.

The IEEE 1451.2 project defines a Transducer Electronic Data Sheet (TEDS) and its data format, along with a 10-wire digital interface and communication protocol between transducers and a microprocessor. The framework of the IEEE 1451.2 interface is shown in Figure 2.10. The TEDS, stored in a non-volatile memory, contains fields that describe the type, attributes, operation,

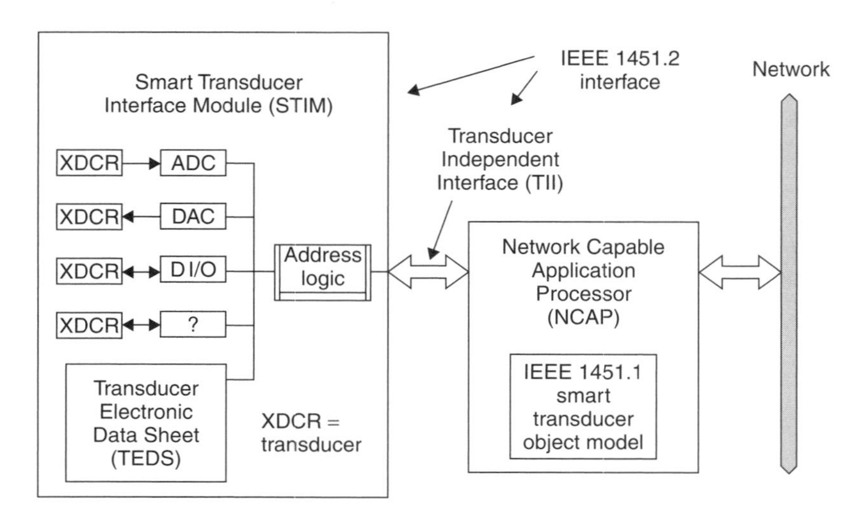

Figure 2.10 Framework of IEEE 1451.1 and 1451.2 interfaces.

and calibration of the transducer. With a requirement of only 178 bytes of memory for the mandatory data, the TEDS is scalable. A transducer integrated with a TEDS provides a feature that makes the self-description of transducers to the network possible. Since the transducer manufacturer data in the TEDS always goes with the transducer, and this information is electronically transferred to a NCAP or host, the human errors associated with manually entering sensor parameters are eliminated. The manufacturer data and the optional calibration data are stored in the TEDS, so losing transducer paper data is not a concern. With the TEDS feature, upgrading transducers with higher accuracy and enhanced capability, and replacing transducers for maintenance purposes, becomes simply 'plug-and-play'. The IEEE 1451.2 interface defines STIM. Up to 255 sensors or actuators of various digital and analog mixes can be connected to a STIM. The STIM is connected to a network node called NCAP through the 10-wire transducer independent interface using a modified Serial Peripheral Interface (SPI) protocol for data transfer.

The IEEE 1451.1 standard defines a common object model for a networked smart transducer and the software interface specifications for each class representing the model. Some of these classes form the blocks, components, and services of the conceptual transducer. The networked smart transducer object model encapsulates the details of the transducer hardware implementation within a simple programming model. This makes programming the sensor or actuator hardware interface less complex by using an input/output (I/O)-driver paradigm. The network services interfaces encapsulate the details of the different network protocol implementations behind a small set of communications methods. The model of the networked smart transducer is shown in Figure 2.11.

During the course of the development of the IEEE 1451.1 and 1451.2 standards, some sensor manufacturers and users recognized the need for a standard interface for distributed multidrop smart-sensor systems. In a distributed system a large array of sensors, in the order of hundreds, needs to be read in a synchronized manner. The bandwidth requirements of these sensors may be relatively high, of the order of several hundred kHz, with time correlation requirements in tens of nanoseconds. IEEE P1451.3 defines the standard specification. The physical representation of the proposed IEEE P1451.3 standard is shown in Figure 2.6. A single transmission line is proposed to supply power to the transducers and to provide the communications between the bus controller and the Transducer Bus Interface Modules (TBIM). A transducer bus is expected to have one bus controller and many TBIMs. A TBIM may contain one or more different transducers. The NCAP contains the controller for the bus and the interface to the network that may support many other buses.

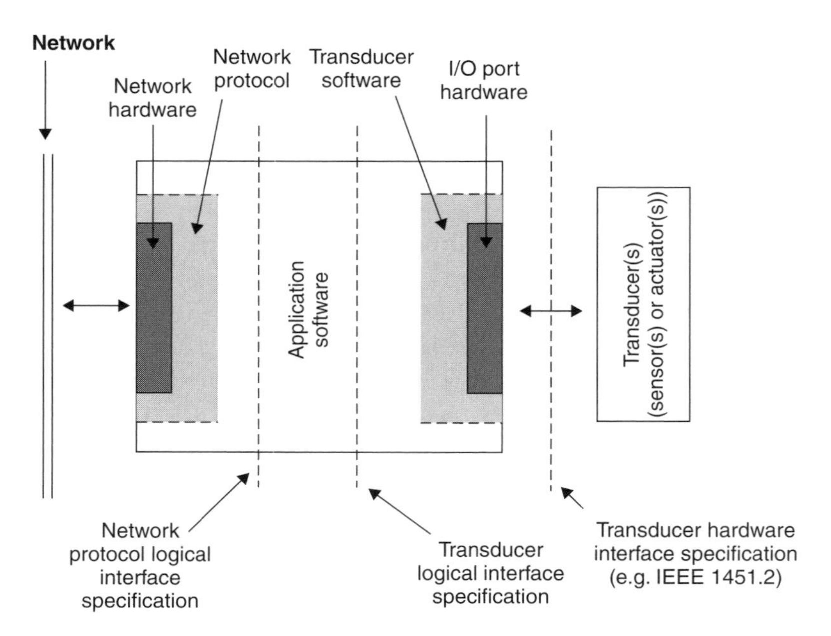

Figure 2.11 Networked smart transducer model.

In the condition-based monitoring and maintenance industry, analog transducers such as piezoelectric, piezoresistive, and accelerometer-based transducers are used with electronics instruments to measure the conditional state of machinery. Transducer measurements are sent to an instrument or computer for analysis. The idea of having small TEDS on analog transducers and the ability to connect transducers to a network is used in the IEEE P1451.4 standard. An IEEE 1451.4 transducer can be a sensor or actuator with, typically, one addressable device, and is referred to as a node-containing TEDS. The IEEE P1451.4 transducer may be used to sense multiple physical phenomena. Each phenomenon sensed or controlled is associated with a node. If more than one node is included in an IEEE 1451.4 transducer, one of the nodes must have a memory block that holds the node list. The node list contains the identifications of the other nodes.

In order to reduce cabling and interfacing costs, a model using different wiring configurations is chosen as a transducer connection interface. If a single wire model is used, the analog transducer signal transmission and communication of the digital TEDS data to an instrument or a network are done on the same wire, but at separate times. If a multiwire model is used, communication of digital data and analog signals can be accomplished simultaneously. The digital communication can be used to read the TEDS information and to configure an IEEE P1451.4 transducer.

The context of the mixed-mode transducer and its interface(s) are shown in Figure 2.12.

A distributed measurement and control system can be easily designed and built based on the IEEE 1451 standards. An application model of IEEE 1451 is shown in Figure 2.13. Three NCAP/STIMs are used for illustration purposes. In scenario one, with sensors and actuators connected to the STIM

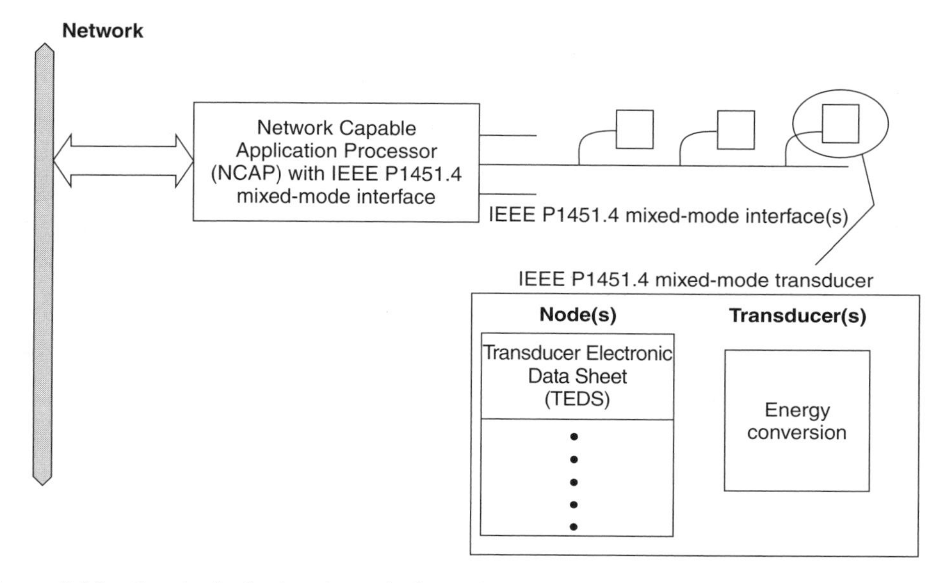

Figure 2.12 Context of mixed-mode transducer and interface.

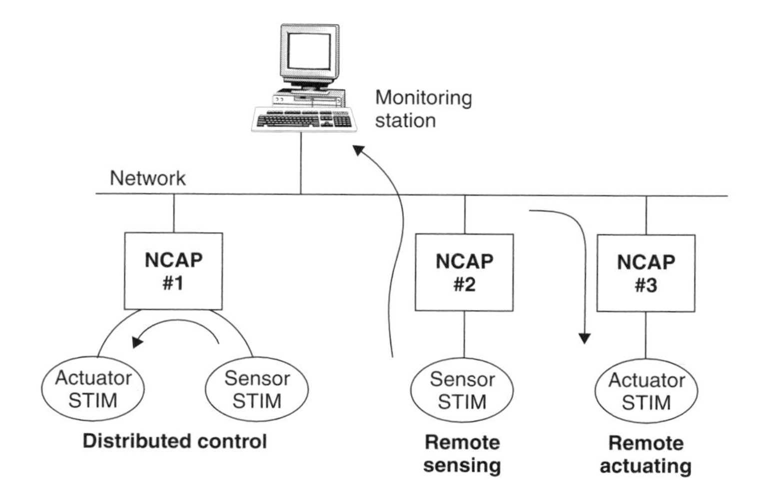

Figure 2.13 An application of IEEE 1451 based sensors on a network.

of NCAP No.1, the application software running in the NCAP can perform a localized control function, for example, maintain a constant temperature of a bath. The NCAP reports measured data, process information, and control status to a remote monitoring station or host. It frees the host from the processor-intensive, closed loop control operation. In the second scenario, NCAP No.2, connected with sensors only, can perform a remote process or condition monitoring function, for instance, to monitor the vibration level of a set of bearings in a turbine. In the third scenario, based on the broadcast data received from NCAP No.2, NCAP No.3 activates an alarm when the vibration level of the bearings exceeds a critical point set.

The Ethernet has been used for networking computers for information and data exchange. The TCP/IP (Transaction Control Protocol/Internet Protocol) enables data transfer between computers across the Internet. An industrial Ethernet NCAP, which is IEEE 1451.2 compatible, can be used to build web-based distributed measurement and control applications which enable the access of sensor information and measurements across the Internet.

2.5. CONTROLLER AREA NETWORK

IEEE 1451 smart transducer standard offers true plug-and-play facilities for connecting sensor and actuator devices to field bus and device-level networks. Although the first implementations of the standard have been developed to allow transducer devices to connect to Ethernet networks, thus creating an industrial Ethernet, the standard can also be applied to CAN (Controller Area Network)-based device level networks. The IEEE 1451 standard describes design and implementation for an IEEE 1451.2 STIM (Smart Transducer Interface Module), involving a software port, onto a standard microcontroller. The IEEE 1451.1 standard defined NCAP (Network Capable Application Processor) can be implemented in a CAN node thus realizing a form of gateway between transducer devices and the CAN network, based on the IEEE 1451 standard.

IEEE 1451 introduces a common interface standard to give a network-independent view of devices. Smart transducers can embed local intelligence to support features such as self-diagnostics, local control and analytical algorithms, and can perform self-declaration to the network based on an electronic data sheet. This self-declaration feature allows transducer devices to be connected to the network in a true plug-and-play way.

The first commercial IEEE 1451 implementations are targeted at the Ethernet networks. Ethernet has traditionally played a role as a LAN (Local Area Network), positioned high up in the CIM (Channel Interface Module) model.

However, as the cost of embedded Ethernet solutions decreases, the Ethernet is applied at the field bus level and even at the device network level, but Ethernet does not support various device profiles, in a formal sense, and IEEE 1451 offers a retrofit solution for Ethernet, defining a device level interface for smart sensors. This solution is referred to as industrial Ethernet.

The IEEE 1451 standard, however, is more than an Ethernet solution. The transducer developers need a network-independent standard for device connection. CAN based networks are good candidates for IEEE 1451 implementations and a number of companies are developing these solutions. The IEEE 1451 standard maps the transducer device to the target network based on an object model defined independently of the network. Each network has an NCAP (Network-Capable Application Processor) which maps to the target network profile.

Along with providing a common-software interface standard for transducer devices, a common-hardware interface is also necessary for network independence. The common hardware interface exists where an architectural difference occurs between the IEEE 1451 standard and the more traditional approach for field bus and device level network interfacing.

The IEEE 1451 standard comprises of four complete sub-standards. Each sub-standard may be used as a stand-alone or as a part of an overall IEEE 1451 family solution. The IEEE 1451.1 and 1451.2 standards have been balloted and accepted by the IEEE. IEEE P1451.3 and P1451.4 are under development, hence the prefix P, which denotes a proposed document. Figure 2.10 shows a block diagram for the IEEE 1451.1 and IEEE 1451.2 solutions.

IEEE 1451.1 defines a network-independent information model, enabling transducers to interface to network-capable application processors (NCAPs). It provides a definition for a transducer and its components using an object-oriented model. The model consists of a set of object classes with specified attributes, actions and behaviors used to provide a clear, comprehensive description of a transducer. The model also provides a hardware independent abstraction for the interface to the sensor and actuator. The model can be mapped onto example networks such as DeviceNet, Ethernet, LonWorks and SDS (Smart Distributed System). This mapping is achieved through a standard API (Application Programming Interface). This standard optionally supports all of the interface module communication approaches taken by the rest of the IEEE 1451 family (i.e. STIM, TBIM, Mixed-mode transducer).

IEEE 1451.2 defines the following:

- a TEDS (Transducer Electronic Data Sheet) and its data format;
- a standard digital interface and the communication protocols used between the transducer(s) and the microprocessor;

- an electrical interface, and
- read and write logic functions to access the TEDS and transducers.

IEEE 1451.2 requires that the TEDS are physically located with the transducers (as part of the STIM) at all times. The TEDS contains information describing the transducers that are embodied within the STIM. The amount of detail held within the TEDS will vary with each specific STIM implementation, but critical information will always be present.

IEEE P1451.3 defines a specification for a standard physical interface for connecting multiple physically separated transducers in a multidrop configuration. This is necessary because in some cases, for example, it is not possible physically to locate the TEDS with the transducers (for instance, due to harsh environments). The IEEE P1451.3 document proposes a bus implementation (known as the Transducer Bus Interface Module, TBIM) that is small and cheap enough to fit easily into a transducer. The network overhead developed is optimized to allow maximum data transfer throughput with a simple control logic interface.

IEEE P1451.4 defines a specification that allows analog transducers (e.g. piezoelectric transducers, strain guages, etc.) to communicate digital information (mixed mode) for the purposes of self-identification and configuration. This standard also proposes that the communication of the digital TEDS data is shared with the analog signal from the transducer with a minimum set of wires, fewer than the 10-wire requirement of the IEEE 1451.2 standard.

IEEE 1451.1 and IEEE 1451.2 together define the specification for networked smart transducers. They provide the framework for the sensor and actuator manufacturers to support multiple networks and protocols easily.

As a whole, the family of IEEE 1451 standard interfaces provides the following benefits:

- enable self-identification of transducers;
- facilitate self-configuration;
- maintain long term self-documentation;
- make for easy transducer upgrade and maintenance;
- increase data and system reliability;
- allow transducers to be calibrated remotely, or to be self-calibrated.

The following components are used in the description of IEEE 1451.2:

XDCR: an abbreviation for transducer, which is a sensor or an actuator.
STIM: Smart Transducer Interface Module (Figure 2.10).

A STIM can range in complexity from a simple single-channel sensor, or actuator, to a product supporting multiple channels of transducers. A transducer channel is denoted smart in this context because:

- it is described by a machine-readable TEDS;
- the control and data associated with the channel are digital;
- triggering, status and control are provided to support the proper functioning of the channel

NCAP: Network Capable Application Processor.
The NCAP mediates between the STIM and a digital network, and may provide local intelligence. The STIM communicates with the network transparently, via the TII that links it to the NCAP.
TII: Transducer Independent Interface.
The TII is a 10-wire serial I/O bus that defines:

- a triggering function that triggers reading and writing from/to a transducer;
- a bit transfer methodology;
- a byte-write data-transport protocol (NCAP to STIM);
- a byte-read data-transport protocol (STIM to NCAP);
- data transport frames.

TEDS: Transducer Electronic Data Sheet.
The TEDS is a data sheet written in electronic format that describes the STIM and the transducers associated with it, such as manufacturer's name, type of transducer, serial number, etc. The TEDS must remain with the STIM for the duration of the STIM's lifetime.

We discuss an example implementation of the IEEE 1451.2 standard to design a STIM. We break down the IEEE 1451.2 standard into its logical parts, and then reorganize these parts into a software model. The resulting software model is shown in Figure 2.14. The 1451.2 STIM contains the following: TEDS; control and status registers; transducer channels; interrupt masks; address and function decoding logic; data transport handling functions; trigger and trigger acknowledge functions for the digital interface to the TII; a TII driver, and a transducer interface. The logical software blocks are shown in Figure 2.15.

The TII contains the physical lines to support data transport, clocking, triggering and acknowledgment. Each STIM must have a TEDS, which consists of eight different subgroupings, which are:

- Meta TEDS [mandatory]

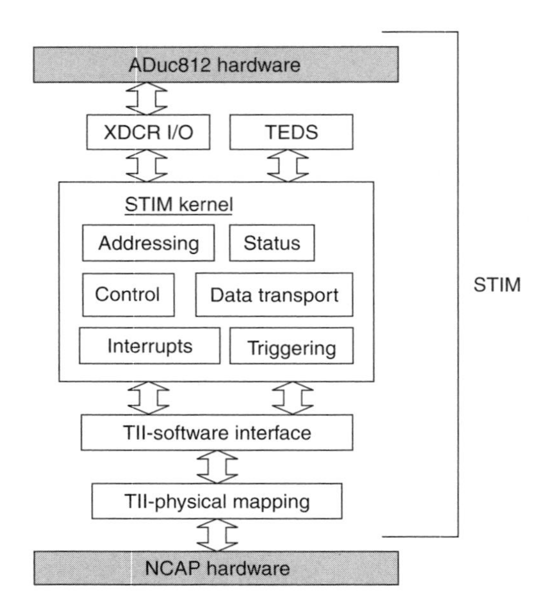

Figure 2.14 STIM software architecture.

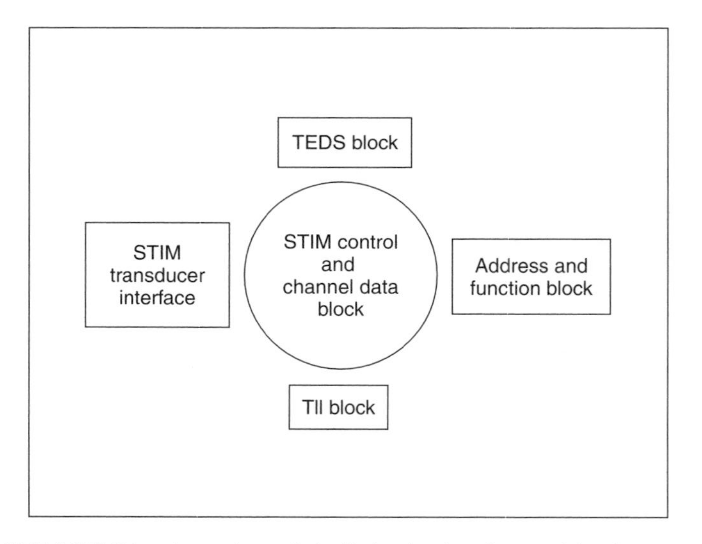

Figure 2.15 IEEE 1451.2 broken down into its logical software blocks.

- makes available, at the interface, all the information needed to gain access to any channel,
- contains information common to all channels,
- information is constant and read-only.

- Channel TEDS [mandatory, one for each channel]
  - makes available, at the interface, all the information concerning the channel being addressed to enable proper operation of that channel,
  - information is constant and read-only.
- Calibration TEDS [optional]
  - makes available, at the interface, all of the information used by the correction engine in connection with the channel being addressed,
  - information may be configured to be read and write capable, or it may be configured as read-only.
- Meta-Identification TEDS [optional]
  - makes available, at the interface, the information needed to identify the STIM,
  - contains any identification information common to all channels,
  - information is constant and read-only.
- Channel-Identification TEDS [optional]
  - makes available at the interface all of the information needed to identify the channel being addressed,
  - information is constant and read-only.
- Calibration-Identification TEDS [optional]
  - makes available at the interface the information describing the calibration of the STIM,
  - information may be configured to be read and write capable, or it may be configured as read-only (it must be the same as for the calibration TEDS).
- End-Users' Application Specific TEDS [optional]
  - contains end-users' writable application-specific data,
  - information is nonvolatile.
- Industry Extensions TEDS [optional]
  - the function of the extension TEDS, the appropriate functional and channel address range where it may reside, and the meaning and type of the data fields will be defined by the creator of the extension.

For the purposes of this implementation, the software was coded in four modules, as follows:

- STIM control, channel data and transducer interface module;
- TII module;
- TEDS module;
- Address and function module.

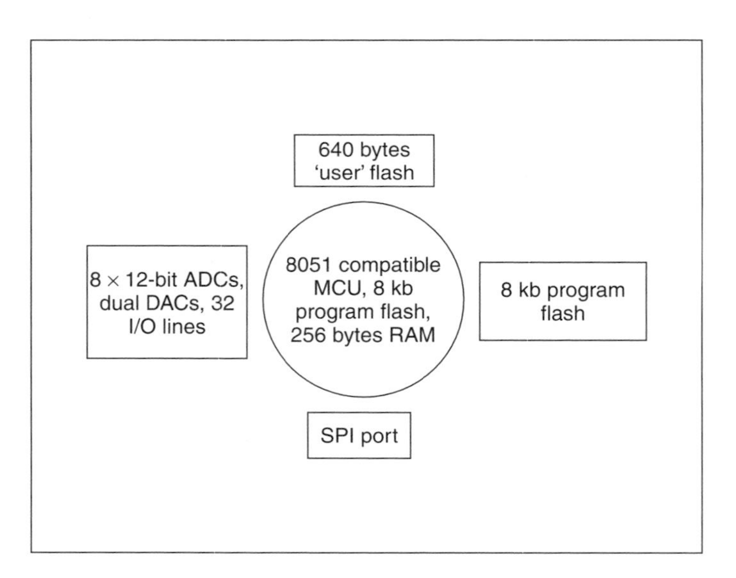

Figure 2.16 A block diagram of the principal features contained on the ADuC812.

This software was built and run on the Analog Devices ADuC812 Microconverter development board. A block diagram of the principal features contained on the ADuC812 is shown in Figure 2.16. The ADuC812 contains an 8051 compatible MCU, 8 kb of program flash/EE, 640 bytes of data flash/EE, 256 bytes of RAM, up to 32 programmable I/O lines, an SPI serial I/O port, dual DACs and an eight-channel true 12-bit ADC. The SPI port is an industry standard four-wire synchronous serial communications interface. It can be configured for master or slave operation, and is externally clocked when in slave mode. The data flash/EE is a memory array which consists of 640 bytes, configured into 160 4-byte pages. The interface to this memory space is via a group of registers that is mapped in the SFR space. The 8-kb program flash/EE will ultimately store and run the end-user's application code.

This board, together with two attached transducers (an AD590 temperature sensor and a digital I/O controlled fan) comprises the STIM. The NCAP used for this implementation was the HP BFOOT 66501. Figure 2.17 shows the mapping between the IEEE 1451.2 software and the ADuC812.

- the STIM is controlled from the program flash/EE, and each channel's transducer data, status and control registers is held in RAM for the duration of the STIM lifetime;
- the transducer interface is mapped onto the ADCs, DACs and I/O lines;
- the TII is a superset of the SPI port (plus some I/O lines);

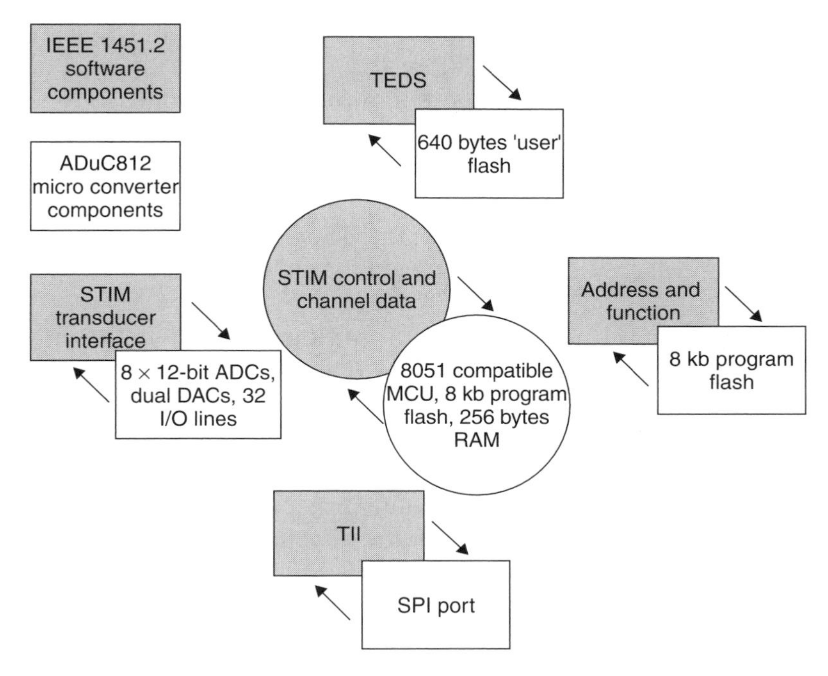

Figure 2.17 Mapping the STIM software onto the microconverter: software blocks overlay onto the ADuC812 feature blocks.

- the TEDS map into the 640 bytes of data flash/EE, and
- the Address and Function block is stored into the program flash/EE.

'STIM Control and Channel Data' and 'Transducer Interface' modules contain the definitions for the transducer channels (there are 256 transducer channels allowed per STIM). One of these channels must be CHANNEL_ZERO, the global channel that is used for globally addressing all of the implemented channels simultaneously. As channels are added, they must be numbered sequentially, starting from the main control flow of the program. Our particular implementation defines two channels: one sensor (an AD590 temperature sensor) and one actuator (a digital-output controlled fan). As with the TII module, this module contains certain definitions that are hardware-coupled (i.e. the physical lines that the sensor and actuator are realized on).

The TII module defines the physical interface to the NCAP. The TII was implemented as a superset of the existing microconverter ADuC812 SPI port. There are 10 lines in the TII, while there are typically only three or four in an SPI implementation. This module must integrate these lines with their hardware interface, and also provide an abstract layer for software interaction.

Viewing the TII functions from the user's perspective, this hardware coupling is transparent. The TII API calls allow for the transport of read and write protocol frames, and for detecting and manipulating the states of the TII lines.

The TEDS module defines the TEDS for the STIM. It defines where the TEDS are physically mapped (e.g. Flash RAM, ROM or EEPROM (Electronically Erasable Programmable read only)), how they are written and stored, how they are retrieved and what they contain. The TEDS supported include the mandatory TEDS (i.e. one meta- and two channel-TEDS) and also the optional meta-ID TEDS.

The 'Address and Function' module implements all of the main functionality that is defined by the IEEE 1451 standards. It takes care of the data transport, control, interrupt, status and trigger functions. Note that each one of the different function types is preceded by a three letter abbreviation representing the grouping to which it belongs, e.g. the function DAT_ReadMetaTEDS() belongs to the data transport function group. The data transport submodule detects activity on the physical transport lines (by calling TII API functions). This submodule controls the transmission and reception of the TEDS information, the status information and the transducer data.

The aim of the software implementation was to create a complete minimal IEEE 1451.2 realization. This work involved dividing the standard into its various parts, then identifying the compulsory and optional sections. The architecture was designed to meet all of the mandatory specifications, and to allow for expansion to support a full and complete specification. As a result, the code was structured, modular and scalable. As far as possible, the architecture abstracted the software into layers, so that all but the most hardware-dependent functions would also be portable.

The program code took just over 5.5 kB of program Flash/EE. The TEDS took up 268 bytes. For every channel that is added, another Channel-TEDS becomes mandatory. Therefore, the 640-byte data Flash/EE on board the ADuC812 will quickly fill up. The TEDS may also be mapped elsewhere should the need arise.

The end product was tested extensively. The NCAP used was HP BFOOT 66501 (which is also a thin web server) was used to test the system, so it was possible to read and display the TEDS over the WWW (World Wide Web). It was possible to track temperature trend remotely over time, using a Java applet to read the AD590 sensor data from the STIM, via the NCAP. It was also possible to control the temperature at the sensor by triggering the fan to actuate when the sensor value exceeded a pre-set threshold. To achieve this, all that needs to be known is the IP address of the NCAP.

For design engineers who are used to developing device profiles and interfaces for transducer devices on CAN based networks, the IEEE 1451

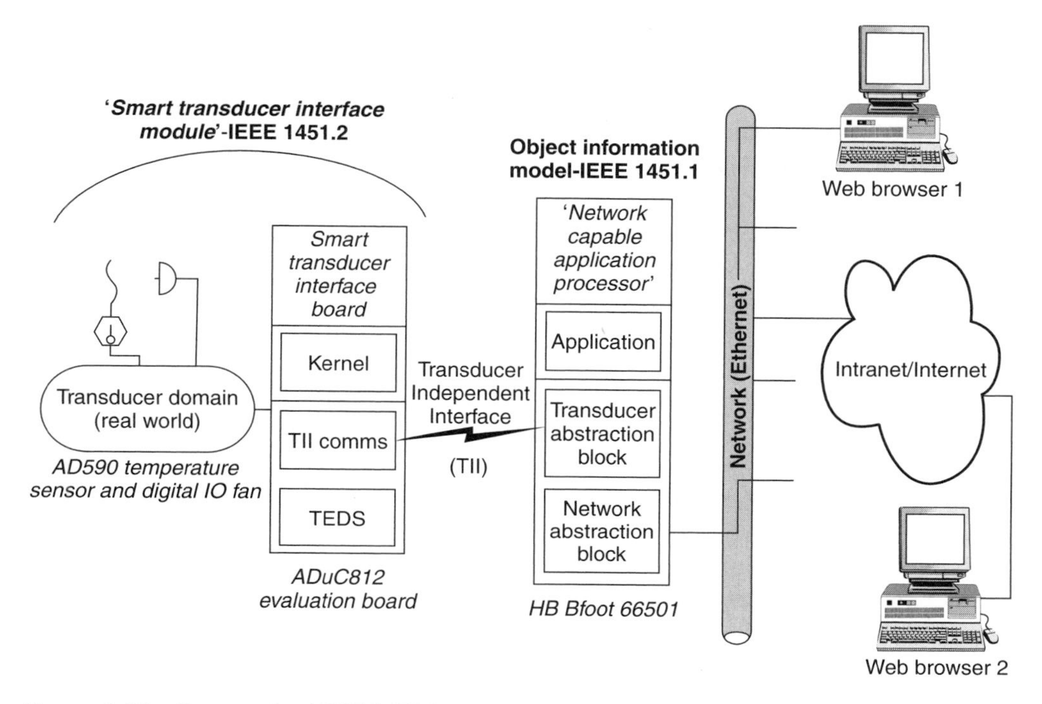

Figure 2.18 Conceptual IEEE 1451 based solution.

approach differs. Figure 2.18 shows a conventional solution where a typical transducer device, comprising some sensors in this example, is interfaced to a CAN-based network, say a DeviceNet solution. Here the sensors are interfaced directly to the network node and a single node processor is used to read and condition the sensor data. Figure 2.18 shows the same sensors interfaced to the DeviceNet network, or some other network, but this time the IEEE 1451 standard is employed. A separate processor is used to implement the NCAP at the node level, where the NCAP interfaces to the network on one side and interfaces to the TII at the other side. A separate processor resides in the STIM module.

It is seen that compared with conventional solutions for transducer interfacing, the IEEE 1451 solution includes an additional serial interface, the TII, and an additional processor for the STIM. These additional hardware features make the implementation appear cumbersome. However, the additional hardware in the IEEE 1451 context allows the transducer manufacturer to develop to a common interface standard, which is independent of any particular network. The transducer developer needs to develop only one product where it is anticipated that the various network developers will provide the

NCAPs, which are in effect gateways between the target network and the STIM. Thus the transducer device manufacturer will develop expertise in STIM and does not need to have knowledge of the NCAP or the networks other than the NCAPs.

Many of the transducer manufacturers are SME (Small to Medium Enterprise) sized companies and it is easier for such companies to develop a single product to suit all networks. Obviously if some particular network or networks are of volume interest to a transducer manufacturer, then the manufacturer may elect to do a dedicated implementation for such a network or networks. However, the STIM solution is more universal, although at an incremental cost.

IEEE 1451 is not just an industrial Ethernet standard but it is a generic standard which can be applied to many field bus or device-level networks.

The IEEE 1451 offers a useful reference for control and data models to describe sensors and actuators. The standard helps to unify the many different models presented in network standards.

The CAN can explore this standard and implement NCAP solutions. The CAN based networks are used to provide device level solutions, and the

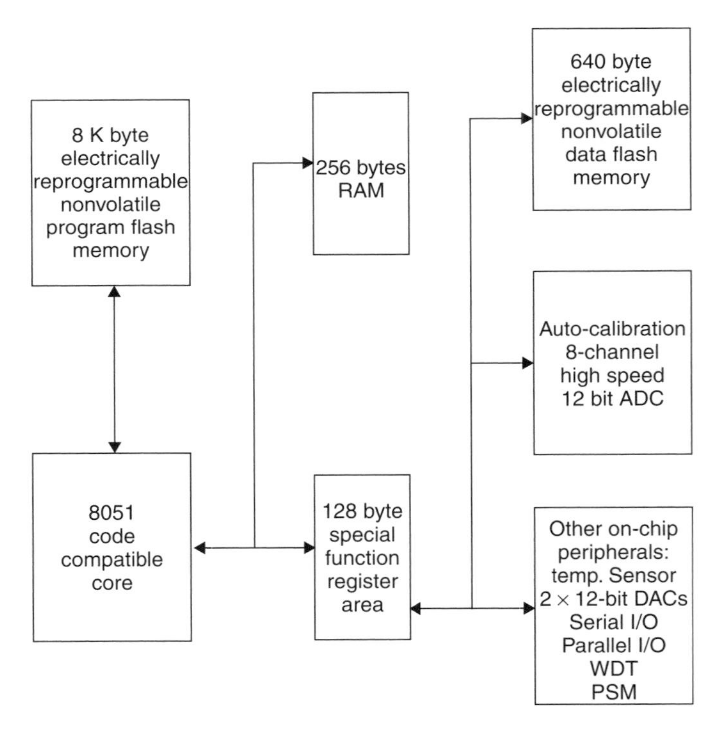

Figure 2.19 The ADuC812 programming model.

IEEE 1451 standard encourages more transducer manufacturers to develop products for networked environments.

Figure 2.19 shows the programming model for the ADuC812. The ADuC812 has separate address spaces for program and data memory. The additional 640 bytes of Flash/EE that is available to the user may be accessed indirectly via a group of control registers in the SFR (Special Function Register) area of the data memory space. The lower 128 bytes of RAM are directly addressable, while the upper 128 bytes may only be addressed indirectly. The SFR area is accessed by direct addressing only and provides an interface between the CPU and all on-chip peripherals.

Figure 2.20 shows how the 1451 implementation makes use of the ADuC812 programming model. The TEDS are located in the 640-byte data Flash/EE, the TII and actuator are directly tied into the peripherals block and the sensor is hanging off of the ADC block. All of these features are accessed and controlled

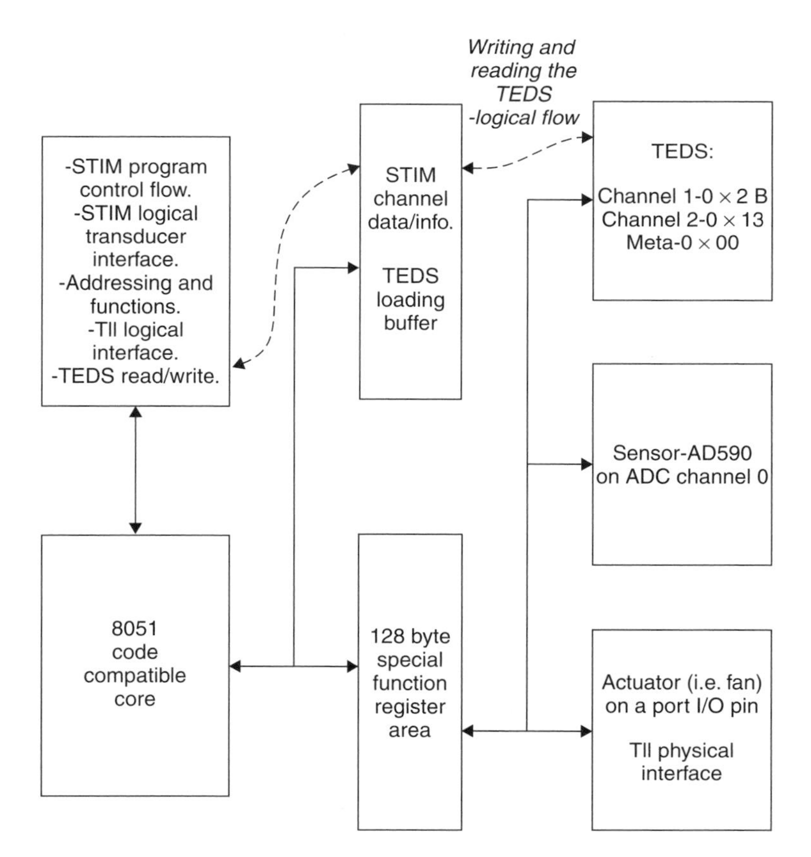

Figure 2.20 Implementation mapped into the ADuC812 programming model.

via the SFR area. The standard data RAM is used for storing the STIM channel transducer data and registers. It must reserve a buffer from which the TEDS may be individually loaded, and into which the TEDS may be read back from the data Flash/EE. Note also that the RAM must contain all local and system variables that are required by the code, and it must also allow for the run-time stack. All of these requirements place limitations on the size of the TEDS buffer, which the end-user should be aware of. All of the programming functions are stored into the program Flash/EE memory. In Figure 2.20, the dotted line shows the logical link between the read/write functions, and the actual writing and reading to/from the TEDS data Flash/EE area. These TEDS_ function calls are designed to be logically transparent to the end-user, and the method of implementation is not important.

2.6. SUMMARY

Smart sensors communicating with their outside world use digital communication. There are many alternative paths along which to develop the potential benefits of an agreed protocol. The proposed IEEE 1451 standard has four levels, three of which focus purely on digital interfaces, while the fourth level, known as P1451.4, defines an interface for mixed mode sensors with analog signals as well as digital information.

In a plug-and-play architecture, sensors and actuators are linked together through a series of common interfaces to modules designed to process the signals and to interface to existing communication networks. In the process control industry, sensors and transducers are connected directly to digital networks over a common interface, and are used in factory automation and closed loop control.

An intelligent tool-condition monitoring method is needed to detect tool wear or damage automatically instead of replacing the tool at regular intervals or discovering defects in material after operation. Direct sensing methods have been developed using multiple sensors to detect vibration, force, acoustic emission, temperature, and motor current. Tool wear is a very complex process and can be detected with sensor fusion, feature extraction, and pattern recognition.

Microprocessors can support smart sensors and devices. With this added capability, it is possible for a smart sensor to communicate measurements to an instrument or a system directly. Networking of transducers (sensors or actuators) in a system can provide flexibility, improve system performance, and make it easier to install, upgrade and maintain systems.

IEEE 1451 defines hardware and software standardized methods to support smart sensor and network connectivity. The standard's specifications place no restrictions on the use of signal conditioning and processing schemes, analog-to-digital converters, microprocessors, network protocols, and network communication media. IEEE 1451 reduces industry's effort to develop and migrate towards networked smart transducers. This standard provides the means to achieve transducer-to-network interchange ability and transducer-to-network interoperability.

The IEEE 1451.1 standard defines a common object model for a networked smart transducer and the software interface specifications for each class representing the model. Some of these classes form the blocks, components, and services of the conceptual transducer. The networked smart transducer object model encapsulates the details of the transducer hardware implementation within a simple programming model. This makes programming the sensor or actuator hardware interface less complex by using an input/output (I/O) driver paradigm. The network services interfaces encapsulate the details of the different network protocol implementations behind a small set of communications methods.

The IEEE 1451.1 defined NCAP (Network Capable Application Processor) can be implemented in a CAN node thus realizing a form of gateway between transducer devices and the CAN network, based on the IEEE 1451 standard. The IEEE 1451 introduces a common interface standard to give a network independent view of devices. Smart transducers can embed local intelligence to support features such as self-diagnostics, local control and analytical algorithms, and can perform self-declaration to the network based on an electronic data sheet. This self-declaration feature allows transducer devices to be connected to the network in a plug-and-play way.

The IEEE 1451 standard maps the transducer device to the target network based on an object model defined independently of the network. Each network has an NCAP (Network Capable Application Processor) which maps to the target network profile.

Along with providing a common software interface standard for transducer devices, a common hardware interface is also necessary for network independence. The common hardware interface exists where an architectural difference occurs between the IEEE 1451 standard and the more traditional approach for field bus and device level network interfacing.

IEEE 1451.1 defines a network-independent information model, enabling transducers to interface to network capable application processors (NCAPs). It provides a definition for a transducer and its components using an object-oriented model.

IEEE 1451.2 defines TEDS (Transducer Electronic Data Sheet) and its data format, a standard digital interface and the communication protocols used between the transducer(s) and the microprocessor, an electrical interface, and read and write logic functions to access the TEDS and transducers.

IEEE P1451.3 defines a specification for a standard physical interface for connecting multiple physically separated transducers in a multidrop configuration. The IEEE P1451.3 document proposes a bus implementation (known as the Transducer Bus Interface Module, TBIM) that is small and cheap enough to fit easily into a transducer.

IEEE P1451.4 defines a specification that allows analog transducers (e.g. piezoelectric transducers, strain gauges, etc.) to communicate digital information (mixed mode) for the purposes of self-identification and configuration.

IEEE 1451.1 and IEEE 1451.2 together define the specification for networked smart transducers. They provide the framework for the sensor and actuator manufacturers to support multiple networks and protocols easily.

PROBLEMS

Learning Objectives

After completing this chapter you should be able to:

- demonstrate understanding of smart sensors;
- explain the role of vibration sensors;
- discuss how smart sensors are applied to condition based maintenance;
- demonstrate understanding of smart transducer networking.

Practice Problems

Problem 2.1: How do smart sensors communicate?
Problem 2.2: What parameters should be included in TEDS data?
Problem 2.3: What does IEEE 1451.1 standard define?
Problem 2.4: What does IEEE 1451.2 standard define?
Problem 2.5: What does IEEE 1451.3 standard define?
Problem 2.6: What does IEEE 1451.4 standard define?
Problem 2.7: What are the benefits of using IEEE 1451 standard interfaces?

Practice Problem Solutions

Problem 2.1:

Smart sensors use digital communication. Smart sensors communicate with their outside world by using the data capture and analysis or a control system.

Problem 2.2:

TEDS data should include identification, e.g. model number; device, e.g. sensor type, sensitivity, and measurement units; calibration, e.g. date of last calibration and correction factors, and application, e.g. channel ID and measurement coordinates.

Problem 2.3:

The IEEE 1451.1 standard defines the Network Capable Application Processor (NCAP).

Problem 2.4:

The IEEE 1451.2 standard specifies Smart Transducer Interface Module (STIM).

Problem 2.5:

The IEEE P1451.3 standard defines Distributed Multidrop System (DMS), a digital interface for connecting multiple physically separated transducers, which allows for time synchronization of data. This transducer bus facilitates communications, data transfer, triggering, and synchronization.

Problem 2.6:

The IEEE P1451.4 standard defines mixed-mode communication protocol and interface to bridge the gap between legacy systems and IEEE 1451 architectures.

Problem 2.7:

The family of IEEE 1451 standard interfaces enables self-identification of transducers, facilitates self-configuration, maintains long-term self-documentation, makes for easy transducer upgrade and maintenance, increases data and system reliability, and allows transducers to be calibrated remotely or to be self-calibrated.

3

Power-Aware Wireless Sensor Networks

3.1. INTRODUCTION

Networks of distributed microsensors are emerging as a solution for a wide range of data gathering applications. Perhaps the most substantial challenge facing designers of small but long-lived microsensor nodes is the need for significant reductions in energy consumption. A power-aware design methodology emphasizes the graceful scalability of energy consumption with factors such as available resources, event frequency, and desired output quality, at all levels of the system hierarchy. The architecture for a power-aware microsensor node highlights the collaboration between software that is capable of energy-quality trade-offs, and hardware with scalable energy consumption.

The energy scalable design methodologies are geared specifically toward microsensor applications. At the hardware level, the unusual energy consumption characteristics are effected by the low duty cycle operation of a sensor node. This design adapts to varying active workload conditions with dynamic voltage scaling. At the software level, energy-agile algorithms for sensor networks, such as adaptive beam forming, provide energy quality trade-offs that are accessible to the user. Power-aware system design encompasses the entire system hierarchy, coupling software that understands the energy-quality trade-off with hardware that scales its own energy consumption accordingly.

Wireless Sensor Network Designs A. Hać

ISBN: 0-470-86736-1

The node's processor is capable of scaling energy consumption gracefully with computational workload. This scalability allows for energy-agile algorithms of scalable computational complexity. Scalability at the algorithm level is highly desirable because a large range of both energy and quality can be achieved. As the energy–quality characteristics of DSP (Digital Signal Processing) algorithms may not be optimal due to data dependencies, it is important to use algorithmic transforms to achieve desirable energy–quality (E–Q) characteristics and accurately model the energy–quality relationship through benchmarking.

Distributed microsensor networks allow a versatile and robust platform for remote environment monitoring. Crucial to long system lifetimes for these microsensors are algorithms and protocols that provide the option of trading quality for energy savings. Dynamic voltage scaling on the sensor node's processor enables energy savings from these scalable algorithms. Dynamic voltage scaling uses a sensor node built of a commercial processor, a digitally adjustable DC–DC regulator, and a power-aware operating system.

A system-level power management technique is used in massively distributed wireless microsensor networks. A power-aware sensor node model enables the embedded operating system to make transitions to different sleep states based on observed event statistics. The adaptive shutdown policy is based on a stochastic analysis and renders desired energy–quality scalability at the cost of latency and missed events. Algorithmic transformations improve the energy–quality scalability of the data gathering network.

Power-aware methodology uses an embedded microoperating system to reduce node energy consumption by exploiting both sleep state and active power management.

The energy dissipated by communication is a key concern in the development of networks of hundreds to thousands of distributed wireless microsensors. To evaluate the dissipation of communication energy in this unique application domain, energy models based on actual microsensor hardware are used for high-density, energy-conscious wireless networks. Assessing and leveraging the energy implications of microsensor hardware and applications is crucial to achieving energy-efficient microsensor network communication.

Power awareness becomes increasingly important in VLSI systems to scale power consumption in response to changing operating conditions. These changes might be brought about by the time-varying nature of inputs, desired output quality, or environmental conditions. Regardless of whether they were engineered for being power aware, systems display variations in power consumption as conditions change.

Low power system design, assuming a worst-case power dissipation scenario, is being supplanted by a more comprehensive philosophy variously termed power-aware, energy-aware or energy–quality-scalable design. The basic idea behind these essentially identical approaches is to allow the system power to scale with changing conditions and quality requirements.

3.2. DISTRIBUTED POWER-AWARE MICROSENSOR NETWORKS

The design of micropower wireless sensor systems has gained increasing importance for a variety of civil and military applications. Advances in microelectromechanical systems (MEMS) technology and its associated interfaces, signal processing, and RF circuitry have enabled the development of wireless sensor nodes. The focus has shifted from limited macrosensors communicating with base stations to creating wireless networks of communicating microsensors, as illustrated in Figure 3.1. Such sensor networks aggregate complex data to provide rich, multidimensional pictures of the environment. While individual microsensor nodes are not as accurate as their expensive macrosensor counterparts, their size and cost enables the networking of hundreds or thousands of nodes in order to achieve high quality, easily deployed, fault-tolerant sensing networks.

A key challenge in the design of a microsensor node is low energy dissipation. A power-aware system design employs a system whose energy consumption adapts to constraints and variations in the environment, onboard resources, or user requests. Power-aware design methodologies offer scalable energy savings that are ideal for the high variabilities of the microsensor environment.

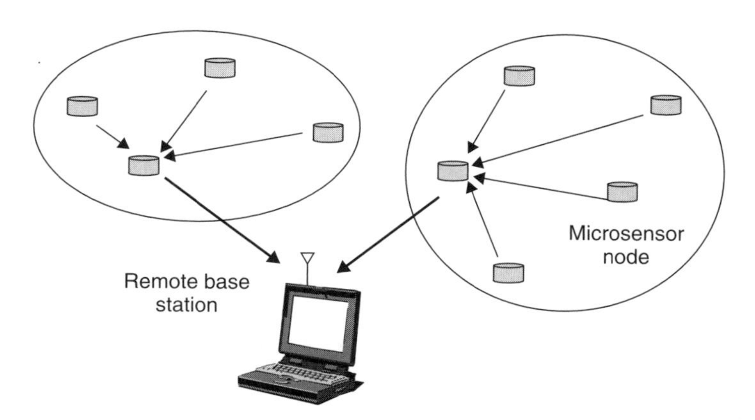

Figure 3.1 Microsensor networks for remote sensing.

Low-power system design allows the system's energy consumption to scale with changing conditions and quality requirements.

There are two main views motivating power-aware design and its emergence as an important paradigm. The first view is to explain the importance of power-awareness as a consequence of the increasing emphasis on making systems more scalable. In this context, making a system scalable refers to enabling the user to trade off system performance parameters as opposed to hard wiring them. Scalability is an important factor, since it allows the end-user to implement operational policy, which often varies significantly over the lifetime of the system. At times, the user of a microsensor network might want extremely high performance (e.g. data with a high signal-to-noise ratio) at the cost of reduced battery lifetime. However, at other times, the opposite might be true, the user may be willing to trade-off quality in return for maximizing battery lifetime. Such trade-offs can only be optimally realized if the system is designed in a power-aware manner. A related motivation for power awareness is that a well-designed system should gracefully degrade its quality and performance as available energy resources are depleted, instead of exhibiting an all-or-none behavior of high-SNR (Signal-to-Noise Ratio) data followed by a network failure.

While the view above argues for power awareness from a user-centric and user-visible perspective, this paradigm can also be motivated in more fundamental, system-oriented terms. With burgeoning system complexity and the accompanying increase in integration, there is more diversity in operating scenarios than ever before. Hence, design philosophies that assume the system to be in the worst case operating state most of the time are prone to yield globally suboptimal results. This naturally leads to the concept of power awareness. For instance, the embedded processor in a sensor node can display tremendous workload diversity depending on activity in the environment. Nodes themselves can also play a variety of roles in the network; a sensor networking protocol may call for the node to act as a data gatherer, aggregator, relay, or any combination of these. Hence, even if the user does not explicitly change quality criteria, the processor can nevertheless exploit operational diversity by scaling its energy consumption as the workload changes.

An example of a sensor node that illustrates power-aware design methodologies is shown in Figure 3.2. This system, the first prototype of μAMPS (micro-Adaptive Multi-domain Power-aware Sensors), is designed with commercial off-the-shelf components for rapid prototyping and modularity.

Power for the sensor node is supplied by a single 3.6-V DC (Direct Current) source. The 5-V supply powers the analog sensor circuitry and A/D (Analog-to-Digital) converter. The 3.3-V supply powers all digital components on the sensor node with the exception of the processor core. The core is powered

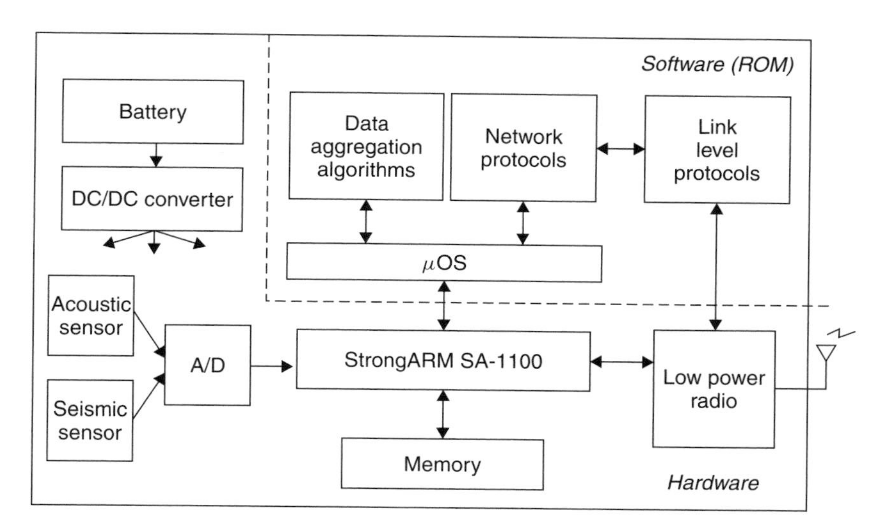

Figure 3.2 μAMPS sensor node hardware and software framework.

by a digitally adjustable switching regulator that can provide 0.9 V to 1.6 V in 20 discrete increments. The digitally adjustable voltage allows the (StrongARM) SA-1100 to control its own core voltage, enabling dynamic voltage scaling techniques.

The node includes seismic and acoustic sensors. The seismic sensor is a MEMS accelerometer capable of resolving 2 mg. The acoustic sensor is an electret microphone with low-noise bias and amplification. The analog signals from these sensors are conditioned with 8th-order analog filters and are sampled by a 12-bit A/D. The high-order filters eliminate the need for over sampling and additional digital filtering in the SA-1100. All components are carefully chosen for low power dissipation; a sensor, filter, and A/D typically requires only 5 mA at 5 volts.

A StrongARM SA-1100 microprocessor is selected for its low power consumption, sufficient performance for signal processing algorithms, and static CMOS (Complementary Metal-Oxide Semiconductor) design. The memory map mimics the SA-1100 Brutus (Config_SA1100_BRUTUS) evaluation platform and thus supports up to 16 Mb of RAM and 512 kb of ROM. The lightweight, multithreaded μOS (microOperating System) running on the SA-1100 is an adaptation of the eCOS (embedded Cygnus operating system) microkernel that has been customized to support the power-aware methodologies. The μOS, data aggregation algorithms, and networking firmware are embedded into ROM.

The radio module interfaces directly to the SA-1100. The radio is based on a commercial single chip transceiver optimized for (Industry Scientific and

Medical) ISM 2.45 GHz wireless systems. The PLL (Phase Lock Loop), transmitter chain, and receiver chain are capable of being shut off under software or hardware control for energy savings. To transmit data, an external voltage controlled oscillator (VCO) is directly modulated, providing simplicity at the circuit level and reduced power consumption at the expense of limits on the amount of data that can be transmitted continuously. The radio module is capable of transmitting up to 1 Mbps at a range of up to 15 meters.

The energy consumption characteristics of the components in a microsensor node provide a context for the power-aware software to make energy quality trade-offs.

Energy consumption in a static (Complementary Metal-Oxide Semiconductor) CMOS-based processor can be classified into switching and leakage components.

While switching energy is usually the more dominant of the two components, the low duty cycle operation of a sensor node can induce precisely the opposite behavior. For sufficiently low duty cycles or high supply voltages, leakage energy can exceed switching energy. For example, when the duty cycle of the StrongARM SA-1100 is 10 %, the leakage energy is more than 50 % of the total energy consumed. Techniques such as dynamic voltage scaling and the progressive shutdown of idle components in the sensor node mitigate the energy consumption penalties of low duty cycle processor operation.

Low duty cycle characteristics are also observable in radio. To power up a radio and transmit a packet of varying length, ideally, the energy consumed per bit would be independent of packet length. At lower data rates, however, the start-up overhead of the radio's electronics begins to dominate the radio's energy consumption. Due to its slow feedback loop, a typical PLL-based frequency synthesizer has a settling time of the order of milliseconds, which may be much higher than the transmission time for short packets. Particular effort is required to reduce transient response time in low-power frequency synthesizers for low data rate sensor systems.

Dynamic Voltage Scaling (DVS) exploits variabilities in processor workload and latency constraints, and realizes this energy quality trade-off at the circuit level. The switching energy of any particular computation is independent of time. Reducing supplied voltage offers savings in switching energy at the expense of additional propagation delay through static logic. Hence, if the workload on the processor is light, or the latency tolerable but the computation is high, the supplied voltage can be reduced with the processor clock frequency to trade-off latency for energy savings. Both switching and leakage energy are reduced by DVS.

Figure 3.3 illustrates the regulation scheme on a sensor node for DVS support. The μOS running on the SA-1100 selects one of the above 11

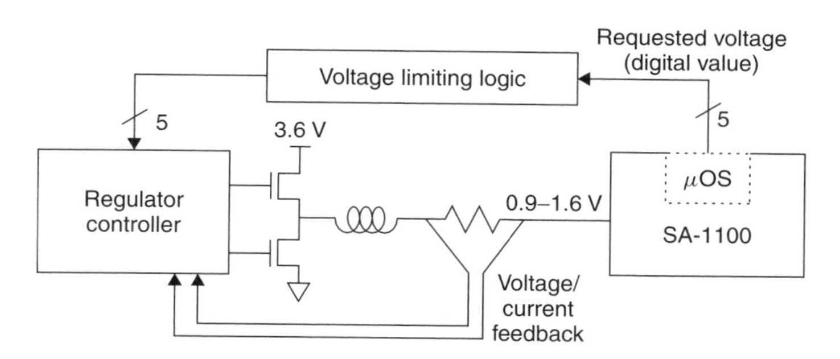

Figure 3.3 Feedback for dynamic voltage scaling.

frequency voltage pairs in response to the current and predicted workload. A 5-bit value corresponding to the desired voltage is sent to the regulator controller, and logic external to the SA-1100 protects the core from a voltage that exceeds its maximum rating. The regulator controller typically drives the new voltage on the buck regulator in under 100 μs (microseconds). At the same time, the new clock frequency is programmed into the SA-1100, causing the on-board PLL to lock to the new frequency. Relocking the PLL requires 150 μs, and computation stops during this period.

The implementation of the above system demonstrates energy–quality trade-offs with DVS. For a fixed computational workload, the latency (the inverse of quality) of the computation increases as the energy decreases. The quality of a FIR (Finite Impulse Response) filtering algorithm is varied by scaling the number of filter taps. The filter quality is sacrificed, and the processor can run at a lower clock speed and thus operate at a lower voltage. The DVS-based implementation of energy quality trade-offs consumes up to 60 % less energy than a fixed-voltage processor.

Algorithmic transformations, such as the most significant first transform, can improve the E–Q characteristics of a particular algorithm by reducing data dependencies. Figure 3.4 shows a testbed of sensors for beam forming, a class of algorithm often used in sensor arrays to make inferences about the environment. In this testbed, an array of six sensors is spaced roughly linearly at intervals of approximately 10 meters, a source moves parallel to the sensor cluster at a distance of 10 meters, and interference exists at a distance of 50 meters. Beam forming on the sensor data with varying numbers of sensors is performed. The energy dissipated on the StrongARM SA-1100 in relation to the number of sensors is measured. The matched filter output (quality) is calculated, and a reliable model of the E–Q relationship is delivered by varying the number of sensors in beam forming.

The E–Q characteristics are compared for two scenarios, the first being traditional beam forming, and the second using a most significant first

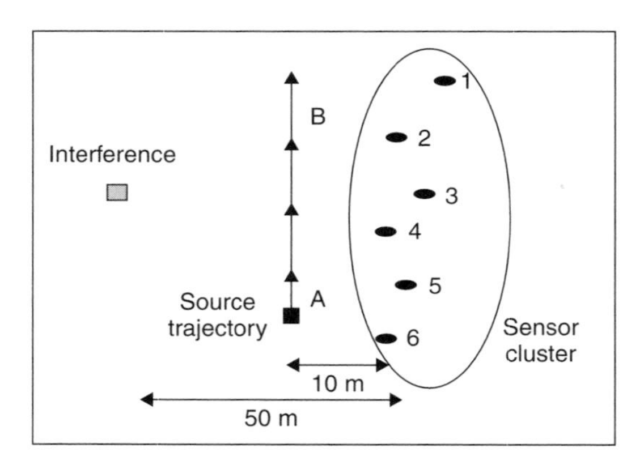

Figure 3.4 A sensor testbed.

transform. In the first scenario, beam forming is simply done in a preset order <1,2,3,4,5,6>. As the source moves from location A to B, the E–Q dependencies change dramatically. When the source is in location A, the beam forming quality is close to maximum because the source is closest to the sensors. However, with the source at B, quality is close to maximum after beam forming only two sensors, thus showing the dependency of the E–Q on the relative source location.

Intelligent data processing can circumvent this dependency. Intuitively, the data should be beam formed from sensors that have higher signal energy to interference energy, or process the most significant first. Figure 3.5 shows a block diagram for applying a most significant first transform to beam forming. To find the desired beam-forming order, each sensor's data energy is estimated. The energies are then sorted using quicksort. The quicksort output determines the desired beam-forming order. By finding the desired beam forming order, the similar E–Q dependencies are achieved even as the source moves with respect to the sensors. The energy cost required to gain this additional scalability is low compared to the energy cost of LMS (Laser Mirror Scanner) beam forming itself: on the SA-1100, the additional computational cost was 8.8 mJ, which is only 0.44 % of the total energy for LMS (Laser Mirror Scanner) beam forming (for the two-sensor case). The incremental refinement characteristics of a sensor's beam-forming algorithm are improved, leading to more uniform and predictably scalable E–Q in the presence of data dependencies.

The energy scalable framework enables the development and implementation of energy-agile applications. It is important that all processing in the sensor node is energy scalable, including link-level protocols, sensor-network protocols, data-aggregation algorithms, and sensor signal processing.

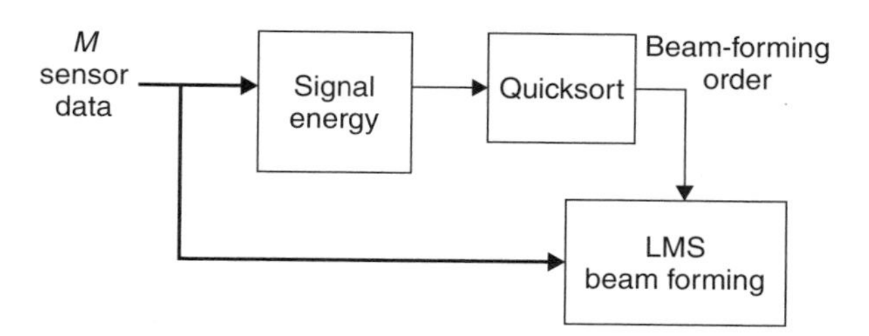

Figure 3.5 The most significant first transform used to improve E–Q characteristics of LMS beam forming.

The μAMPS sensor node prototype demonstrates the effectiveness of power-aware system design methodologies. Inefficiencies of low-duty cycle operation are countered with a focus on leakage current and start-up time reduction, and variations in processor workload are exploited by dynamic voltage scaling. Variations in incoming data rate and volume are exploited by energy-agile algorithms whose computational complexity scales with the arrival statistics of the data, allowing switching energy savings in the hardware. Close collaboration between the hardware and software of a microsensor node results in dramatic energy savings.

3.3. DYNAMIC VOLTAGE SCALING TECHNIQUES

Distributed microsensor networks are emerging as a compelling new hardware platform for remote environment monitoring. Researchers are considering a range of applications including remote climate monitoring, battle-field surveillance, and intramachine monitoring. A distributed microsensor network consists of many small, expendable, battery-powered wireless nodes. Once the nodes are deployed throughout an area, they collect data from the environment and automatically establish dedicated networks to transmit their data to a base station. The nodes collaborate to gather data and extend the operating lifetime of the entire system. Compared with larger macrosensor based systems, microsensor networks offer a longevity, robustness, and ease of deployment that is ideal for environments where maintenance or battery replacement may be inconvenient or impossible.

The μAMPS (micro-Adaptive Multi-Domain Power-Aware Sensors) use the enabling technologies for distributed microsensor networks. Microsensor networks are composed of hundreds to thousands of small, inexpensive, and homogeneous nodes. Once deployed, the nodes periodically organize themselves into clusters, based on the selection and location of cluster head

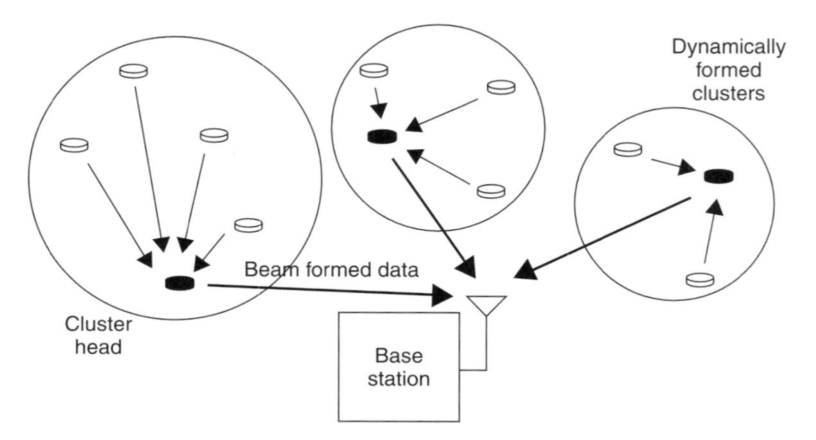

Figure 3.6 Operation of the μAMPS distributed microsensor network.

nodes. Cluster heads receive data from the nodes in their cluster, aggregate the data locally through beam forming, and transmit the result to a base station. Figure 3.6 illustrates this organization and data flow. The use of a two-hop routing protocol and data fusion at the cluster head substantially reduces total transmission energy in the network, thus prolonging system lifetime.

The μAMPS sensor node is designed as a highly integrated solution. To demonstrate enabling technologies, the prototypes are integrated with commercial, off-the-shelf components, for example, the StrongARM SA-1100 microprocessor.

Attaining months or years of useful life from a distributed microsensor network requires system design for power awareness. An essential component of power aware systems is the ability to make intelligent trade-offs between energy and quality, and energy scalability in microsensor networks.

In algorithms where additional computation incrementally refines a result, the energy of computation can be traded off for quality. There are many examples in microsensor networks. For instance, transmitted data can be encrypted with a key of varying length, allowing trade-offs between computational energy and the security of the transmission. A beam-forming algorithm can fuse data from a varying number of sensors, with the mean square error of the result decreasing as data from more sensors are combined. The number of taps in a FIR filter can be varied; longer impulse responses yield a more powerful filter.

We can assume that energy can scale gracefully with these variations. A variable voltage microprocessor can reduce the energy consumed during low workload periods through dynamic voltage scaling.

Variations in the quality of an algorithm appear to a processor as variations in processor utilization, which affect the number of clock cycles and the total switched capacitance for the computation. When a processor is idle due to a light workload, clock cycles and energy are wasted. Gating the clock during idle cycles reduces the switched capacitance of idle cycles. Reducing the clock frequency during periods of low workload eliminates most idle cycles altogether. Neither approach, however, affects the total switched capacitance for the computation and the supplied voltage for the actual computation, or substantially reduces the energy lost to leakage. Reducing the supply voltage in conjunction with the clock frequency achieves energy savings for the actual computation. Scaling the frequency, and supply voltage together results in a nearly quadratic savings in energy and reduces leakage current.

Dynamic voltage scaling (DVS) is the active adjustment of the supply voltage and the clock frequency in response to fluctuations in a processor's utilization. A voltage scheduler, running in tandem with an operating system's task scheduler, can adjust voltage and frequency in response to *a priori* knowledge or predictions of the system's workload. DVS has been successfully applied to custom chip sets.

Dynamic voltage scaling capabilities can be demonstrated on the SA-1100, the processor chosen for the μAMPS wireless microsensor prototype. A DC–DC converter circuit with a digitally adjustable voltage delivers power to the SA-1100 core and is controlled by a multithreaded, power-aware operating system.

The SA-1100 operates at a nominal core supply voltage of 1.5 volts and is capable of on-the-fly clock frequency changes in the range of 59 MHz to 206 MHz. Each frequency change incurs a latency of up to 150 s while the SA-1100's on-board PLL (Phase Lock Loop) locks to the new frequency. The SA-1100 is a completely static component and thus facilitates energy savings with DVS as discussed above.

A PCB (Printed Circuit Board) containing a custom DC–DC converter circuit provides a dynamically adjustable voltage to the SA-1100's core. Figures 3.7 and 3.8 illustrate the operation of this circuit. A buck regulator composed of discrete components is driven by a commercial step-down switching regulator controller. This controller is programmed with a 5-bit digital value to regulate one of 32 voltages between 0.9 and 2.0 volts. The operating system running on the SA-1100 commands the core voltage as a 5-bit digital value that is passed to the regulator controller. External programmable logic between the SA-1100 and the regulator controller prevents the regulator from delivering a voltage beyond the SA-1100 core's rated maximum.

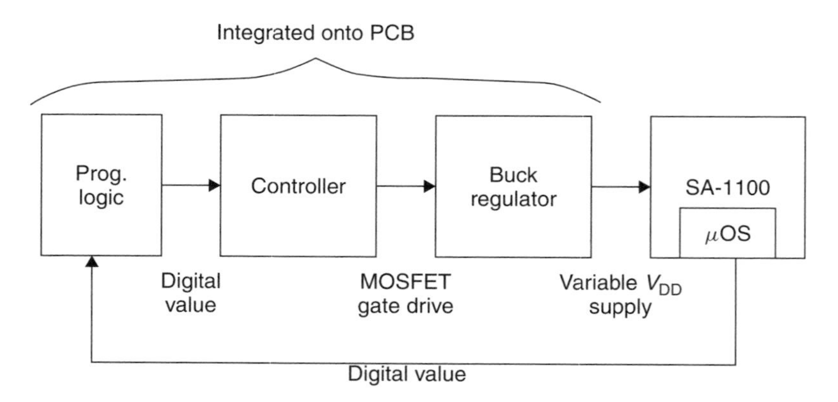

Figure 3.7 Overview of the adjustable DC-DC converter.

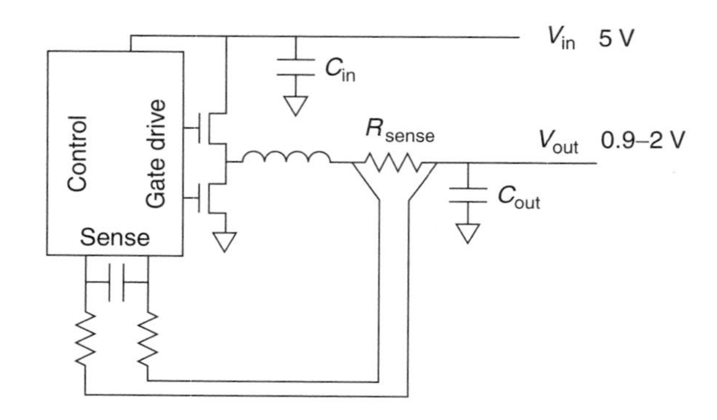

Figure 3.8 Simplified schematic for the buck regulator.

The power-aware operating system is based on the embedded Cygnus operating system (eCOS) kernel. This μOS supports preemptive multitasking, allowing threads such as data fusion, data packetization, network protocol handling, and voltage scheduling to operate simultaneously within a microsensor node. The μOS resides within a bootable instruction ROM.

In the implementation, the μOS monitors load on the processor and adjusts the clock frequency and supply voltage together to meet throughput requirements imposed by the tasks. Though the majority of throughput requirements and real-time deadlines on a microsensor network are known *a priori*, more sophisticated load prediction algorithms may be needed for more optimal voltage scheduling.

DVS can be used on the system in two scenarios, each with and without voltage scaling. The first scenario trades energy for computational latency on a fully loaded processor, and the second trades energy for output quality on a constant throughput FIR filter. All energy measurements are based on measured current into the regulator circuit and therefore include losses in the variable voltage regulator.

The SA-1100 is running the μOS with multiple threads and no idle time. Reducing the clock frequency without altering the voltage does not decrease the energy per operation. Supply voltage scaling, on the other hand, can reduce the energy cost of an operation dramatically. An ideal system with DVS would operate at the lowest voltage possible for each supported frequency.

The energy per operation is greater for lower frequencies since the energy lost to leakage is distributed over fewer computations. Voltage scaling exhibits up to 60 % in energy savings over fixed-voltage approaches. A regulator controller optimized for low loads would provide immediate improvements; all of the commercial options were designed for a higher voltage and much higher average currents than those typically used by the SA-1100.

To evaluate the energy consumption of energy-scalable algorithms on this system, a FIR filter is run under the μOS, adjusting the number of filter taps to alter the quality of the filter.

The FIR filter is run at a constant throughput, and the impulse response length is varied. The μOS dynamically adjusts the core voltage and frequency to meet the throughput requirement with the lowest possible energy.

3.4. OPERATING SYSTEM FOR ENERGY SCALABLE WIRELESS SENSOR NETWORKS

Massively distributed, dedicated, wireless microsensor networks have gained importance in a wide spectrum of civil and military applications. Advances in MEMS technology, combined with low power, low cost DSP (Digital Signal Processing) and RF (radio frequency) circuits have resulted in cheap and wireless microsensor networks becoming feasible. A distributed, self-configuring network of adaptive sensors has significant benefits. These networks can be used for remote monitoring of inhospitable and toxic environments. A large class of benign environments also require the deployment of a large number of sensors, such as intelligent patient monitoring, object tracking, assembly line sensing, etc. Their massively distributed nature provides wider resolution, as well as increased fault tolerance, than would a single sensor node.

A wireless microsensor node is typically battery operated and is thus energy constrained. To maximize the lifetime of the sensor node after its deployment, all aspects, including circuits, architecture, algorithms and protocols, have to be made energy efficient. Once the system has been designed, additional energy savings can be obtained by using dynamic power management concepts whereby the sensor node is shut down if no interesting events occur. Such event-driven power consumption is critical in obtaining maximum battery life. In addition, it is highly desirable that the node has a graceful energy–quality (E–Q) scalability such that, if the application so demands, the user is able to extend the mission lifetime at the cost of sensing accuracy. Energy scalable algorithms and protocols are required for such energy constrained situations.

Sensing applications present a wide range of requirements in terms of data rates, computation, average transmission distance, etc., as protocols and algorithms have to be tuned to each application. Therefore, embedded operating systems and software are critical in such microsensor networks as programmability is a necessary requirement. Operating system directed power-management technique can improve the energy efficiency of the sensor nodes. Dynamic Power Management (DPM) is an effective tool for reducing system power consumption without significantly degrading performance. The basic idea is to shut down devices when they are not needed and wake them up when necessary. DPM in general is a nontrivial problem. If the energy and performance overheads in transitioning to sleep states were negligible, then a simple greedy algorithm, which makes the system go into the deepest sleep state as soon as it is idle, would be perfect. However, in reality, transitioning to a sleep state has the overhead of storing the processor state and shutting off the power supply. Waking up too takes a finite amount of time. Therefore, implementing the right policy for transitioning to the sleep state is critical for the success of DPM. The algorithm can be used to provide desirable E–Q characteristics in sensing applications. In energy-scalable algorithms the principal notion is that computation is done in such way that a drop in energy availability results in minimum possible quality degradation.

The fundamental idea in distributed sensor applications is to incorporate sufficient processing power in each node such that they are self-configuring and adaptive. Figure 3.9 illustrates the basic sensor node architecture. Each node consists of the embedded sensor, A/D converter, a processor with memory (which in this case will be the StrongARM SA-1100 processor) and the RF circuits. Each of these components are controlled by the microOperating System (μOS) through microdevice drivers. An important function of the μOS is to enable Power Management (PM). Based on event statistics, the μOS decides which devices to turn off or on.

This network consists of homogeneous sensor nodes distributed over a rectangular region R with dimensions $W \times L$ with each node having a visibility radius of ρ (shown by the region C_k). Three different communication

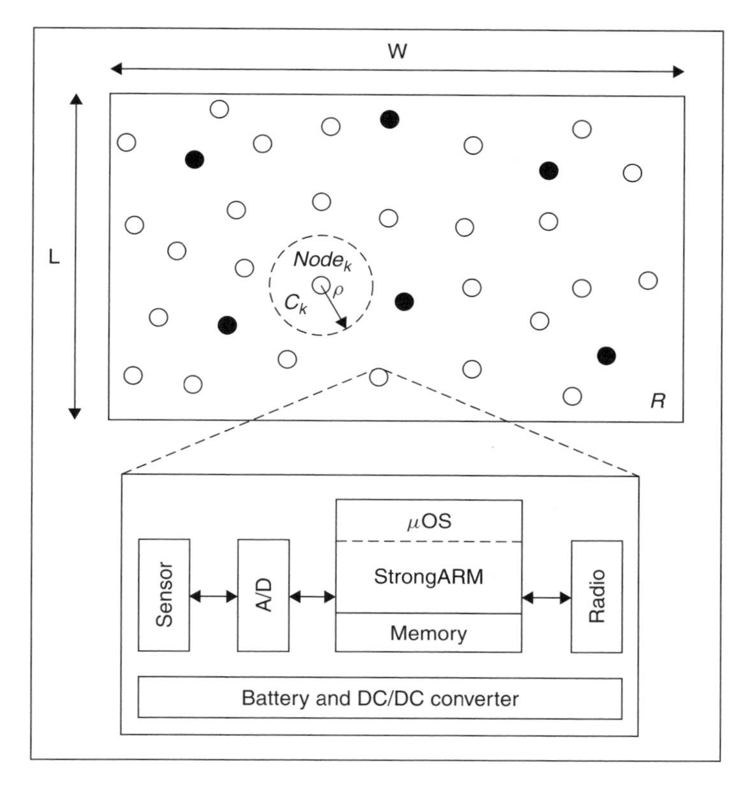

Figure 3.9 Sensor network and node architecture.

models can be used for such a network:

- direct transmission (every node directly transmits to the base station);
- multihop (data is routed through the individual nodes towards the base station), and
- clustering.

If the distance between the neighboring sensors is less than the average distance between the sensors and the user or the base station, transmission power can be saved if the sensors collaborate locally. Further it is likely that sensors in local clusters share highly correlated data. Some of the nodes elect themselves as cluster heads (as depicted by nodes in black) and the remaining nodes join one of the clusters based on a minimum transmit power criteria. The cluster head then aggregates and transmits the data from the other cluster nodes. Such application-specific network protocols for wireless microsensor networks have been developed. A clustering scheme is an order of magnitude more energy efficient than a simple direct transmission scheme.

A power-aware sensor-node model essentially describes the power consumption in different levels of node sleep state. Every component in the node can have different power modes, e.g. the StrongARM can be in active, idle or sleep mode; the radio can be in transmit, receive, standby or off mode. Each node sleep state corresponds to a particular combination of component power modes. In general, if there are N components labeled $(1, 2, \ldots, N)$ each with k_i number of sleep states, the total number of node sleep states are $\prod_{i=1}^{N} k_i$. Every component power mode is associated with a latency overhead for transitioning to that mode. Therefore each node sleep mode is characterized by a power consumption and a latency overhead. However, from a practical point of view not all the sleep states are useful. Table 3.1 enumerates the component power modes corresponding to five different useful sleep states for the sensor node. Each of these node sleep modes corresponds to an increasingly deeper sleep state and is therefore characterized by an increasing latency and decreasing power consumption. These sleep states are chosen based on actual working conditions of the sensor node, e.g. it does not make sense to have the A/D in the active state and everything else completely off. The design problem is to formulate a policy of transitioning between states based on observed events so as to maximize energy efficiency. The power aware sensor model is similar to the system power model in the Advanced Configuration and Power Interface (ACPI) standard. An ACPI compliant system has five global states. SystemStateS0 (working state), and SystemStateS1 to SystemStateS4 corresponding to four different levels of sleep states. The sleep states are differentiated by the power consumed, the overhead required in going to sleep and the wake up time. In general, the deeper the sleep state, the less the power consumption, and the longer the wake-up time. Another aspect of similarity is that in ACPI the Power Manager (PM) is a module of the μOS.

Table 3.1 Useful sleep states for the sensor node.

Sleep state	StrongARM	Memory	Sensor, analog–digital converter	Radio
S_0	Active	Active	On	Tx, Rx
S_1	Idle	Sleep	On	Rx
S_2	Sleep	Sleep	On	Rx
S_3	Sleep	Sleep	On	Off
S_4	Sleep	Sleep	Off	Off

Tx = transmit, Rx = receive.

3.5. DYNAMIC POWER MANAGEMENT IN WIRELESS SENSOR NETWORKS

While shutdown techniques can yield substantial energy savings in idle system states, additional energy savings are possible by optimizing the sensor node performance in the active state. Dynamic Voltage Scaling (DVS) is an effective technique for reducing CPU (Central Processing Unit) energy. A block diagram of a DVS processor system is shown in Figure 3.10. Most microprocessor systems are characterized by a time varying computational load. Simply reducing the operating frequency during periods of reduced activity results in linear decreases in power consumption, but does not affect the total energy consumed per task. Reducing the operating voltage implies greater critical path delays, which, in turn, compromise peak performance.

Significant energy benefits can be achieved by recognizing that peak performance is not always required and, therefore, the processor's operating voltage and frequency can be dynamically adapted according to the instantaneous processing requirement. The goal of DVS is to adapt the power supply and operating frequency to match the workload, so the visible performance loss is negligible. The problem is that future workloads are often nondeterministic.

The rate at which DVS is carried out also has a significant bearing on performance and energy. A low update rate implies greater workload averaging, which results in lower energy use. The update energy and performance cost is also amortized over a longer time frame. On the other hand, a low update rate also implies a greater performance hit since the system will not respond to a sudden increase in workload.

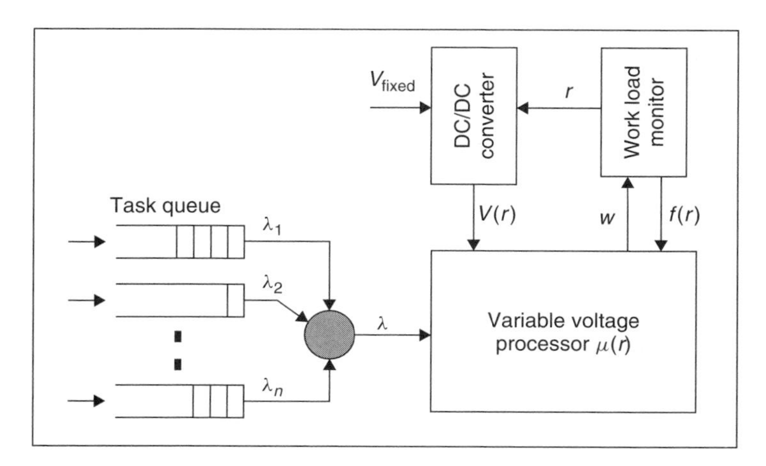

Figure 3.10 Block diagram of a DVS processor system.

A workload prediction strategy based on adaptive filtering of the past workload profile is used and analyzed with several filtering schemes. A performance hit metric is used to judge the efficacy of these schemes.

An event occurs when a sensor node picks up a signal with power above a predetermined threshold. For analytical tractability, every node is assumed to have a uniform radius of visibility, r. In real applications, the terrain might influence the visible radius. An event can be static (such as a localized change in temperature/pressure in an environment monitoring application), or can propagate (such as signals generated by a moving object in a tracking application).

In general, events have a characterizable (possibly nonstationary) distribution in space and time. There are three distinct classes of events:

- the events occur as stationary points;
- the event propagates with fixed velocity (such as a moving vehicle); and
- the event propagates with fixed speed but random direction (such as a random walk)

The processor must watch for preprogrammed wake up signals. The CPU programs these signal conditions prior to entering the sleep state. To wake up on its own, the node must be able to predict the arrival of the next event. An optimistic prediction might result in the node waking up unnecessarily, while a pessimistic strategy results in some events being missed. There are two possible approaches:

(1) Completely disallow the state that results in missed events as the node is not alerted.

If the sensing task is critical and events cannot be missed this state must be disabled.

(2) Selectively disallow the state that results in missed events as the node is not alerted.

This technique can be used if events are spatially distributed and not all critical. Both random and deterministic approaches can be used. In the clustering protocol, the cluster heads can have a disallowed state while normal nodes can transition to this state. Alternatively, the scheme can be more homogeneous. Every node that satisfies the sleep threshold condition for the selectively disallowed node, enters sleep with a system defined by a probability for a certain time duration.

The advantage of the algorithm is that efficient energy trade-offs can be made with event detection probability. By increasing this probability, the system energy consumption can be reduced, while the probability of missed events will increase and vice versa. Therefore, the overall shut down policy is governed by two implementation specific probability parameters.

3.6. ENERGY-EFFICIENT COMMUNICATION

A distributed network of thousands of collaborating microsensors promises a maintenance-free, fault-tolerant platform for gathering rich, multidimensional observations of the environment. Microsensor networks is a specialized class of dedicated networks with several distinguishing characteristics: high node density, low data rate, and an unprecedented attention to energy consumption. Unlike laptop or palmtop devices, microsensor nodes are expected to operate from 5 to 10 years from an amount of energy equivalent to a battery cell, requiring innovative design methodologies to eliminate energy inefficiencies that would have been overlooked in the past. An area potentially ripe with inefficiencies is microsensor communication. Building an energy-efficient protocol stack for microsensors requires a thorough investigation of the interactions among the sensor application, network protocol, MAC (Media Access Control) layer, and radio. Energy consumption characteristics that are unique to this domain of wireless systems must be addressed and exploited for maximally energy-efficient communication.

Communication protocols for traditional dedicated networks generally employ multihop routing to ameliorate the high path losses incurred by radio transmission. Two general routing methodologies, source routing and distance vector approaches, are analogous to their counterparts in wired networks. Source routing specifies complete, hop-by-hop paths for each packet, while distance vector protocols maintain only next-hop information to each destination. These protocols are typically intended for wireless IP (Internet Protocol) applications rather than microsensor networks.

Protocols have been specifically designed for energy-constrained sensor networks. Directed diffusion relies on local interactions among nodes to create efficient paths for data flow. No global routing state is kept anywhere in the system; rather, each node chooses its own source(s) from which to receive data, leading to reasonably efficient data propagation at a global level. LEACH (Low-Energy Adaptive Clustering Hierarchy) forms rotating clusters of adjacent nodes, within which nodes transmit to a single cluster head that bears the burden of a long-distance transmission. Clustering explicitly encourages data aggregation to reduce further the transmission burden on the network.

With the increasing interest in battery-powered wireless systems, energy consumption has become a primary metric of wireless communication protocols, alongside traditional metrics such as throughput and fault tolerance. However, little prior work has characterized the energy consumption of wireless network protocols with realistic hardware models and behavioral characteristics of microsensors. Transmission energy is often modeled in μJ/bit, a model that fails to consider hardware and protocol overheads. Physical and MAC layer models have almost universally adopted energy consumption and performance characteristics from IEEE 802.11b, whose high power consumption and complexity are unsuitable for wireless microsensors.

Reducing the energy of communication in wireless microsensors demands that each aspect of communication, such as the protocol and MAC layers, is tailored to the application.

A wireless node in an dedicated network traditionally seeks its nearest neighbors as candidates for next hop transmission. However, with the high node densities that enable the high robustness and resolution of microsensor networks, the paradigm of routing through nearest neighbors must be reconsidered.

In a model for the power consumed by multi-hop transmission as the distance of the required transmission increases, it becomes advantageous to increase the number of hops. However, it is clear that there is a large range of distances for which direct transmission is more energy-efficient than multi-hop transmission.

That model has assumed ideal multi-hop communication with no overhead. The energy characterization must account for protocol and MAC overhead, suboptimal node spacing, and the fact that radio receivers, not being omniscient to packet arrivals, must occasionally poll for packets. These sources of overhead introduce the additional power overhead into multi-hop routing. For thousand-node microsensor networks, these overheads are substantial. The communication energy is a function of both transmission distance and the transmit duty cycle.

Multi-hop routing can be beneficial is some cases. In the receive-dominant regime, the hardware and protocol overhead of receiving packets outweighs the energy savings of shorter radio transmissions. In the transmit-dominant regime, both the transmission distance and the number of bits transmitted are sufficiently large that multi-hop routing is beneficial. As intuition would suggest, both the transmission distance and quantity of transmitted bits determine the break point between the two regimes, a result concealed by the simpler model. If transmit duty cycles are sufficiently small, then protocol-free direct transmission is more energy efficient even at very large distances. With overhead accounted for, the transmission distance at which multi-hop transmission becomes advantageous over direct transmission is much greater than the total transmission distance of 30 m.

This observations holds two noteworthy implications for microsensor networks:

- As most microsensor networks utilize a small mean distance between nodes, nearest neighbors are often the wrong candidates for energy-efficient next-hops.
- Large classes of applications exist for which the entire network diameter is in the receive-dominant regime.

For these classes of networks, such as those completely enclosed within a room, machine, or small lawn, transmission is most efficient with no multi-hop protocol at all. In these situations, it is increasingly important to focus on the energy dissipation characteristics of the hardware.

Presuming that the network is operating in a transmit-dominated regime, the techniques for reducing power overhead for multi-hop wireless microsensor communication are considered. Exploiting the microsensor network's architecture and application-specific characteristics allows for an energy-conscious optimization in the protocol stack.

Reduction of radio receive energy is primary concern. A radio receiver that is on and idle consumes a substantial amount of power – often as much as transmission. Given that many nodes will likely be in the receiving range of any node's transmission, it is desirable to shut down the radio receiver in the majority of idle nodes. Unfortunately, most wireless protocol and MAC layers utilize unique addresses to route packets to specific destinations, with the expectation that these destinations are actively listening for packets. With radio shutdown, this assumption no longer holds, and routing tables, whether they contain source routes or next-hop information, suddenly become very unstable.

At the application level, communication in a microsensor network is one-way, from the observer nodes to a base station. There are many data sources and relays, but few actual sinks. As the individual relays have no need for the data they are relaying, the entire notion of addressing a packet to a specific relay node is unnecessary. The concern is that packets move progressively closer to a base station.

Microsensor nodes may not utilize explicit addresses at the protocol or MAC level, but rather a metric of their approximate distance to the nearest base station. This metric can be propagated across the network by flooding, with the base stations initially broadcasting a zero metric and each node adding a constant factor to the smallest value heard.

A node with a packet destined for the base station simply broadcasts its packet with its current distance metric. Nodes that receive the packet compare their own distance metric to that of the packet. To minimize the number of hops, the receiving node that is closest to the base station and farthest from the originating node would relay the packet onward. Such behavior could be implemented, for instance, with a delay timer proportional to the difference between the packet's and relay node's distance metrics, or simply the RSSI (Receiver Signal Strength Indication). The node with the lowest delay would forward the packet, and the others, hearing the forward, would drop their respective copies.

Address-free forwarding allows any active node, rather than one that is specifically addressed, to relay a packet. This enables flexible, protocol-independent radio receiver shutdown.

The energy consumption of a microsensor network is discussed for both the transmit-dominant and receive-dominant regimes.

The μAMPS (micro-Adaptive Multidomain Power-Aware Sensors) develop the enabling technologies for energy-efficient microsensor networks. The μAMPS-1 node is the basis for the hardware energy consumption models. The μAMPS-1 node consists of sensing, processing, and radio subsystems. The sensing system consists of an acoustic sensor and low-power A/D converter. Data processing, as well as some network functions, are carried out by a StrongARM SA-1110 microprocessor. The radio transmits and receives at 1 Mbps at half-duplex in the 2.4 GHz range.

The measured energy consumption and performance of μAMPS-1 node form the basis of the hardware model, in which node antennas are at ground level model. The radio path loss is modeled with an empirical r^3 rather than the conventional two-ray ground wave propagation model.

When packets are extremely short, the energy required for radio startup exceeds the energy of the actual transmission. By buffering packets at the local nodes and transmitting many short packets in a single transmission, the number of energy-consuming startups is reduced. By sending ten packets at 10-second intervals, for instance, less than half the total energy of immediate transmission is used. As individual packets grow larger in size, the impact of radio startup energy on total system energy is inherently reduced.

The trade-off is the increased latency of the buffered observations. Hence, for applications such as short-distance, low-rate environmental sensors, exposing the number of buffered packets as a dynamically adjustable quantity provides an effective energy-quality trade-off. For instance, a node could choose to transmit observations immediately if an observed parameter fell outside a normal range, but buffer them otherwise.

3.7. POWER AWARENESS OF VLSI SYSTEMS

Power awareness can be enhanced by using a systematic technique. This technique is illustrated by applying it to VLSI systems at several levels of the system hierarchy: multipliers, register files, digital filters, dynamic voltage scaled processing and data-gathering wireless networks. The power awareness of these systems can be significantly enhanced leading to increases in battery lifetimes.

There are two main aims in motivating power-aware design and its emergence as an important paradigm. The first is to explain the importance of power awareness as a consequence of the increasing emphasis on making systems more scalable. In this context, making a system scalable refers to enabling the user to trade off system performance parameters as opposed to hard-wiring them. Scalability allows the end user to implement operational policy, which often varies significantly over the lifetime of the system. For example, consider the user of a portable multimedia terminal. At times, the user might want extremely high performance (for instance, high quality video) at the cost of reduced battery lifetime. At other times, the opposite might be true, the user might want bare minimum perceptual quality in return for maximizing battery lifetime. Such trade-offs can only be optimally realized if the system was designed in a power-aware manner. A related motivation for power-awareness is that a well designed system must gracefully degrade its quality and performance as the available energy resources are depleted. Continuing the video example, this implies that as the expendable energy decreases, the system should gracefully degrade video quality (seen by the user as increased blockiness, for instance) instead of exhibiting a cliff-like, all-or-none behavior (perfect video followed by no video).

While the above argues for power-awareness from a user-centric and user-visible perspective, one can also motivate this paradigm in more fundamental, system-oriented terms. With burgeoning system complexity and the accompanying increase in integration, there is more diversity in the operating scenarios than ever before. Hence, design philosophies that assume the system to be in the worst-case operating state most of the time are prone to yield suboptimal results. In other words, even if there is little explicit user intervention, there is an imperative to track operational diversity and scale power consumption accordingly. This naturally leads to the concept of power-awareness. For instance, the embedded processor that decodes the video stream in a portable multimedia terminal can display tremendous workload diversity depending on the temporal correlation of the incoming video bit stream. Hence, even if the user does not change quality criteria, the processor must exploit this operational diversity by scaling its power as the workload changes.

Since low energy and low power are intimately linked to power awareness, it is important and instructive to provide a first-cut delineation of these concepts. Power awareness as a metric and design driver does not devolve to traditional worst-case centric low-power/low-energy design. As preliminary evidence of this, consider the system architect faced with the task of increasing the power awareness of the portable multimedia terminal alluded to above. While the architect can claim that certain engineering reduces worst-case dissipation and/or overall energy consumption of the terminal and so on, these traditional measures still fall short of answering the related but different questions:

- How well does the terminal scale its power with user or data or environment dictated changes?
- What prevents it from being arbitrarily proficient in tracking operational diversity?
- How can we quantify the benefits of such proficiency?
- How can we systematically enhance the system's ability to scale its power?
- What are the costs of achieving such enhancements?

The process of formally understanding power awareness uses a multiplier as an example. The basic power-awareness formalisms using a simple system, a 16×16 bit array multiplier, is developed. Consider a given system H that performs a certain set of operations F while obeying a set of constraints C. For the illustrative system, H is the given implementation of a 16×16 bit array multiplier. While the set F ideally contains all m-bit by n-bit multiplications, where $m, n \in F$ is restricted to be set of all m-bit by m-bit multiplications instead. The constraint may be simply one of fixed latency (i.e. H cannot take more than a given time, t, to perform F).

We discuss how well the energy of a system H scales with changing operating scenarios. Note that the energy rather than power in the statement above is used, because energy allows us seamlessly to include latency constraints. Next, observe that the understanding of power awareness can only be as exact as the understanding of operating scenarios. These scenarios can be characterized with arbitrarily high detail. For instance, in the case of the multiplier, the scenario can be defined by the precision of the current multiplicands, the multiplicands themselves, or even the current multiplicands and the previous multiplicands, since the power dissipation depends on those factors. To simplify this approach, the set of scenarios S is characterized by the precision of the multiplicands. This needs a two tuple since there are two multiplicands. However, in this case, F is only one number (the precision of the two identical bit-width multiplicands), which characterizes the scenario. Hence, H can find itself in one of 16 scenarios. We denote henceforth, a scenario by s and the set of 16 scenarios by S.

After defining the scenarios, the first step is to characterize the power awareness of H by tracing its energy behavior as it moves from one scenario to the other. For a 16-bit multiplier, a large number of different scenarios is executed, and the energy consumed is measured in each scenario.

The multiplier has a natural degree of power awareness even though it was not explicitly designed for it. This is because the lower precision vectors lead to lesser switched capacitance than do higher precision ones.

The perfectly power-aware system is a system $H_{perfect}$, which is defined as the most power-aware system if and only if for every scenario in S, $H_{perfect}$ consumes only as much energy as its current scenario demands. More formally, $H_{perfect}$ is the most power-aware system if and only if for every scenario in S, $H_{perfect}$ consumes only as much energy as demanded by its current operation $\in F$ executing in the current scenario under constraints C. In the multiplier example, S is constructed such that it has a one-to-one correspondence with F and hence, the energy of a scenario executing on H is discussed.

We need formally to capture the concept of only as much energy as a scenario demands. To derive this energy for a given scenario, s_1, a system H_{s_1} is designed to execute this and only this scenario. The reasoning is that a given system H can never consume lesser energy in a scenario compared with H_{s_1}, a dedicated system that was specially designed to execute only that scenario. We often refer to the H_{s_i}s as point systems because of their focused construction to achieve low energy for a particular scenario (or point) in the energy dependency. Hence, in the context of power awareness, the energy consumed by H_{s_1} is in a sense, the lower bound on the dissipation of H while executing scenario s_1. Generalizing this statement, the bounds on efficiency of tracking scenarios are discussed.

The energy consumed by a given system H while executing a scenario s_i cannot be lower than that consumed by the a dedicated system H_{s_i} constructed to execute only that scenario s_i as efficiently as possible.

This leads to the next definition of $H_{perfect}$ as the perfectly power aware system. The perfect system, $H_{perfect}$, is as energy efficient as H_{s_i} while executing scenario s_i for every $s_i \in S$.

The energy of a perfect system is denoted by $E_{perfect}$. From a system perspective, the perfect system behaves as if it contains a collection of dedicated point systems, one for each scenario. When $H_{perfect}$ has to execute a scenario s_i, it routes the scenario to the point system H_{s_i}. After H_{s_i} has finished processing, the result is routed to the common system output. This abstraction of $H_{perfect}$ as an ensemble of point systems is illustrated in Figure 3.11.

The task of identifying the scenario by looking at the data input is carried out by the scenario-determining block. Once this block has identified the scenario, it configures the mux (multiplexer) and de-mux (de-multiplexer) blocks such that data is routed to, and results routed from, the point system that corresponds to the current scenario. Note that if the energy costs of identifying the scenario, routing to and from a point system, and activating the right point system are zero, then the energy consumption of $H_{perfect}$ will indeed be equal to that of H_{s_i} for every scenario s_i. Since these costs are never zero in real systems, this implies that $H_{perfect}$ is an abstraction and does not correspond to a physically realizable system. Its function is to provide a nontrivial lower bound for the energy dependency.

To construct the $E_{perfect}$ dependency for the 16-bit multiplier, the ensemble of points construction outlined above is emulated. The point systems in this example were 16 dedicated point multipliers: 1×1-bit, 2×2-bit, ..., 16×16-bit, corresponding to H_{s_1} to $H_{s_{16}}$. When a pair of multiplicands with precision

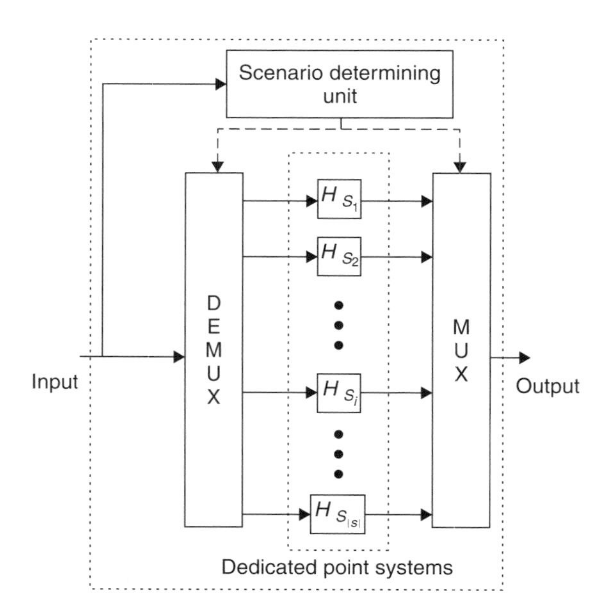

Figure 3.11 The perfect system ($H_{perfect}$) can be viewed as an ensemble of point systems.

i came by, we diverted them to H_{s_i} (i.e. the ixi-bit multiplier). Since $E_{perfect}$ is derived, only the energy consumed by the H_{s_i}s is taken into account.

Note that $E_{perfect}$ scales extremely well with precision since the scenarios are being executed on the best possible point systems that can be constructed. It is essential to note that the $E_{perfect}$ really depends on the kind of point system allowed. In the case of the multiplier, any ixi-bit multiplier is allowed. The set of point systems allowed is henceforth denoted by P. This set captures the resources available to engineer a power-aware system. Like the scenario and constraint sets, it can be specified with increasing rigor and detail. This new formalism, P has two key purposes:

- P gives a more fundamental basis to $E_{perfect}$. While it is not possible to talk about the best possible energy, it is indeed possible to talk about the best possible energy dependency for a specified P.
- P is also important when enhancing the power awareness of H. In that context, P specifies exactly which building blocks are available for such an enhancement.

Enhancing the power awareness of a system is composed of two well defined steps:

(1) Engineering the best possible point systems;

(2) Engineering the desired system using the point systems constructed in step (1) such that power awareness is maximized.

In the context of a power aware multiplier, the first task involves engineering $1 \times 1, 2 \times 2, \ldots, 16 \times 16$ bit multipliers that are as efficient as possible while performing $1 \times 1, 2 \times 2, \ldots, 16 \times 16$-bit multiplications respectively. The second task of engineering a system using point systems is illustrated by the multiplier shown in Figure 3.12.

Note the overall similarity between this figure and the abstraction of $H_{perfect}$ in Figure 3.11. The ensemble of point systems is used as an abstract concept in the context of explaining energy dependency of $H_{perfect}$. A physical realization of a system based on this concept is illustrated. The basic idea is to detect the precision of the incoming operands using a zero detection circuit and then route them to the most suitable point system. In the case of $H_{perfect}$, the matching is done trivially, and multiplier operands which need a minimum precision of i-bits are directed to a ixi-bit multiplier. Similarly, the output of the chosen multiplier is multiplexed to the system output. However, $H'_{perfect}$ has significant overheads. Even if the area cost of having 16-point

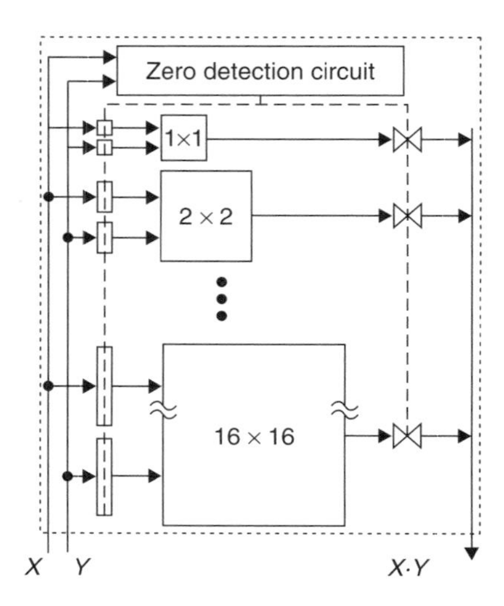

Figure 3.12 The $H'_{perfect}$ system mimics the abstract $H_{perfect}$ system by using an ensemble of 16 dedicated point multipliers and a zero-detection circuit as the scenario-detector.

multipliers was ignored, and the focus was solely on the power awareness, the energy dependency of $H'_{perfect}$ would not be the same as $E_{perfect}$. This is because, while the scenario execution itself is the best possible, the energy costs of determining the scenario (the zero detection circuit), routing the multiplicands to the right point system and routing the result to the system output (the output mux) can be nontrivial.

A system that uses a less aggressive ensemble in an effort to reduce the energy overhead of assembling point systems is shown in Figure 3.13. The basic operation of this multiplier ensemble is the same. The precision requirement of the incoming multiplicand pair is determined by the zero detection circuitry. Unlike the previous 16-point ensemble, this four-point ensemble is not complete and hence mapping scenarios to point systems is not one–one. Rather, precision requirements of:

(1) $\leqslant 9$ bits are routed to the nine-point multiplier;
(2) 10, 11 bits are routed to the 11-point multiplier;
(3) 12–14 bits are routed to the 14-point multiplier;
(4) 15, 16 bits are routed to the 16-point multiplier.

Similarly, the results are routed back from the activated multiplier to the system output. While scenarios are no longer executed on the best possible

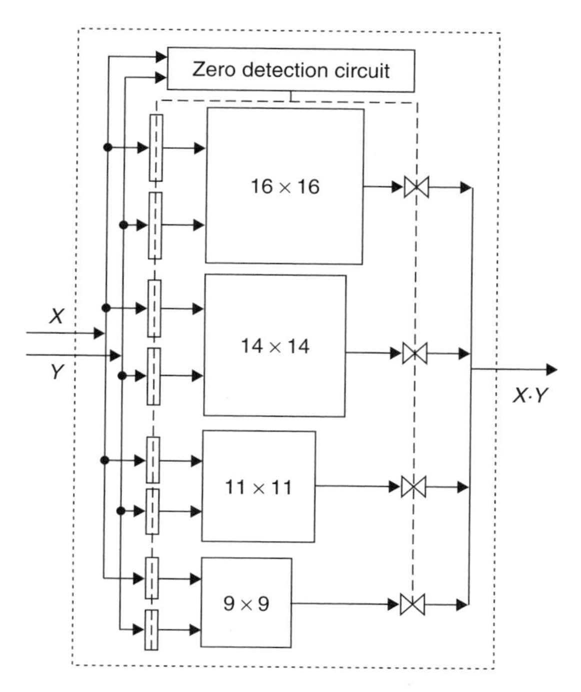

Figure 3.13 The four-point ensemble multiplier system.

point system (with the exception of 16-, 14-, 11- and 9-bit multiplications), this ensemble has the advantage that energy overheads of routing are significantly reduced over $H'_{perfect}$. Also, while the scenario-to-point system mapping of the four-point ensemble is not as simple as the one–one mapping, it is important to realize two things. First, the energy dissipated by the extra gates needed for the slightly more involved mapping in the four-point ensemble is low relative to that dissipated in the actual multiplication. Second, only four systems have to be informed of the mapping decision compared with 16 earlier. This reduction further offsets the slight increase in scenario mapping.

It is not difficult to see the basic trade-off at work here. Increasing the number of point systems decreases the energy needed for the scenario execution itself but increases the energy needed to coordinate these point systems. Hence, it is intuitively reasonable to assume the existence of an optimal ensemble of point systems that strikes the right balance.

We discuss whether a system $H_{optimal}$ can be constructed as an ensemble of point systems drawn from P such that $H_{optimal}$ is unconditionally more power aware than any other such constructed system. The unconditional power awareness only leads to partial ordering. Hence, the existence of a unique

$H_{optimal}$ cannot be guaranteed. While it is possible to present a set of solutions that are unconditionally more power aware than all other solutions, there is no guarantee that this set will have only one member. In fact, this last condition is highly unlikely to occur in practice, unless routing costs are very low or very high compared with scenario execution costs (in which cases the optimal ensembles would be the complete and single-point solutions respectively). Hence, in general, it is futile to search for an optimal ensemble of point systems that is unconditionally better than all other ensembles.

We discuss whether a system $H_{optimal}$ can be constructed as an ensemble of a point systems drawn from P such that $H_{optimal}$ is more power aware than any other such constructed system for a specified scenario d_{given}. Since a specified scenario distribution d_{given} imposes a total ordering on the power awareness of all possible subsets of P, it is easy to prove the existence of an optimal system. Note that the proof based on total ordering is nonconstructive, i.e. it only tells us that $H_{optimal}$ exists but does not help us determine what it is. This is unfortunate because a brute-force search of the optimal subset of P would require an exponential number of operations in $|P|$–a strategy that takes unacceptably long even for the modestly large P.

To see if there are algorithms that can find $H_{optimal}$ in nonexponential runtimes the problem is defined more formally by using practical illustrations of enhancing power awareness.

Enhancing power awareness by constructing ensembles of point systems carefully chosen from P is a general technique that can be used not just for multipliers but other systems as well. We illustrate how this ensemble idea can be applied to enhance the power awareness of multiported register files, digital filters, and a dynamic voltage scaled processor. In each case, the problem is described in terms of the framework developed above and characterizes the power awareness of the system. Then an ensemble construction is used to enhance power awareness. It is interesting to note that these applications cover not just spatial ensembles, but purely temporal (processor example) and spatial–temporal hybrid ensembles (register files and adaptive digital filters) as well.

Architecture and VLSI technology trends point in the direction of increasing energy budgets for register files. The key to enhancing the power awareness of register files is the observation that over a typical window of operation, a microprocessor accesses a small group of registers repeatedly, rather than the entire register file. This locality of access is demonstrated by the 20 benchmarks comprising the SPEC92 (Systems Performance Evaluation Consortium) benchmark suite that were run on a MIPS (Million Instructions Per Second) R3000. More than 75 % of the time, no more than 16 registers were accessed by the processor in a 60-instruction window. Equally importantly, there was strong locality from window to window. More than 85 % of the time, less than five registers changed from window to window.

The number of registers the processor typically needs over a certain instruction window is considered a scenario. The smaller files have lower costs of access because the switched bit-line capacitance is lower. Hence, from a power awareness perspective, over any instruction window, as small as possible a file is used.

There are significant motivations for investigating power aware filters. As an example, consider the adaptive equalization filters that are ubiquitous in communications ASICs (application specific integrated circuits). The filtering quality requirements depend strongly on the channel conditions (line lengths, noise and interference), the state of the system (training, continuous adaptation, freeze, etc.), the standard dictated specifications, and the quality of service (QoS) desired. All these considerations lead to tremendous scenario diversity which a power-aware filtering system can exploit.

The three examples show power aware subsystems, multipliers, register files, and digital filters. The power awareness at the next level of the system hierarchy is a power-aware processor that scales its energy with workload. Unlike previous examples, however, this one illustrates how an ensemble can be realized in a purely temporal rather than a spatial manner.

It is well known that processor workloads can vary significantly and it is highly desirable for the processor to scale its energy with the workload. A powerful technique that allows such power awareness is dynamic frequency and voltage scaling. The basic idea is to reduce energy in nonworst-case workloads by extending them to use all available time, rather than simply computing everything at the maximum clock speed and then going into an idle or sleep state. This is because using all available time allows one to lower the frequency of the processor which, in turn, allows scaling down of the voltage leading to significant energy savings. In terms of the power-awareness framework, a scenario would be characterized by the workload. The point systems would be processors designed to manage a specific workload. As the workload changes, we would ideally want the processor designed for the instantaneous workload to execute it. It is clear that implementing such an ensemble spatially is meaningless and must be done temporally using a dynamic voltage scaling system.

Increased levels of integration and advanced low power techniques, are enabling dedicated, wireless networks of microsensor nodes. Each node is composed of a sensor, analog preconditioning circuitry, A/D, processing elements (DSP, RISC, FPGA, etc.) and a radio link, all powered by a battery. Replacing high quality macrosensors with such networks has several advantages: robustness and fault tolerance, autonomous operation for years,

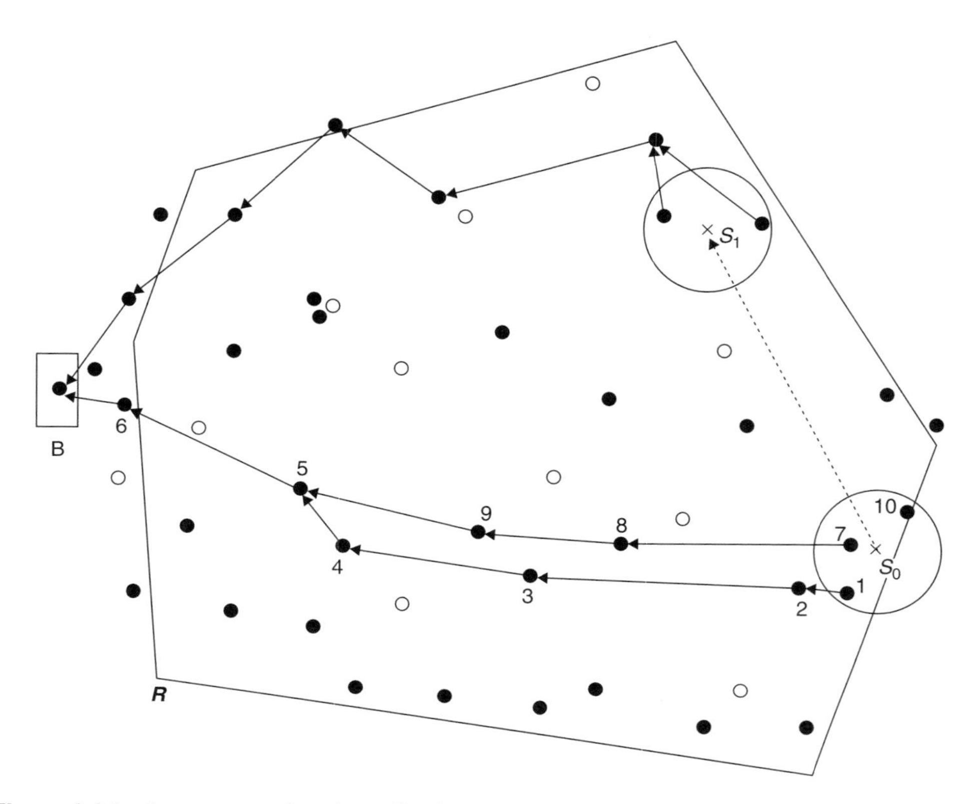

Figure 3.14 A sensor network gathering data from a circularly observable source (denoted by ×) residing in the region *R*. Live nodes are denoted by ● and dead ones by ○. The base station is marked *B*. In this example we require that at least two nodes sense the source. When the source is at S_0, nodes 1 and 7 assume the role of sensors and nodes 2, 3, 4, 5, and 6 form the relay path for data from node 1 while nodes 7, 8, 9, 5, and 6 form the relay path for data from node 7. Data might be aggregated into one stream at node 5. This is not the only feasible role assignment that allows the source to be sensed. For instance, node 10 could act as the second sensor instead of node 7 and 10, 7, 8, 4, 5, and 6 could form the corresponding relay path. Also, node 6 might aggregate the data instead of node 5, etc. The sensor, aggregator, and relay roles must change as the source moves from S_0 to S_1.

enhanced data quality, and optimal cost-performance. Such data gathering networks are expected to find wide use in remote monitoring applications, intrusion detection, smart medicine, etc. An illustrative data gathering network is shown in Figure 3.14. The network is live as long as it can guarantee that any source in region *R* will be sensed and the data relayed back to a fixed base station. To accomplish this objective, different nodes take on different roles over the lifetime of the network as seen in the figure. A noteworthy point is that nodes must often change roles even if the source does not move. This is to enable energy drain to be spread throughout the network, which leads to increased lifetimes. An assignment of roles to nodes that leads to data gathering is termed a feasible role assignment We also require feasible role assignments to be nonredundant, i.e. data from a sensor should not be routed via multiple links. A data-gathering strategy or collaborative strategy can be completely characterized by specifying a sequence of feasible role assignments and the time for which the assignment is sustained.

3.8. SUMMARY

The power-awareness philosophy captures more than just energy savings. Inherent to power-awareness is an adaptability to changing environmental conditions and resources, as well as the versatility to prioritize either system lifetime or output quality at the user's request. Such flexibility and adaptability are essential characteristics of a microsensor node, a system subjected to far more resource, workload, and input variability than most electronic devices. As continuing developments in VLSI (Very Large Scale Integration) technology reduce the size and increase the functionality of microsensor nodes, the power-aware design methodology becomes the dominant enabler for a practical, energy-efficient microsensor node.

The energy savings are offered by dynamic voltage scaling on an unmodified commercial microprocessor using an adjustable DC–DC regulator and a power-aware μOS. Energy scalable algorithms running under this implementation consume up to 60 % less energy with DVS than with a fixed supply voltage. Refinements to the regulator circuit and the addition of voltage scheduling algorithms to the μOS further increase energy savings. The DVS enabled system, when incorporated into a complete prototype sensor node for μAMPS, enables the energy–quality trade-offs inherent to power-aware algorithms for distributed microsensor networks.

Wireless distributed microsensor networks have gained importance in a wide spectrum of civil and military applications. Advances in MEMS (MicroElectroMechanical Systems) technology, combined with low-power, low-cost, digital signal processors (DSPs) and radio frequency (RF) circuits, have resulted in the feasibility of inexpensive and wireless microsensor networks. A distributed, self-configuring network of adaptive sensors has significant benefits. They can be used for remote monitoring of inhospitable and toxic environments. A large class of benign environments also requires the deployment of a large number of sensors such as for intelligent patient monitoring, object tracking, and assembly-line sensing. The massively distributed nature of these networks provides increased resolution and fault tolerance as compared with a single sensor node.

The severe energy constraints on distributed microsensor networks demand the utmost attention to all aspects of energy consumption, and the use of energy models that are suited to the task of evaluating high-density, energy-conscious microsensor networks. For multi-hop communication, the high node density of microsensor networks demands receiver shutdown, which is enabled by a distance-metric addressing scheme that takes advantage of the network's one-way communication from nodes to the base station. Nodes with sufficiently little or very short-distance transmissions are most energy efficient with direct transmission to the base station. In this regime, inefficiencies of the hardware, such as radio startup is the concern, rather than protocols.

While a stratified communication architecture that separates the physical, MAC protocol, and application layers is convenient for abstraction and instruction, energy efficiency is clearly gained by each level's awareness of the characteristics of the others. For microsensors, where energy matters most, the energy is conserved by tailoring the MAC and protocol architectures to the specific characteristics of the radio and application domain.

A key challenge in unlocking the potential of data-gathering networks is attaining a long lifetime despite the severely energy-constrained nature of the network. For example, networks composed of ultracompact nodes carrying less than 2 J of battery energy might be expected to last for 5–10 years. It is possible to address these challenges by power aware design. Data gathering networks can be aware of the desired quality of gathered data, of changing source behavior, of the changing state of the network, and finally of the environment in which they reside. We focus on this last aspect, i.e. the problem of designing a power-aware, data-gathering network that tracks changes in the environment to maximize energy efficiency. It is well known that the transmit power can be scaled with changing noise power to maintain the same SNR (signal-to-noise ratio) and hence the same link performance. A more holistic approach is to view environmental variations as affecting changes in the energy needed to process a bit (i.e. carry out some computation on it) versus the energy needed to communicate it. A power-aware network is then simply one that can track changes in the computation-to-communication energy ratio. For large ratios, i.e. high computation costs, the network will favor unaggregated or raw sensor streams. Conversely, for low ratios, i.e. high communication costs, aggregation will be favored. Hence, the challenge in power-aware data gathering is to determine and execute the collaborative strategy that assigns roles optimally for a specified computation-to-communication energy ratio.

PROBLEMS

Learning Objectives

After completing this chapter you should be able to:

- demonstrate understanding of power-aware wireless sensor networks;
- discuss what is meant by distributed power-aware microsensor networks;
- explain the dynamic voltage scaling techniques for distributed microsensor networks;
- demonstrate understanding of an operating system for energy scalable wireless sensor networks;
- discuss what is meant by design integration;
- explain what dynamic power management in wireless sensor networks is;
- demonstrate understanding of energy-efficient communication for dedicated wireless-sensor networks;
- discuss power awareness of VLSI systems.

Practice Problems

Problem 3.1: What is the power-aware system design?

Problem 3.2: How is energy consumption in a static CMOS-based processor classified?

Problem 3.3: What is the role of dynamic voltage scaling (DVS)?

Problem 3.4: Where does the energy scalable framework apply?

Problem 3.5: How do microsensor networks compare with the larger macrosensor based systems?

Problem 3.6: How can the energy of computation be traded off for quality in microsensor networks?

Problem 3.7: What are the requirements in sensing applications?

Problem 3.8: What are the requirements for sensor node in a distributed sensing application?

Problem 3.9: How can the sensor node's lifetime be maximized?

Problem 3.10: Why are embedded operating systems and software critical for microsensor networks?

Problem 3.11: How can energy savings be achieved with the power supply and operating frequency?

Problem 3.12: When is it advantageous to increase the number of hops in a wireless sensor network?

Practice Problem Solutions

Problem 3.1:

A power-aware system design employs a system where energy consumption adapts to constraints and variations in the environment, on board resources, or user requests. Power-aware design methodologies offer scalable energy savings that are ideal for the high variabilities of the microsensor environment.

Problem 3.2:

Energy consumption in a static (Complementary Metal-Oxide Semiconductor) CMOS-based processor can be classified into switching and leakage components.

Problem 3.3:

Dynamic voltage scaling (DVS) exploits variabilities in processor workload and latency constraints, and realizes this energy quality trade-off at the circuit level. The switching energy of any particular computation is independent of time. Reducing supplied voltage offers savings in switching energy at the expense of additional propagation delay through static logic. Hence, if the workload on the processor is light, or the latency tolerable but the computation is high, the supplied voltage and the processor clock frequency together can be reduced to trade-off latency for energy savings. Both switching and leakage energy are reduced by DVS.

Problem 3.4:

The energy scalable framework enables the development and implementation of energy-agile applications. It is important that all processing in the sensor node is energy scalable, including link level protocols, sensor network protocols, data aggregation algorithms, and sensor signal processing.

Problem 3.5:

Compared with larger macrosensor based systems, microsensor networks offer longevity, robustness, and ease of deployment that are ideal for environments where maintenance or battery replacement may be inconvenient or impossible.

Problem 3.6:

In algorithms where additional computation incrementally refines a result, the energy of computation can be traded off for quality. In microsensor networks, for instance, transmitted data can be encrypted with a key of varying length, allowing trade-offs between computational energy and the security of the transmission.

Problem 3.7:

Sensing applications present a wide range of requirements in terms of data rates, computation, average transmission distance, etc., as protocols and algorithms have to be tuned to each application.

Problem 3.8:

The fundamental idea in distributed sensor applications is to incorporate sufficient processing power in each node such that they are self-configuring and adaptive.

Problem 3.9:

A wireless microsensor node is typically battery operated and therefore energy constrained. To maximize the sensor node's lifetime after its deployment, other aspects including circuits, architecture, algorithms, and protocols have to be energy efficient. Once the system has been designed, additional energy savings can be attained by using dynamic power management (DMP) where the sensor node is shut down if no events occur. Such event driven power consumption is critical to maximum battery life. In addition, the node should have a graceful energy quality scalability, so that the mission lifetime can be extended if the application demands, at the cost of sensing accuracy.

Problem 3.10:

Sensing applications present a wide range of requirements in terms of data rates, computation, and average transmission distance. Protocols and algorithms have to be tuned for each application. Therefore embedded operating systems and software are critical for such microsensor networks because programmability is a necessary requirement.

Problem 3.11:

Significant energy benefits can be achieved by recognizing that peak performance is not always required and therefore the processor's operating voltage and frequency can be dynamically adapted, based on instantaneous processing requirement. The goal of DVS is to adapt the power supply and operating frequency to match the workload so the visible performance loss is negligible.

Problem 3.12:

There is a large range of distances for which direct transmission is more energy-efficient than multi-hop transmission. However, in a model for the power consumed by multi-hop transmission, as the distance of the required transmission increases, it becomes advantageous to increase the number of hops.

4

Routing in Wireless Sensor Networks

4.1. INTRODUCTION

Sensor networks are dense wireless networks of heterogeneous nodes collecting and disseminating environmental data. There are many scenarios in which such networks might be used, for example, environmental control in office buildings, robot control and guidance in automatic manufacturing environments, interactive toys, the smart home providing security, identification, and personalization, and in interactive museums.

Networking a large number of low-power mobile nodes involves routing, addressing, and support for different classes of service at the network layer.

Self-configuring wireless sensor networks consist of hundreds or thousands of small, cheap, battery-driven, spread-out nodes bearing a wireless modem to accomplish a monitoring or control task jointly. An important concern is the network lifetime: as nodes run out of power, the connectivity decreases and the network can finally be partitioned and become dysfunctional. The concept of altruistic nodes can be applied to the routing protocol of the pico radio (the Energy Aware Routing protocol, EAR).

The EAR protocol is built on the principle of attribute-based addressing. EAR and directed diffusion belong to the class of reactive routing protocols, where the routing information between nodes is set up only on demand and maintained only as long as it is needed. This eliminates the need to maintain permanent routing tables. Hence, before any communication

Wireless Sensor Network Designs A. Hać

ISBN: 0-470-86736-1

can take place, a route discovery has to be performed. Furthermore, the consumers of data (called sinks) initiate the route discovery. In the other types of routing protocols for sensor networks, for example, the SPIN (Sensor Protocol Information via Negotiation) protocol, the data producer (called source) advertises its data.

The database generic query interface for data aggregation can be applied to dedicated networks of sensor devices. Aggregation is used as a data reduction tool. Networking approaches have focused on application specific solutions. The network aggregation approach is driven by a general purpose, SQL (Structured Query Language)-style interface that can execute queries over any type of sensor data while providing opportunities for significant optimization.

Deployment of large networks of sensors requires tools for collecting and querying data from these networks. Of particular interest are aggregates whose operations summarize current sensor values in a part of, or the entire, sensor network. For example, given a dense network of a thousand sensors querying temperature, users want to know temperature patterns in relatively large regions encompassing tens of sensors, as individual sensor readings are of little value.

4.2. ENERGY-AWARE ROUTING FOR SENSOR NETWORKS

Crucial to the success of ubiquitous sensor networks is the availability of small, lightweight, low cost network elements, called pico nodes. These nodes must be smaller than one cubic centimeter, weigh less than 100 grams, and cost substantially less than 1 dollar (US). Even more important, the nodes must use ultra-low power to eliminate frequent battery replacement. A power dissipation level below 100 microwatts would enable self-powered nodes using energy extracted from the environment, an approach called energy scavenging or harvesting.

Routing protocols are low power, and can be scalable with the number of nodes, and fault tolerant to nodes that go up or down, or move in and out of range. A more useful metric for routing protocol performance is network survivability. The protocol should ensure that connectivity in a network is maintained for as long as possible, and the energy status of the entire network should be of the same order. This is in contrast to energy optimizing protocols that find optimal paths and then burn the energy of the nodes along those paths, leaving the network with a wide disparity in the energy levels of the nodes, and eventually disconnected subnets. If nodes in the network burn energy more equitably, then the nodes in the center of the network continue to provide connectivity for longer, and the time to network partition increases. This leads to a more graceful degradation of the network, and is the idea of survivability of networks.

EAR protocol ensures the survivability of low energy networks. It is a reactive protocol such as Ad Hoc On Demand Distance Vector Routing (AODV) and directed diffusion; however, the protocol does not find a single optimal path to use for communication. EAR keeps a set of good paths and chooses one using a probability. This means that instead of a single path, a communication uses different paths at different times, thus any single path does not deplete energy. EAR is quick to respond to nodes moving in and out of the network, and has minimal routing overhead. The network performance improves by using this method, and the network lifetime increases.

The main functions in a sensor network are sensing, controlling and actuating. These functions can be placed on separate nodes or located on the same physical node. Each physical node has a logical repeater function that helps in multihop routing. Three types of nodes are sensors, controllers, and actuators. Based on the system description, most of the sensors and actuator nodes remain static. Controllers, on the other hand, can be mobile, but their speed is low, of the order of 1 to 5 m/s.

The bit rates in sensor networks are fairly low, at about a few hundred bits per second per node. The peak bit rate supported is about 10 kb/s, which enables simple voice messaging, but not in real time. Sensor data is highly redundant, which means that end-to-end reliability is not required for most data packets.

Most of the communication is fairly periodic in nature, and sensor values are sent at regular intervals to the controllers. The network can be optimized for such recurrent communication, while loosely optimizing for less often, one-time cases.

The three main layers for designing the pico node are the physical, Media Access Control (MAC), and network layers.

The physical layer handles the communication across the physical link, which involves modulating and coding the data so that the intended receiver can optimally decode it in the presence of channel interference.

The MAC layer's primary functions are to provide access control, channel assignment, neighbor list management, and power control. The MAC layer has a location subsystem that computes the x, y and z-coordinates based on the received signal strength of neighboring nodes, and the presence of certain anchors in the network that know their exact positions.

The MAC coordinates channel assignment such that each node gets a locally unique channel for transmission, while the channels are globally reused. There is also a global broadcast channel that is used for common control messages and for waking up nodes. Each node has two radio receivers, one of which runs at 100 % duty cycle, but has a very low bit rate and consumes very little power. The second radio runs at a very low duty cycle (about 1 %) and is switched on only when the node needs to receive data. This is a higher rate radio (about 10 kbps) and consumes more power.

To send data, the MAC layer sends a wake-up signal on the broadcast channel. The address of the node to which it needs to send data is modulated with the wake-up. Access to the broadcast channel is CSMA/CA (Carrier Sense Multiple Access/Collision Avoidance). On receiving the message, the node to which this message is addressed powers on its main radio and communication begins. Since each node has a locally unique channel, no collisions occur during data transmission. Thus the MAC layer enables deep sleep of the nodes, leading to substantial power savings.

The MAC layer also keeps a list of its neighbors and metrics such as the neighbor's position and the energy needed to reach it. This list is used heavily by the network layer to take decisions regarding packet routing. The MAC layer also performs power control to ensure power savings and maintain an optimal number of neighbors.

The network layer has two primary functions: node addressing, and routing.

Traditional network addressing assigns fixed addresses to nodes, for example, on the Internet. The advantage of such schemes is that the addresses can be made unique. However, there is a high cost associated with assigning and maintaining these addresses. The problem occurs in mobile networks where the topology information keeps changing. Packet routing becomes difficult if the node address does not provide any information about the direction in which to route the packet. There are two approaches to solving this problem. One is to maintain a central server that keeps up-to-date information about the position of every node. Another way is to take the mobile IP approach, where every node has a home agent that handles all the requests for the node and redirects those requests to the current position of the node.

There is an important property of information flow that can be used in sensor networks. Most of the communication in sensor networks is in the form of, for example, 'give me the temperature in the room'. This way, the nodes can be addressed based on their geographical position. This information is very useful for the routing protocol to forward the information in the right direction. A class based addressing is used in pico radio. The address is a triplet in the form of $<$ location, node type, node subtype $>$. Location specifies a particular point or region in space that is of interest; node type defines which type of node is required, such as sensor, controller or actuator. The node subtype further narrows down the scope of the address, such as temperature sensor, humidity sensor, etc. Class-based addressing defines the type of node in the region of space. A class-based addressing is assumed within the network layer.

The dedicated routing protocols have to contend with the wireless medium, i.e. low bandwidth, high error rates and burst losses, as well as the limitations imposed by these networks, such as frequently changing topology and low power devices. These protocols have to scale well with a large number of nodes in the network. The dedicated protocols can be categorized into proactive and reactive.

Proactive routing protocols have the distinguishing characteristic of attempting to maintain consistent up-to-date routing information from each node to every other node in the network. Every node maintains one or more routing tables that store the routing information, and topology changes are propagated throughout the network as updates, so that the network view remains consistent. The protocols vary in the number of routing tables maintained and the method by which the routing updates are propagated.

The Destination Sequenced Distance Vector Routing protocol (DSDV) is proactive protocol based on the Bellman–Ford algorithm for shortest paths, which ensures that there is no loop in the routing tables. Every node in the network maintains the next hop and distance information to every other node in the network. Routing table updates are periodically transmitted throughout the network to maintain table consistency.

Link State Routing (LSR) is a proactive protocol in which each node floods the cost of all the links to which it is connected throughout the network. Every node then calculates the cost of reaching every other node using shortest path algorithms. The protocol works correctly even if unidirectional links are present, whereas DSDV assumes bidirectional links.

In contrast to proactive routing protocols, reactive routing protocols create routes only when desired. An explicit route discovery process creates routes only on demand. These routes can be either source initiated or destination initiated. Source-initiated routing means that the source node begins the discovery process, while destination-initiated routing occurs when the destination begins discovery protocol. Once a route is established, the route discovery process ends, and a maintenance procedure preserves this route until the route breaks down or is no longer desired.

Ad Hoc On Demand Distance Vector Routing (AODV) is a routing protocol based on the distance vector algorithm similar to DSDV, with the difference that AODV is reactive. It is a source-initiated protocol, with the source node broadcasting a Route Request (RREQ) when it determines that it needs a route to a destination and does not have one available. This request is broadcast until the destination or an intermediate node with a route to the destination is located. Intermediate nodes record the address of the neighbor from which the first copy of the broadcast packet is received, in their route tables, thus establishing a reverse path.

Dynamic Source Routing (DSR) is a reactive protocol that is source initiated and based on the concept of source routing, in which the source specifies the entire route to be taken by a packet, rather than just the next hop. If the source node does not have a route, it floods the network with a Route Request (RREQ). Any node that has a path to the destination can reply with a Route Reply (RREP) to the source. This reply contains the entire path recorded in the RREQ packet. The entire path is added to the header of every packet to the destination.

Directed diffusion is a communication paradigm specifically for sensor networks. It is a destination-initiated reactive protocol that is data centric and application aware. Diffusion works well for sensor networks where queries, for instance, 'give me the temperature in a particular area, and query responses', are the most common form of communication. A destination node (controller) requests data by sending interests for data. This interest is flooded over the network, but each node knows only the neighbor from which it received the request, and the node sets up a gradient to send data to the neighbor. In this process, the interest reaches the source node (sensor), but each node knows only its neighbor(s) who asked for the data, and does not know the consumer of the data. If each node receives the same interest from more than one neighbor, the data will travel to the controller node along multiple paths. Among these paths, one high-rate path is defined, and the remaining paths are low rate. This is achieved by sending out positive reinforcements to increase the rate of a particular path. There is also a mechanism for negative reinforcements to change high-rate paths to low rate, which are used when a more efficient path emerges.

The potential problem in routing protocols is that they find the lowest energy route and use it for all communication. This is not favorable for the network lifetime. Using a low-energy path frequently leads to energy depletion in the nodes along that path, and may lead to network partition.

The basic idea of EAR is to increase the survivability of networks, which may lead to using suboptimal paths. This ensures that the optimal path does not get depleted and the network degrades gracefully and does not become partitioned. To achieve this, multiple paths are found between the source and destinations, and each path is assigned a probability depending on the energy metric. Every time the data is sent from the source to destination, a path is randomly chosen, depending on the probability. None of the paths is used all the time, which prevents energy depletion. Different paths are tried continuously, improving tolerance to nodes moving around the network.

EAR is a reactive routing protocol, and destination initiated protocol. The consumer of data initiates the route request and maintains the route subsequently. Multiple paths are maintained from source to destination. EAR uses only one path at all times whereas diffusion sends data along all the paths at regular intervals. Due to the probabilistic choice of routes, EAR can continuously evaluate different routes and choose the probabilities accordingly. EAR protocol has three phases:

- Set-up phase or interest propagation, in which the localized flooding occurs to find all the routes from source to destination and their energy costs. This occurs when routing (interest) tables are built up.
- Data communication phase or data propagation in which data is sent from source to destination, using the information from the earlier phase. This occurs when paths are chosen probabilistically according to the energy costs calculated earlier.
- Route maintenance, which is minimal. Localized flooding is performed infrequently from destination to source to keep all the paths alive.

The set-up phase is performed as in the following seven steps.

(1) The destination node initiates the connection by flooding the network in the direction of the source node. It also sets the cost field to zero before sending the request.

$$Cost(N_D) = 0$$

(2) Every intermediate node forwards the request only to the neighbors that are closer to the source node than itself and farther away from the destination node. Thus at a node N_i, the request is sent only to a neighbor N_j which satisfies:

$$d(N_i, N_S) \geqslant d(N_j, N_S)$$

$$d(N_i, N_D) \leqslant d(N_j, N_D)$$

where $d(N_i, N_j)$ is the distance between N_i and N_j.

(3) On receiving the request, the energy metric for the neighbor that sent the request is computed and is added to the total cost of the path. Thus, if the request is sent from node N_i to node N_j, N_j calculates the cost of the path as:

$$C_{N_j,N_i} = Cost(N_i) + Metric(N_j, N_i)$$

(4) Paths that have a very high cost are discarded and not added to the forwarding table. Only the neighbors N_i with paths of low cost are added

to the forwarding table FT_j of N_j.

$$FT_j = \{i | C_{N_j,N_i} \leqslant \alpha \times (\min_k C_{N_j,N_k})\}$$

(5) Node N_j assigns a probability to each of the neighbors N_i in the forwarding table FT_j, with the probability inversely proportional to the cost.

$$P_{N_j,N_k} = (C_{N_j,N_i})^{-1} \Big/ \sum_{k \in FT_j} (C_{N_j,N_k})^{-1}$$

(6) Thus, each node N_j has a number of neighbors through which it can route packets to the destination. N_j then calculates the average cost of reaching the destination using the neighbors in the forwarding table.

$$Cost(N_j) = \sum_{i \in FT_j} P_{N_j,N_i} C_{N_j,N_i}$$

(7) This average cost, $Cost(N_j)$ is set in the cost field of the request packet and forwarded along towards the source node as in (2).

The data communication phase is performed in the following steps:

(1) The source node sends the data packet to any of the neighbors in the forwarding table, with the probability of the neighbor being chosen being equal to the probability in the forwarding table.
(2) Each of the intermediate nodes forwards the data packet to a randomly chosen neighbor in its forwarding table, with the probability of the neighbor being chosen being equal to the probability in the forwarding table.
(3) This continues until the data packet reaches the destination node.

The energy metric used to evaluate routes is a very important component of the protocol. Depending on the metric, the characteristics of the protocol can change substantially. This metric can include information about the cost of using the path, energy health of the nodes along the path, topology of the network, etc. EAR uses the metric;

$$C_{ij} = e_{ij}^{\alpha} R_i^{\beta}$$

where C_{ij} is the cost metric between nodes i and j, e_{ij} is the energy used to transmit and receive on the link, and R_i is the residual energy at node i normalized to the initial energy of the node. The weighting factors α and β can be chosen to find the minimum energy path, the path with nodes having the highest energy, or a combination of these. This metric has a deep impact on the protocol performance and needs to be thoroughly evaluated.

4.3. ALTRUISTS OR FRIENDLY NEIGHBORS IN THE PICO RADIO SENSOR NETWORK

Self-configuring wireless sensor networks consist of hundreds or thousands of small, cheap, battery-driven, spread-out nodes bearing a wireless modem to accomplish a monitoring or control task jointly. An important concern is the network lifetime: as nodes run out of power, the connectivity decreases and the network can finally be partitioned and become dysfunctional. The concept of altruistic nodes can be applied to the routing protocol of the pico radio (the Energy Aware Routing protocol, EAR). The concept of altruists is a lightweight approach for exploiting differences in the node capabilities. The altruist approach can achieve significant gains in terms of network lifetime over the already lifetime optimized EAR protocol of pico radio.

Example applications of sensor networks are microclimate control in buildings, environmental monitoring, home automation, distributed monitoring of factory plants or chemical processes, interactive museums, etc. The ultimate goal is to make sensor network nodes so small that they can be just thrown out somewhere, or smoothly woven into other materials such as wallpapers.

Sensor nodes are typically battery driven, and the batteries are too small, too cheap and too numerous to consider replacing or recharging. Hence, their energy consumption is a major concern, imposing a design constraint of utmost importance. An immediate consequence is that the transmitting power of the nodes should be restricted to a few meters. Furthermore, nodes should go into sleep mode as often as possible. In sleep mode, a node switches off its radio circuitry and other subsystems. The restricted transmit power leads to the necessity of multihop communications: if the distance between two communicating nodes is too large, then the intermediate nodes have to relay the packets, which in turn drains the batteries of the relaying nodes.

To achieve a maximum network lifetime, it is mandatory to optimize the energy consumption in all layers of the protocol stack, from the physical layer to the application layer. The approach of jointly designing the application and the communication related layers can be effective. The data transmitted over the network is specified in terms of the application, for example, 'what is the temperature in the neighboring room?' To respond to this request, it is not necessary to use general purpose routing protocols. Instead, the routing process can explicitly take geographical information into account to perform location based routing. The routing functionality is application aware. The needs of the application layer and the routing protocol also influence the design of link layer, MAC layer, and physical layer. This way, the protocol architecture for sensor networks differs from other networks.

The example of temperature sensors helps to explain the notion of network lifetime: a monitor station that wants to get the temperature is not concerned about which sensor delivers the temperature in the room. If there are many of these sensors in this room, at least one of them will send its data, and this value is probably similar to those of the other sensors. As long as there are intermediate nodes to forward data packets to the monitor, the network can function. However, as a node becomes depleted and dies, the possible forwarding routes may be eliminated. The network may split into two or more clusters with no connectivity, and will no longer serve its purpose.

The ultimate design goal of pico radio is an ultra-low-power wireless sensor network with cheap nodes (substantially less than 1 dollar), which are small (less than 1 cubic centimeter), do not weigh much, and are battery driven. The hardware and the software/firmware design is targeted for a power dissipation level of below 100 microwatts, whereas a Bluetooth radio consumes more than 100 milliwatts. The protocol stack for pico radio is designed with the assumption that all nodes have the same capabilities (battery, processor power) and all protocols work in a decentralized manner.

There are applications, however, where this is not necessarily true. If the network has actuator nodes (e.g. a small motor controlling a room), the latter will likely be connected to a regular power line, since operation of these nodes requires significant amounts of energy. Making use of these asymmetries causes the more capable stations to perform as altruists: they announce their capabilities to their neighbors, and may use their services. The altruist approach is a light-weight approach as compared with clustering schemes. The altruist approach is applied to the data forwarding stage of the pico radio network layer protocol, which is the EAR protocol. Specifically, a simple process to implement altruistic add-on to EAR is considered. The experiment presented by Willig *et al.* (2002) compares different performance metrics for the unmodified EAR protocol and the altruist scheme, and shows that the altruist scheme can achieve significant gains in terms of network lifetime.

The pico radio is a sensor network of ultra-low-powered nodes, called pico nodes. There are three types of pico node: sensor nodes, actuator nodes, and monitor nodes. The sensor nodes acquire data (using a built-in sensor facility), which is typically processed by monitor nodes. The resulting output (control actions) is sent to the actuator nodes.

Pico nodes use two channels in the 1.9-GHz band. An on–off keying scheme is employed as the modulation scheme, providing a data rate of 10 kbit/sec per channel. One channel is used for data packets, the other for packet management. Pico nodes can also use a Bluetooth radio.

Pico nodes MAC layer uses a combination of CSMA with a cycled receiver scheme, where a node goes into sleep mode periodically. Communication only takes place when a node is awake. A promising solution for power savings is the wake-up radio. We assume that a node A spends most of the time in a sleep mode. When another node, B, wants to transmit a packet to A, it sends a wake-up signal on the wake-up radio channel, a dedicated, very low bit rate and very low-power channel. The wake-up signal carries the address of A. Upon reception of the signal, A wakes up, participates in the packet exchange, and goes back to sleep when finished. Transmission and reception on the wake-up radio channel consumes less power than on the data and management channels. The wake-up radio is always on.

Besides the MAC layer, several different functions are performed between physical layer and network layer:

- Allocation subsystem helps nodes to discover their geographical position in terms of (x, y, z) coordinates within the network, using the help of so called anchor nodes, which know their position *a priori* (configured during network set-up). Nodes can determine their position using signal strength measurements to nodes of a known position, or they infer their position from the hop-count distance between their immediate neighbors to the anchor nodes.
- A local address assignment protocol determines locally unique node addresses. 'Locally unique' means that no node has two neighbors with the same address x, but x can be reused in more distant parts of the network. (In sensor networks globally unique node addresses like Ethernet MAC address or IP addresses, have disadvantages, since address assignment involves complex management (e.g. address resolution protocols), specifically in the presence of mobile nodes).
- A power control and topology control algorithm is responsible for adjusting the transmit powers of the pico nodes in order to find a proper network topology. The goal is to find a well connected graph, and to avoid too many neighbors per node. It is necessary to restrict the transmit power in order to reduce the interference imposed by a node on its neighbors.
- A neighbor-list management facility maintains a table of currently reachable neighbors of a node and their (x, y, z) coordinates. The information is obtained directly from the topology control algorithm.

4.3.1. Energy-Aware Routing

In EAR and directed diffusion, the routing is data centric and takes the application-layer data into account. A sink generates an interest specification (ispec), which specifies the type of data it is interested in and the geographical location or area where this data is expected. Location is specified by using (x, y, z) coordinates. To enable more user-centric descriptions, for instance, the left window in the next room, another level of indirection is needed, which maps these descriptions to spatial coordinates. It is assumed that every node knows its own position (from the locationing subsystem) and its (type, subtype) tuple, where the type can be sensor, actuator or monitor, and a subtype can be a temperature sensor, light sensor, or pressure sensor. Furthermore, every node has a locally unique node address, as determined by the local address assignment protocol.

EAR and directed diffusion schemes distinguish between route-discovery phase and data-transmission phase. The route discovery is initiated by the sink. A flooding scheme (e.g. directional flooding) is used to find the source(s). Flooding approaches tend to find not only a single route, but all the routes. A difference between directed diffusion and EAR is that the directed diffusion introduces a reinforcement phase, where among the several possible routes between the source and the sink the most energy efficient route is selected by using control messages issued by the sink. The consequence is that for a longer lasting communication between the source and the sink, all data packets take the same route, which may quickly deplete the node power along that route. In contrast to this, the EAR approach keeps most of the possible routes, and only the very inefficient routes are discarded. In the data-transmission phase the packet route is chosen randomly from the available routes. This reduces the load for a fixed intermediate node and increases battery lifetime.

The energy aware routing scheme (EAR) works as follows:

- The sink generates an interest message. The interest message contains (amongst others) an interest specification (ispec), and a cost field, initialized with 0. The sink also includes its own node specification (position, (type, subtype) tuple, abbreviated as node spec). The interest message is sent to those of its neighbors, which are closer to the target area (of the ispec).
- When node i receives the interest message from an upstream node j, it takes the following actions:
  - the ispec is inserted into an interest cache, along with j's node address, the received cost field and the sink's node spec. If there is already an entry with the same ispec and node spec in the interest cache, the node does not forward the interest message any more, in order to bound the number of interest packets. (By taking both ispec and node spec into account, a single sink node can issue different interests at the same time). When the received cost is already very high, the node may choose to drop the interest.

- When the ispec matches node i, it starts generating the requested data. In addition, i broadcasts the interest message locally in order to propagate it to neighboring nodes of the same type (which are potential data sources, too).
- If the ispec does not match node i, the interest message is forwarded. The first step is to update the cost field:

$$new\_cost\_field = cost\_field + metric(i, j),$$

where $metric(i, j)$ represents the costs for node i to transmit a data packet to node j. There are many different ways to use this field: e.g. setting $metric(i, j) = const$, is equivalent to a hop-count *metric*, and setting $metric(i, j)$ to the inverse of node i's remaining energy assigns a costly route to a node with reduced energy. This way, a node with reduced energy is less likely to be selected as the next data forwarder.
- The final forwarding step for node i is to send a copy of the interest message to those neighboring nodes that are geographically closer to the source and farther away from the sink. To do this, i uses neighborhood information collected by the MAC layer. This information includes the neighbor's geographical position.

- When an intermediate node k receives a data packet not destined to itself, k has to forward this packet towards the sink. To do so, it looks up all the interests in the interest cache to which the data fits: the data packet contains type and subtype fields describing the data and the node spec (position) of the source, which are compared to the respective values of the ispecs stored in the interest cache. The matching cache entries differ only in the stored cost field and the node addresses of the upstream nodes. Among the possible upstream nodes one is randomly chosen, and the probabilities are assigned proportionally to the respective cost values.

This scheme is different from the EAR scheme in two respects. In the original EAR scheme:

- intermediate nodes do not filter out the second and following copies of an interest packet from the same sink, and
- for every forwarded interest packet they set the cost field to the mean value of the costs of all routes known so far.

Hence, the EAR scheme tends to produce more copies of interest messages, while the scheme presented above propagates only the costs of the path with the minimum delay and the number of hops between sink and intermediate node. Currently some further alternatives can be considered:

- After getting the first copy of an interest message, an intermediate node waits a certain amount of time for further packets. After this time it forwards only one packet with the average cost. However, this approach tends to increase the delays. In addition, it is hard to find good values for the waiting time. These should be suitable for intermediate nodes close to the sink or far away from the sink.
- The first copy of an interest message is immediately sent out. Further copies are sent, when the accumulated average cost value differs significantly from the last sent value.

4.3.2. Altruists or Friendly Neighbors

The altruist approach explores asymmetries in node capabilities. In a sensor network, not all nodes are the same type. When there are actuator or monitor nodes, these are probably attached to a permanent power source, or have more powerful processors and more memory than other nodes. Sensors either have batteries or a permanent power supply. While the battery driven sensors are spread out, sensors with a permanent power supply are placed carefully to increase the network lifetime. These asymmetries can be discussed at different levels:

- *Application level*: some nodes can perform data aggregation and concentration or data filtering. As a simple example, all temperature sensors in a small geographical area deliver similar temperature values to a monitor station. If the packets traverse the same intermediate node, it can accumulate a number of packets, calculate a average temperature and forward only a single packet with the average value to the monitor.
- *Network level*: restrict data forwarding to stations with more energy.
- *MAC layer and link layer*: A more capable node can act as a central station in centralized MACs by scheduling transmissions to its associated nodes. A node that has no outstanding transmission can go into a sleep mode. This approach is explored in the IEEE 802.11 PCF (Point Coordination Function) for power saving.

Two different approaches to exploit asymmetries are clustering schemes and altruist schemes. In clustering schemes, the network is partitioned into clusters. Each cluster has a cluster head, which does most of the work. Each node is associated with, at most, one cluster head and all communications are relayed through the cluster head. These schemes typically require protocols for cluster-head election and node association. In the presence of mobile nodes, both functions have to be carried out frequently enough to maintain a consistent network state.

In the altruist or friendly neighbor approach, a node simply broadcasts its capabilities to its neighbors, along with its position, i.e. node address, and a lifetime value. In this broadcast, the altruist uses an altruist announcement packet. The lifetime value indicates how long the altruist node is willing to do more work in the soft-state approach. The other nodes in the altruist's neighborhood can freely decide whether they use the service offered by the altruist or not. Altruist protocols are lightweight as compared with cluster approaches, since they only involve an occasional altruist announcement packet, whereas cluster approaches need cluster-head election and association protocols (typically implemented with two-way or three-way handshake, e.g. in the IEEE 802.11 standard).

The altruist approach can be used in the data packet forwarding stage of the EAR protocol. We make an assumption that only nodes with access to a power line send altruist announcement packets, hence, a node has some facility to query the type of its power supply. Other schemes are possible, where the probability of a node becoming an altruist can depend on its remaining energy, the number of altruists in its neighborhood, the time elapsed since it was last an altruist, etc. Every node that receives an altruist announcement packet stores the issuing node address in an altruist cache, and starts a timer for this cache entry according to the indicated lifetime. (The size of the altruist cache and the number of parallel timers is limited by the node's number of neighbors). If the timer expires, the entry is removed from the altruist cache. When an arbitrary node receives a data packet and has to decide about the next data forwarder, it first looks up all the possible upstream nodes j and their respective costs c_j from the interest cache. The costs c_j of those upstream nodes j that are currently altruists (according to the altruist cache) are reduced by a fixed factor $0 \geqslant \alpha \geqslant 1$ (called cost reduction factor):

$$c'_j = \begin{cases} \alpha \times c_j & \text{if } \textit{node } j \textit{ is an altruist} \\ c_j & \text{if } \textit{node } j \textit{ is not an altruist} \end{cases}$$

This increases the probability that an altruist is chosen as the next data forwarder. The EAR protocol with the altruist scheme is denoted as EAR+A.

Note that EAR+A works somewhat in opposition to the original idea of EAR to distribute the forwarding load as smoothly as possible over all available routes. In fact, depending on α, the EAR+A protocol favors altruistic nodes. The problem with this is that the nodes behind the altruists also experience an increased forwarding load as compared to EAR. This is a positive aspect as long as these nodes are also altruists. Otherwise, these nodes are potentially depleted faster than the nodes in EAR. Furthermore, the altruist scheme tends to increase the mean number of hops taken by a data packet.

In many applications network reliability is a critical issue. The altruist scheme as described above is basically a soft-state scheme, since the altruist announcements have only a limited lifetime. Furthermore, the operation of the network does not depend critically on the altruists. If an altruist node dies for some reason, its neighbors have inaccurate state information for a time no longer than the announced lifetime. If this time expires, the network returns to its normal mode. Hence, the network designer can choose whether to accept the inaccuracy for longer lifetimes and less overhead by altruist announcement packets.

4.3.3. Analysis of Energy Aware and Altruists Routing Schemes

The experiment was designed to gain insight into the following questions:

- Does the presence of power unconstrained stations have an impact on network lifetime for both the unmodified EAR and the EAR plus altruist routing schemes?
- Does the altruist scheme have an effect on network lifetime and is there a dependence on the percentage of altruistic nodes or on the load patterns?

The model is divided into a node model describing the internal structure of a single pico node, and a channel model, which defines the physical channel and the channel error behavior. The model is built with a steady state assumption: the network initialization (localization algorithm, topology control, local address assignment, neighbor-list determination) is already done and is not a part of the model, furthermore, there are no mobile stations.

A node model consists of a MAC layer, a network layer, an application layer and a node controller:

- The application layer of sink nodes generates interests for other nodes (randomly chosen). The interests are artificially restricted to match a single node position; the more common case of an interest specifying a larger geographical area is foreseen but not used. A sink can issue several different interests at the same time. When an interest matches a source node, the source-node application layer generates data packets at a certain rate for a certain duration. Interest (data) packets have a length of 288 (176) bits.

- The network layer implements the EAR protocol and the EAR+A scheme on top of it. The cost metric *metric*(i, j) is inversely proportional to node's i remaining energy r_i, i.e., metric $(i, j) = 1/r_i$.
- On the MAC layer the experiment used a simple nonpersistent CSMA (Carrier Sense Multiple Access) protocol where the backoff times are drawn uniformly from a fixed interval (0 to 100 ms). The carrier sense operation is assumed to indicate a carrier when at the node position the composite signal level from other node transmissions is above a certain threshold.
- The node controller is essentially an abstraction of a node's energy supply. For battery driven nodes the experiment uses negligible computation costs as compared to the cost of transmitting or receiving packets. A node spends energy on transmitting a packet and on receiving a packet destined for it (i.e. with its own node address or the broadcast address). The latter assumption corresponds to the wake-up radio scheme. Transmitting needs 4 milliwatts, and receiving needs 3 milliwatts in the pico nodes. If a battery powered node i has less than 1 % remaining energy r_i, it is considered dead and does not communicate any more. A certain percentage of nodes has infinite power.

The channel model considers only mutual interference, which is computed by a simple path loss model: for an isotropic antenna, a transmit power of P_T, and a distance of d meters to the destination node, the received power at the destination node is given by

$$P_R = P_T \times g \times d^{\gamma}$$

where g is a scaling factor (incorporating antenna gains and wavelength) and is the path loss exponent. The experiment uses the optimistic assumption of $\gamma = 2$ (this exponent varies typically between 2 and 5, from ideal free-space propagation to attenuation on obstacles). For $d < 1$ meter, the experiment takes $P_R = P_T \times g$. Beyond a certain distance depending on P_T and g the signal is below a prespecified threshold and is considered undetectable. The channel model computes the overall signal level at some geographical location by adding the received power coming from all ongoing transmissions at this point. This computation is invoked at the start and end of packet transmissions.

The channel model is also responsible for generating packet errors. The strategy is simple: it marks a packet as erroneous, if the ratio of the packet's signal strength at the receiver as compared with all the interference is below some threshold, called minimum signal to interference ratio (SIR). The experiment uses an SIR of 10 so that parallel transmissions can be successful, if their distance is large enough. Only the data channel is used, the management channel is not modeled. The data channel has a bit rate of 10 kbit/sec.

An experiment with the Large-Scale Office Scenario (LSOSC) is meant to resemble a microclimate control application in a large scale office (20 × 30 m as shown in Figure 4.1). The node placement is nonuniform; close to the windows the density is much higher than in the middle of the room (there is a total of 121 nodes). A single monitor station in a corner is the only sink in this network. Only the monitor node generates interests for randomly

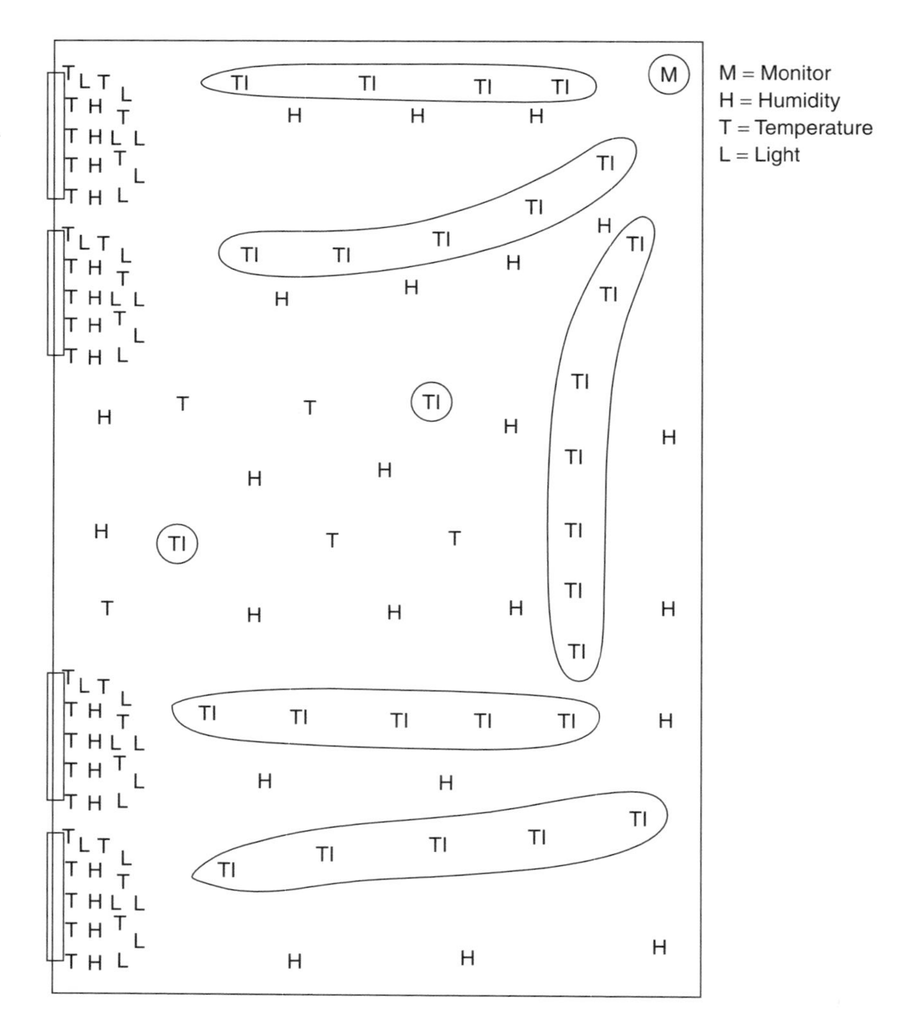

Figure 4.1 Large-scale office scenario.

chosen sensors, at most one interest is active at any time. Besides the monitor, some strategically placed temperature sensors have an infinite power supply (marked as TI in Figure 4.1). A node can transmit over 6 meters. For the altruist announcements, $\alpha = 0.01$ and a lifetime of 10 000 s. A source generates data packets every 3 seconds.

The focus is on the network lifetime. This experiment is based on the time that 50 % out of the total number of nodes needs to die due to energy depletion. (A similar measure, the 50 % lethal dose LD_{50} is used in medicine to assess the efficiency of toxins). Other possible measures are the time until the first node dies or the time before the network of the alive nodes is partitioned the first time (i.e. loses its full connectivity). However, the 50 % metric can be applied to both scenarios, since in LSOSC the network cannot get disconnected (due to the chosen placement of nodes with infinite power supply and the transmission range of 6 meters), as opposed to USC.

Results for uniform scenario discuss the mean 50 % nodes dead time versus the percentage of nodes with unconstrained power supply.

- Both EAR and EAR+A protocols can take advantage of nodes with unconstrained power supply, even despite the fear that nodes in the neighborhood of unconstrained nodes get depleted faster. Due to the energy metric used (costs inversely proportional to remaining energy), packets tend to go more and more over the unconstrained nodes, when the other nodes run out of energy. This takes the forwarding burden from the other nodes.
- The EAR+A scheme gives in the main some advantage over EAR, and the gain increases with the percentage of unconstrained nodes. However, the altruist scheme is not always better, since with fixed unconstrained node percentage there are some random seeds for which EAR gives better network lifetime.

The mean interest answer time (as taken over all interests and all needs) is defined as the time between a node issuing an interest and getting the first data packet. For EAR+A the mean interest answer time is higher than for EAR. This can be attributed to the tendency of EAR+A to favor altruists, which may well not be the shortest path.

The comparably high mean interest-answer times are determined by contributions with high values, primarily from the first phase of the network lifetime, when all nodes are alive. We consider the time needed for an interest to reach the source node. The source node immediately starts generating data packets. The problem here lies in the combination of directional flooding and the nonpersistent CSMA protocol: for a single interest a large number of interest packets is generated successively as the interest moves towards the source (flooding). At the time when the first interest packet reaches the source, many copies of the same interest are stored in upstream nodes. This means that at the time the first interest packet reaches the source, the area around the source is congested by additional interest packets. Data packets have to penetrate this congested area, which may take a long time due to the CSMA operation and lack of packet priorities. The MAC throughput does not increase by reducing the backoff window size.

In the LSOSC, the experiment uses the mean 50 % nodes dead time against the interest lifetime and data generation period. This corresponds to varying the ratio between data packets and interest packets. The larger the interest lifetime, the more data packets are transmitted per single interest.

- For both EAR and EAR+A, the network lifetime increases with an increased interest lifetime. This permits the conclusion that actual interest propagation (which uses directional flooding) is expensive as compared with data transmission.
- The altruist scheme significantly increases the network lifetime. The altruist scheme is specifically designed to improve the data transmission phase while not affecting the interest propagation phase.

The high relative costs of the interest propagation phase can be explained by the comparably large number of interest packets a single node receives. This number is for the LSOSC scenario directly reflected by the interest cache length, which varies typically between 10 and 20. Hence, a power constrained node burns energy for between 10 and 20 packet receptions and one packet transmission per interest, while not involved in the corresponding data transmission phase. This suggests that the altruist concept can also be applied to the interest propagation phase.

The altruistic nodes can be used in the pico radio sensor network. This scheme is applied to the data forwarding stage of the pico radio EAR protocol, and it shows up that significant improvements in network lifetime can be achieved as compared to the already lifetime optimized EAR protocol. This holds true specifically for the case where much more bandwidth is spent on data transmission than on interest propagation.

4.4. AGGREGATE QUERIES IN SENSOR NETWORKS

The database generic query interface for data aggregation can be applied to dedicated networks of sensor devices. Aggregation is used as a data reduction tool. Networking approaches have focused on application specific solutions. The network aggregation approach is driven by a general purpose, SQL (Structured Query Language)-style interface that can execute queries over any type of sensor data while providing opportunities for significant optimization.

Advances in computing technology have led to the production of a new class of computing device: the wireless, battery-powered, smart sensor. Unlike traditional sensors deployed throughout buildings, laboratories, and equipment everywhere, these new sensors are not merely passive devices that modulate a voltage based on some environmental parameter: they are fully fledged computers, capable of filtering, sharing, and combining sensor readings.

Small sensor devices are called motes. Motes are equipped with a radio, a processor, and a suite of sensors. An operating system makes it possible to deploy dedicated networks of sensors that can locate each other and route data without any detailed knowledge of network topology.

Of particular interest are aggregates, whose operations, as stated earlier, summarize current sensor values in part of, or the entire, sensor network, and when users want to know temperature patterns in relatively large regions, individual sensor readings are of little value.

Sensor networks are limited in external bandwidth, i.e. how much data they can deliver to an outside system. In many cases the externally available bandwidth is a small fraction of the aggregate internal bandwidth. Thus computing aggregates in-network is also attractive from a network performance and longevity standpoint: extracting all data over all time from all sensors will consume large amounts of time and power as each individual sensor's data is independently routed through the network. Studies have shown that aggregation dramatically reduces the amount of data routed through the network, increasing throughput and extending the life of battery-powered sensor networks as less load is placed on power-limited radios.

Networking research considered aggregation to be application specific technique that can be used to reduce the amount of data that must be sent over a network. Database community views aggregates as a generic technique that can be applied to any data, irrespective of the application. The system provides a generic aggregation interface that allows aggregate queries to be posed over networks of sensors. The benefits of this approach over the traditional network solution are as follows:

- By defining the language that users use to express aggregates, we can significantly optimize their computation.

- The same aggregation language can be applied to all data types, thus, the programmers can issue declarative, SQL (Structured Query Language) style queries rather than implement custom networking protocols to extract the data they need from the network.

Basic database aggregates (COUNT, MIN, MAX, SUM, and AVERAGE) can be implemented in special networks of sensors. This generic approach leads to significant power savings. Sensor network queries can be structured as time series of aggregates, and adapted to the changing network structure.

Motes are equipped with a 4-MHz Atmel microprocessor with 512 bytes of RAM (random access memory) and 8 kb of code space, a 917 MHz RFM (Radio Frequency Module) radio running at 10 kb/s, and 32 kb of EEPROM (Electronically Erasable Programmable read only). An expansion slot accommodates a variety of sensor boards by exposing a number of analog input lines as well as chip-to-chip serial buses. The sensor options include: light, temperature, magnetic field, acceleration (and vibration), sound, and power.

The radio hardware uses a single channel, and on–off keying. It provides an unbuffered bit-level interface; the rest of the communication stack (up to message layer) is implemented by operating-system software. Like all single-channel radios, it offers only a half duplex channel. This implementation uses a CSMA (Carrier Sense Multiple Access) media access protocol with random backoff scheme. Message delivery is unreliable by default, though applications can build up an acknowledgement layer. Often, a message acknowledgement can be obtained.

Power is supplied via a free hanging battery pack or a coin-cell attached through the expansion slot.

The effective lifetime of the sensor is determined by its power supply. The power consumption of each sensor node is dominated by the cost of transmitting and receiving messages, including processor cost, where sending a single bit of data requires about 4000 nJ of energy, whereas a single instruction on a 5-mW processor running at 4 MHz consumes only 5 nJ. Thus, in terms of power consumption, transmitting a single bit of data is equivalent to 800 instructions. This energy trade-off between communication and computation implies that many applications will benefit by processing the data inside the network rather than simply transmitting the sensor readings.

The operating system provides a number of services greatly to simplify writing programs that capture and process sensor data and transmit messages over the radio. The API (application programming interface) can send and receive messages and read from sensors. The messaging and networking aspects of the operating system and wireless sensors are the most relevant to aggregation.

Radio is a broadcast medium, where a sensor within hearing distance can hear any message, irrespective of whether or not this sensor is the intended recipient. The radio links are typically symmetric: if sensor α can hear sensor β, we assume sensor β can also hear sensor α. Note that this may not be a valid assumption in some cases: if α's signal strength is higher, because its batteries are fresher or its signal is more amplified, β will be able to hear α but not to reply to it.

Each 30-byte message type has a message id (identifier) that distinguishes it from other types of messages. Sensor programmers write message-ID specific handlers that are invoked by the operating system when a message of the appropriate ID is heard on the radio. Each sensor has a unique sensor ID that distinguishes it from other sensors. All messages specify their recipient (or broadcast, meaning all available recipients), allowing sensors to ignore messages not intended for them, although nonbroadcast messages must still be received by all sensors within range, unintended recipients simply drop messages not addressed to them.

The sensors route data by building a routing tree. This is one of many possible techniques that can be used. A tree can be built and maintained efficiently in the presence of a changing network topology. One sensor is dedicated as a root. The root is the point from which the routing tree will be built, and upon which aggregated data will converge. Thus, the root is typically the sensor that interfaces the querying user with the rest of the network. The root broadcasts a message asking sensors to organize into a routing tree; in that message it specifies its own ID and its level, or distance from the root, which is zero. Any sensor that hears this message assigns its own level to be the level in the message plus one, if its current level is not already less than or equal to the level in the message. The sensor also chooses the sender of the message as its parent, through which it will route messages to the root. Each of these sensors then rebroadcasts the routing message, inserting their own identifiers and levels. The routing message floods down the tree, with each node rebroadcasting the message until all nodes have been assigned a level and a parent. Nodes that hear multiple parents choose one arbitrarily. Multiple parents can be used to improve the quality of aggregates. These routing messages are periodically broadcast from the root, so that the process of topology discovery goes on continuously. This constant topology maintenance makes it relatively easy to adapt to network changes caused by mobility of certain nodes, or to the addition or deletion of sensors: each sensor simply looks at the history of received routing messages, and chooses the best parent, while ensuring that no routing cycles are created with that decision.

This approach makes it possible to route data efficiently towards the root. When a sensor wishes to send a message to the root, it sends the message to its parent, which in turn forwards the message on to its own parent, and so on, eventually reaching the root. This application does not address point-to-point routing. Flooding aggregation request, and routing replies up the tree to the root, is acceptable. As data is routed towards the root, it can be combined with data from other sensors so as to combine routing and aggregation efficiently.

Aggregation in SQL (Structured Query Language)-based database systems is defined by an aggregate function and a grouping predicate. The aggregate function specifies how a set of values should be combined to compute an aggregate; the standard set of SQL (Structured Query Language) aggregate functions is COUNT, MIN, MAX, AVERAGE, and SUM. These compute the obvious functions; for example, the SQL (Structured Query Language) statement:

SELECT AVERAGE (temp) FROM sensors

computes the average temperature from a table of sensors, which represents a set of sensor readings that have been read into the system. Similarly, the COUNT function counts the number of items in a set, the MIN and MAX functions compute minimal and maximal values, and SUM calculates the total of all values. Additionally, most database systems allow user-defined functions (UDFs) that specify more complex aggregates.

Grouping is also a standard feature of database systems. Rather than merely computing a single aggregate value over the entire set of data values, a grouping predicate partitions the values into groups based on an attribute. For example, the query:

SELECT TRUNC (temp/10), AVERAGE (light)

FROM sensors

GROUP BY TRUNC (temp/10)

HAVING AVERAGE (light) ⩾ 50

partitions sensor readings into groups according to their temperature reading and computes the average light reading within each group. The HAVING clause excludes groups whose average light readings are less than 50.

Users are often interested in viewing aggregates as sequences of changing values over time. The user is stationed at a desktop class PC (Personal Computer) with ample memory. Despite the simple appearances of this architecture, there are a number of difficulties presented by the limited capabilities of the sensors.

Throughout the following analyses, the focus is on reducing the total number of messages required to compute an aggregate; this is because message transmission costs typically dominate energy consumption of sensors, especially when performing only simple computation such as the five standard database aggregates.

4.4.1. Aggregation Techniques

A possible implementation of sensor network aggregation would be to use a centralized, server-based approach where all sensor readings are sent to the host PC, which then computes the aggregates. However, a distributed, in-network approach where aggregates are partially or fully computed by the sensors themselves as readings are routed through the network towards the host PC, can be considerably more efficient. The in-network approach, if properly implemented, has the potential of both lower latency and lower power than the server-based approach.

To illustrate the potential advantages of the in-network approach, a simple example of computing an aggregate over a group of sensors arranged is shown in Figure 4.2. Dotted lines represent connections between sensors, solid lines

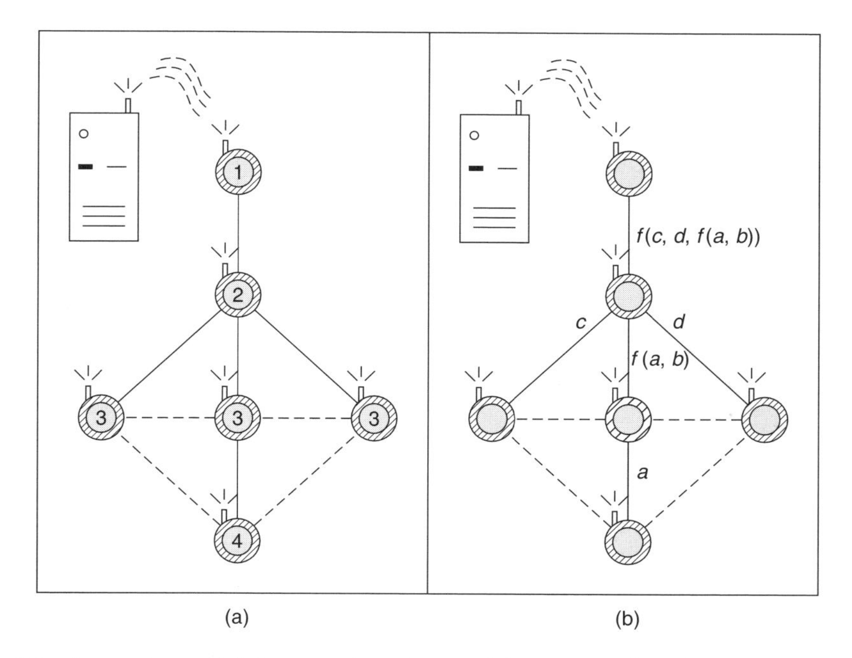

Figure 4.2 Server-based (a) versus in-network (b) aggregation. In (a), each node is labeled with the number of messages required to get data to the host PC: a total of 16 messages is required. In (b), only one message is sent along each edge as aggregation is performed by the sensors.

represent the routing tree imposed on top of this graph (as discussed above) to allow sensors to propagate data to the root along a single path. In the centralized approach, each sensor value must be routed to the root of the network; for a node at depth n, this requires $n - 1$ messages to be transmitted per sensor. The sensors in Figure 4.2(a) have been labeled with their distance from the root; adding these numbers gives a total of 16 messages required to route all aggregation information to the root. The sensors in Figure 4.2(b), with no children, simply transmit their readings to their parents. Intermediate nodes (with children) combine their own readings with the readings of their children via the aggregation function and propagate the partial aggregate, along with any extra data required to update the aggregate, up the tree.

The amount of data transmitted in this solution depends on the aggregate. In the AVERAGE function at each intermediate node n, the sum and count of all children's sensor readings are needed to compute the average of sensor readings of the subtree rooted at n. We assume that, in the case of AVERAGE, both pieces of information will easily fit into a single 30-byte message. Thus, a total of five messages needs to be sent for the average function. In the case of the other standard SQL (Structured Query Language) aggregates, no additional state is required: COUNT, MIN, MAX, and SUM can be computed by a parent node, given sensor or partial aggregate values at all of the child nodes.

A class of aggregation predicates is particularly well suited to the in-network regime. Such aggregates can be expressed as an aggregate function over the sets a and b such that:

- The basic SQL (Structured Query Language) aggregates all exhibit the above property, and the problems with this substructure map easily onto the underlying network.
- Aggregation queries are pushed down into a sensor network and the results are returned to the user. We assume that aggregate queries do not specify groups.

Computing an aggregate consists of two phases: a propagation phase, in which aggregate queries are pushed down into sensor networks, and an aggregation phase, where the aggregate values are propagated up from children to parents. The most basic approach to propagation works just like the network discovery algorithm described above, except that leaf nodes (nodes with no children) must discover that they are leaves and propagate singular aggregates up to their parents. Thus, when a sensor p receives an aggregate a, either from another sensor or from the user, it transmits a and begins listening. If p has any children, it will hear those children retransmit to their children, and will know it is not a leaf. If, after some time interval t, p has heard no children, it concludes it is a leaf and transmits its current sensor value up the routing tree. If p has children, it assumes they will all report within time t, and so after time t it computes the value of a applied to its own value and the values of its children and forwards this partial aggregate to its parent.

Notice that choosing too short a duration for t can lead to missed reports from children, and also that the proper value of t varies depending on the depth of the routing tree. We assume that t is set to be long enough for the message to have time to propagate down to all leaves below and back, or, numerically:

$$t = 2 \times (d_p - d_{tree}) \times (t_{xmit} + t_{process})$$

where t_{xmit} is the time to send a message and $t_{process}$ is the time to process an aggregation request. Empirical studies suggest that $(t_{xmit} + t_{process})$ needs to be 200 or more milliseconds. The time to transmit a 30-byte message on a 10-kbit radio is about 50 ms: each nibble must be DC balanced (have the same number of ones and zeros), costing extra bits, and simple forward error correction is used, meaning that for every byte, 18 bits must be transmitted; 18×30 bytes/10 000 bits/sec $= 50$ ms. Computation time is small, but significantly more than 50 ms must be allocated per hop to account for differences in clock synchronization between sensors, and random collision detection back-off in which those sensors engage. Thus, for a deep sensor network, computing a single aggregate can take several seconds. The unreliable communication inherent to sensor networks, coupled with such long computation times, makes this simple in-network approach undesirable.

Sensor networks are inherently unreliable: individual radio transmission can fail, nodes can move, and so on. Thus, it is very hard to guarantee that a significant portion of a sensor network was not detached during a particular aggregate computation. For example, what happens when a sensor, p, broadcasts a and its only child, c, somehow misses the message (perhaps because it was garbled during transmission). P will never hear c rebroadcast, and will assume that it has no children and that it should forward only its own sensor value. The entire network below p is thus excluded from the aggregation computation, and the end result is probably incorrect. Indeed, when any subtree of the graph can fail in this way, it is impossible to give any guarantees about the accuracy of the result.

One solution to this problem is to double check aggregates by computing them multiple times. The simplest way to do this would be to request the aggregate be computed multiple times at the root of the network; by observing the common case value of the aggregate, the client could make a reasonable guess as to its true value. The problem with this technique is that it requires retransmitting the aggregate request down the network multiple times, at a significant message overhead, and the user must wait for the entire aggregation interval for each additional result.

Pipelined aggregates are propagated into the network as described above. However, in the pipelined approach, time is divided into intervals of duration i. During each interval, every sensor that has heard the request to aggregate transmits a partial aggregate by applying a to its local reading and to the values its children reported during the previous interval. Thus, after the first interval, the root hears from sensors one radio-hop away. After the second, it hears aggregates of sensors one and two hops away, and so on. In order to include sensors which missed the request to begin aggregation, a sensor that hears another sensor reporting its aggregate value can assume that it too should begin reporting its aggregate value.

In addition to tending to include nodes that would have been excluded from a single pass aggregation, the pipelined solution has a number of interesting properties: first, after aggregates have propagated up from leaves, a new aggregate arrives every i seconds. Note that the value of i can be quite small, about the time it takes for a single sensor to produce and transmit a sensor reading, versus the value of t in the simple multiround solution proposed above, which is roughly $depth_{tree}$-times larger. Second, the total time for an aggregation request to propagate down to the leaves and back to the root is roughly t, but the user begins to see approximations of the aggregate after the first interval has elapsed; in very deep networks, this additional feedback may be a useful approximation while waiting for the true value to propagate out and back. These two properties provide users with a stream of aggregate values that changes as sensor readings and the underlying network change. As discussed above, such continuous results are often more useful than a single, isolated aggregate, as they allow users to understand how the network performs over time. Figure 4.3 illustrates a simple pipelined aggregate in a small sensor network.

The most significant drawback with this approach is that a number of additional messages is transmitted to extract the first aggregate over all sensors. In the example shown in Figure 4.3, 22 messages are sent, since each aggregating node transmits once per time interval. The comparable nonpipelined aggregate requires only 10 messages, one down and one back along each edge. In this example, after the initial 12-message overhead, each additional aggregate arrives at a cost of only five messages and at a rate of one update per time interval. Still, it is useful to consider optimizations to reduce this overhead. One option is that sensors could transmit only when the value of the aggregate computed over their subtree changes, and parents could assume that their children's aggregate values are unchanged unless

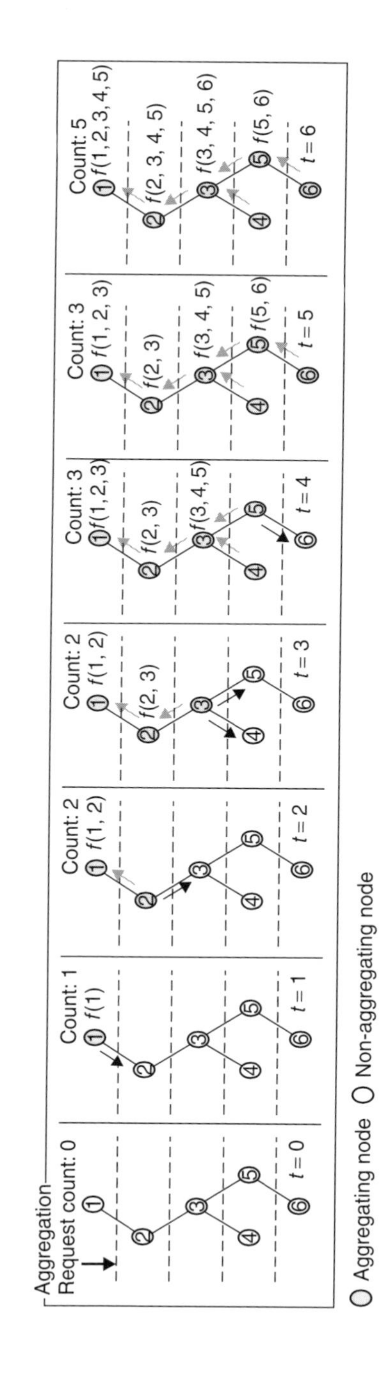

Figure 4.3 Pipelined computation of aggregates.

they hear differently. In such a scheme, far fewer messages will be sent, but some of the ability to incorporate nodes that failed to hear the initial request to aggregate will also be lost, as there will be fewer aggregate reports for those nodes to snoop on.

A hybrid pipeline scheme significantly improves the robustness of aggregates by tending to incorporate nodes that lose initial aggregation requests. Pipelining also improves throughput, which can be important when a single aggregate requires seconds to compute.

In aggregation algorithms, sensors communicate over a shared radio channel. Every message is effectively broadcast to all other sensors within range, which enables a number of optimizations that can significantly reduce the number of messages transmitted, and hence increase the accuracy of aggregates in the face of transmission failures.

A shared channel can be used to increase message efficiency when a sensor misses an initial request to begin aggregation: it can initiate aggregation even after missing the start request by snooping on the network traffic of nearby sensors. When it sees another sensor reporting an aggregate, it can assume that it too should be aggregating.

This technique is not only beneficial for improving the number of sensors participating in any aggregate; it also substantially reduces the number of messages that must be sent when using the pipelined aggregation scheme. Because nodes assume they should begin aggregation any time they hear an aggregate reported, a sensor does not need to explicitly tell its children to begin aggregation. It can simply report its value to its parents, which its children will also hear. The children will assume they missed the start request and initiate aggregation locally. For the simple example in Figure 4.3, none of the messages associated with black arrows actually need to be sent. This reduces the total messages required to compute the first full aggregate of the network from 22 to 17, a total saving of 23 %.

Of course, for later rounds in the aggregation, when no messages are sent from parents to children, this saving is no longer available. Snooping can, however, be used to reduce the number of messages sent for certain classes of aggregates. Consider computing a maximum over a group of sensors: if a sensor hears a peer reporting a maximum value greater than its local maximum, it can elect to not send its own value and be assured of not affecting the value of the final aggregate.

In addition to reducing the number of messages that must be sent, the inherently broadcast nature of radio also offers communications redundancy, which improves reliability. Consider a sensor with two parents: instead of sending its aggregate value to just one parent, it can send it to both parents. It is easy for a node to discover that it has multiple parents, since it can simply build a list of nodes it has heard that are one step closer to the root. Of course, for aggregates other than MIN and MAX, sending to multiple parents has the undesirable effect of causing the node to be counted several times. The solution to this is to send part of the aggregate to one parent and the rest to the other. Consider a COUNT; a sensor with $(c - 1)$ children and two parents can send a COUNT of $c/2$ to both parents instead of a count of c to a single parent. A simple statistical analysis reveals the advantage of doing this: assume that a message is transmitted with probability p, and that losses are independent, so that if a message m from sensor s is lost in transition to parent P_1, it is no more likely to be lost in transit to P_2. (Although failure independence is not always a valid assumption, it will occur when a hidden node garbles communication to P_1 but not to P_2, or when one parent is forwarding a message and another is not). First, consider the case where s sends c to a single parent; the expected value of the transmitted count is $p \times c$ [0 with probability $(p - 1)$ and c with probability p], and the variance is $c^2 \times p \times (1 - p)$, since these are standard Bernoulli trials with a probability of success multiplied by a constant c. For the case where s sends $c/2$ to both parents, linearity of expectation allows the expected value to be the sum of the expected value through each parent, or $2 \times p \times c/2$. Similarly, the sum of variances through each parent computes:

$$var = 2 \times (c/2)^2 \times p \times (1 - p) = c^2/2 \times p \times (1 - p)$$

Thus, the variance of the multiple parent COUNT is much less, although its expected value is the same. This is because it is much less likely (assuming independence) for the message to both parents to be lost, and a single loss will less dramatically affect the computed value. Note that the probability that no data is lost is actually lower with multiple parents (p^2 versus p), suggesting that this may not always be a useful technique. However, since losses are almost assured of happening occasionally when aggregating, the users will prefer that their aggregates be closer to the correct answer more often than exactly right.

This technique applies equally well for SUM and AVERAGE aggregates or for any aggregate which is a linear combination of a number of values. For rank-based aggregates, like mode and median, this technique cannot be applied.

The efficiency of aggregates can be increased by rephrasing aggregates as hypotheses dramatically to reduce the number of sensors required to respond to any aggregate.

Although the above techniques offer significant gains in terms of the number of messages transmitted and robustness with respect to naive approaches, these techniques still require input from every node in a network in order to compute an aggregate. We only need to hear from a particular sensor if that sensor's value will affect the end value of the aggregate. For some aggregates, this can significantly reduce the number of nodes that need to report.

When computing a MAX or MIN, a sensor can snoop on the values its peers report and omit its own value if it knows it cannot affect the final value of the aggregate. This technique can be generalized to an approach called hypothesis testing. If a node is presented with a guess as to the proper value of an aggregate, either by snooping on another sensor's aggregate value or by explicitly being presented with a hypothesis by the user or root of the network, it can decide locally whether contributing its reading and the readings of its children will affect the value of the aggregate.

For MAX, MIN and other top n aggregates, this technique is directly applicable. There are a number of ways it can be applied and the snooping approach is one. As another example, the root of the network seeking a MIN sensor value might compute the value of the aggregate over the top k levels of the network (using the pipelined approach described above), and then abort the aggregate and issue a new request asking for only those sensor values less than the minimum observed in the top k levels. In this approach, leaf nodes will be required to send no message if their value is greater than the minimum observed over the top levels (intermediate nodes must forward the request to aggregate, so they must still send messages.) If the sensor values are independent and randomly distributed (a big assumption!), then a particular leaf note must transmit with probability $(1/b^k)$, where b is the branching factor of the tree and b^k is the number of sensors in the top k levels, which is quite low for even small values of k. Since, in a balanced tree, half the nodes are in the bottom-most level, this can reduce the total number of messages that must be sent almost by a factor of two.

For other aggregates that accumulate a total, such as SUM and COUNT, this technique will never be applicable. For the third class of statistical aggregates, such as AVERAGE or variance, this technique can reduce the number of messages, although not as drastically. To obtain any benefit from such aggregates, the user must define an error bound to be tolerated over the value of the aggregate. Given this error bound, the same approach as for top n aggregates can be applied. Consider the case of an average: any sensor that is within the error bound of the approximate answer need not answer, its parent can assume that its value is the same as the approximate answer and count it accordingly (this scheme requires parents to know how many children they have.) The total computed average will not deviate from the actual average by more than the error bound, and leaf sensors with values close to the average will not be required to report. Obviously, the value of this scheme depends greatly on the distribution of sensor values. If values are uniformly distributed, the fraction of leaves that does need to report will approximate the size of the error bound. If values are normally distributed, a much larger percentage of leaves will not report. Thus, the value of this scheme depends on the expected distribution of values and the tolerance of the user to inaccurate error bounds.

In-network aggregation is used to compute aggregates. By pipelining aggregates, the throughput is increased and smoothed over intermittent losses inherent in radio communication. This basic approach is improved with several other techniques: snooping over the radio to reduce message load and improve accuracy of aggregates, and hypothesis testing to invert problems and further reduce the number of messages sent.

4.4.2. Grouping

Grouping computes aggregates over partitions of sensor readings. The basic technique for grouping is to push down a set of predicates that specify group membership, ask sensors to choose the group they belong to, and then, as answers flow back, update the aggregate values in the appropriate groups.

Group predicates are appended to requests to begin aggregation. If sending all predicates requires more storage than will fit into a single message, multiple messages are sent. Each group predicate specifies a group id (identifier), a sensor attribute (e.g. light, temperature), and a range of sensor values that defines membership in the group. Groups are assumed to be disjointed and defined over the same attribute, which is typically not the attribute being aggregated. Because the number of groups can be so large that information about all groups does not fit into the RAM of any one sensor, sensors pick the group they belong to as messages defining group predicates flow past and discard information about other groups.

Messages containing sensed values are propagated just as in the pipelined approach described above. When a sensor is a leaf, it simply tags the sensor value with its group number. When a sensor receives a message from a child, it checks the group number. If the child is in the same group as the sensor, it combines the two values just as above. If it is in a different group, it stores the value of the child's group along with its own value for forwarding in the next interval. If another child message arrives with a value in either group, the sensor updates the appropriate aggregate. During the next interval, the sensor will send out the value of all the groups about which it collected information during the previous interval, combining information about multiple groups into a single message as long as the message size permits. Figure 4.4 shows an example of computing a query, grouped by temperature, that selects average light readings. In this snapshot, data is assumed to have filled the pipeline, such that results from the bottom of the tree have reached the root.

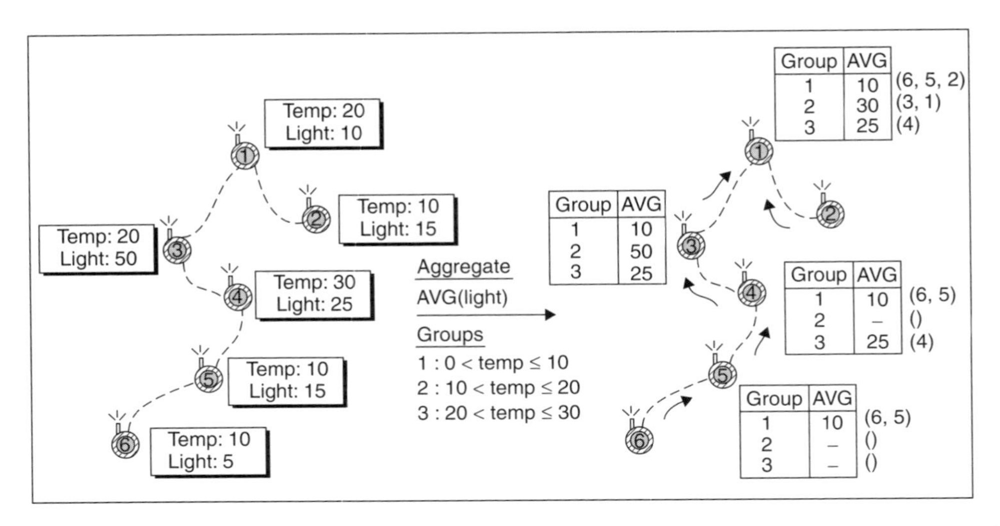

Figure 4.4 A sensor network (left) with an in-network, grouped aggregate applied to it (right). Parenthesized numbers represent the sensors that contributed to the average, but the sensors do not track this information.

Recall that SQL (Structured Query Language) queries also contain a HAVING clause that constrains the set of groups in the final query result by applying a filtration predicate to each group's aggregate value. This predicate may be passed into the network along with partitions. The predicate is only sent into the network if, potentially, it can be used to reduce the number of messages that must be sent: for example, if the predicate is of the form MAX(attr) $< x$, then information about groups with MAX(attr) $> x$ need not be transmitted up the tree, and so the predicate is sent down into the network. However, other HAVING predicates, such as those filtering AVERAGE aggregates, or of the form MAX(attr) $> x$, cannot be applied in the network because they can only be evaluated when the final group aggregate value is known.

Because the number of groups can exceed available storage on any one sensor, a way to evict groups is needed. Once an eviction victim is selected, it is forwarded to the sensor's parent, which may choose to hold on to the group or continue to forward it up the tree. Because groups can be evicted, the user workstation at the top of the network may be called upon to combine partial groups to form an accurate aggregate value. Evicting partially computed groups is known as partial preaggregation, as described in the database literature.

There are a number of possible policies for choosing which group to evict. The policies which incur a significant storage overhead (more than a few bits per group) are undesirable because they reduce the number of groups that can be stored and increase the number of messages that must be sent. Evicting groups with low membership is probably a good policy, as those are the groups that are least likely to be combined with other sensor readings, and so are the groups that benefit the least from in-network aggregation.

Evicting groups forces information about the current time interval into higher level nodes in the tree. In the standard pipelined scheme, the aggregates are computed over values from the previous time interval, which presents an inconsistency, but does not dramatically effect the aggregates.

This method shows how to partition sensor readings into a number of groups and properly compute aggregates over those groups, even when the amount of group information exceeds available storage in a sensor.

With respect to aggregation, the semantics used here are largely a part of the SQL (Structured Query Language) standard. The partial preaggregation techniques used to enable group eviction were proposed as a technique to deal with very large numbers of groups to improve the efficiency of hash joins and other bucket-based database operators.

When computing multiple simultaneous aggregates over a single sensor network, it should be possible for sensors to accommodate multiple queries (just as they handle multiple groups) up to some small number of queries. There may be an eviction option, as with grouping, but there may also be a point at which the in-network approach is so slow that the server-based approach again becomes viable. The implementation issues associated with simultaneous aggregates must be explored before these in-network approaches can be implemented in a database system that supports concurrent queries.

This approach offers the ability to query arbitrary data in a sensor network without custom building applications. By pipelining the flow of data through the sensor network, the aggregates are robustly computed while providing rapid and continuous updates of their value to the user. Finally, by snooping on messages in the shared channel and applying techniques for hypothesis testing, the performance of basic approach is improved.

SQL (Structured Query Language), as it has developed over many years, has proven to work well in the context of database systems. A similar language, when properly applied to sensor networks, will offer similar benefits to SQL: ease of use, expressiveness, and a standard on which research and industry can build.

4.5. SUMMARY

EAR is a reactive routing protocol, and destination initiated protocol. The consumer of data initiates the route request and maintains the route subsequently.

Multiple paths are maintained from source to destination. EAR uses only one path at all times whereas diffusion sends data along all the paths at regular intervals. Due to the probabilistic choice of routes, EAR can continuously evaluate different routes and choose the probabilities accordingly.

The pico radio is a sensor network of ultra low powered nodes, called pico nodes. There are three types of pico node: sensor nodes, actuator nodes, and monitor nodes. The sensor nodes acquire data (using a built in sensor facility), which is typically processed by monitor nodes. The resulting output (control actions) is sent to the actuator nodes.

The altruistic nodes can be used in the pico radio sensor network. This scheme is applied to the data forwarding stage of the pico radio EAR protocol, and it shows up that significant improvements in network lifetime can be achieved as compared to the already lifetime-optimized EAR protocol. This holds true specifically for the case where much more bandwidth is spent for data transmission than for interest propagation.

In-network aggregation is used to compute aggregates. By pipelining aggregates, the throughput is increased and smoothed over intermittent losses inherent in radio communication. This basic approach is improved with several other techniques: snooping over the radio to reduce message load and improve accuracy of aggregates, and hypothesis testing to invert problems and further reduce the number of messages sent.

The partial preaggregation techniques can be used to deal with very large numbers of groups to improve the efficiency of the database operators.

The implementation issues associated with simultaneous aggregates must be explored before the in-network approaches discussed can be implemented in a database system that supports concurrent queries. By pipelining the flow of data through the sensor network, the aggregates can be robustly computed and provide rapid and continuous updates to the user. Snooping on messages in the shared channel and applying techniques for hypothesis testing can improve the performance of the basic approach.

SQL has proven to work well in the context of database systems and a similar language, could offer similar benefits.

PROBLEMS

Learning Objectives

After completing this chapter you should be able to:

- demonstrate understanding of routing in wireless sensor networks;
- discuss what is meant by energy aware routing for sensor networks;

- explain what altruists or friendly neighbors in the pico radio sensor network are;
- explain the role of aggregate queries in sensor networks;
- demonstrate understanding of aggregation techniques.

Practice Problems

Problem 4.1: What is a useful metric for the performance of a routing protocol?
Problem 4.2: What are the main layers for designing the pico node?
Problem 4.3: What is the function of the physical layer in the pico node?
Problem 4.4: What is the function of the MAC layer in the pico node?
Problem 4.5: What is the function of the network layer in the pico node?
Problem 4.6: What are the categories of dedicated protocols?
Problem 4.7: What is the characteristic of proactive routing protocols?
Problem 4.8: What is the characteristic of reactive routing protocols?
Problem 4.9: What is the potential problem of using the lowest energy route for communication?
Problem 4.10: How the maximum network lifetime can be achieved?
Problem 4.11: How does the altruist or friendly neighbor approach work?
Problem 4.12: What are motes?

Practice Problem Solutions

Problem 4.1:

A useful metric for routing protocol performance is network survivability. The protocol should ensure that connectivity in a network is maintained for as long as possible, and the energy status of the entire network should be of the same order.

Problem 4.2:

The three main layers for designing the pico node are the physical, media access control (MAC), and network layers.

Problem 4.3:

The physical layer in the pico node handles the communication across the physical link, which involves modulating and coding the data so that the intended receiver can optimally decode it in the presence of channel interference.

Problem 4.4:

The MAC layer's primary functions in the pico node are to provide access control, channel assignment, neighbor list management, and power control.

Problem 4.5:

The network layer in the pico node has two primary functions: node addressing, and routing.

Problem 4.6:

The dedicated protocols are categorized into proactive and reactive.

Problem 4.7:

Proactive routing protocols have the distinguishing characteristic of attempting to maintain consistent up to date routing information from each node to every other node in the network. Every node maintains one or more routing tables that store the routing information, and topology changes are propagated throughout the network as updates so that the network view remains consistent. The protocols vary in the number of routing tables maintained and the method by which the routing updates are propagated.

Problem 4.8:

Reactive routing protocols create routes only when desired. An explicit route discovery process creates routes initiated only on demand. These routes can be either source initiated or destination initiated. Source-initiated routing means that the source node begins the discovery process, while destination-initiated routing occurs when the destination begins discovery protocol. Once a route is established, the route discovery process ends, and a maintenance procedure preserves this route until the route breaks down or is no longer desired.

Problem 4.9:

The potential problem in routing protocols is that they find the lowest energy route and use it for all communication. This is not favorable for the network lifetime. Using a low-energy path frequently leads to energy depletion in the nodes along that path, and may lead to network partition.

Problem 4.10:

To achieve a maximum network lifetime, it is mandatory to optimize the energy consumption in all layers of the protocol stack, from the physical layer to the application layer. The approach of jointly designing the application and the communication related layers can be effective.

Problem 4.11:

In the altruist or friendly neighbor approach, a node simply broadcasts its capabilities to its neighbors, along with its position, i.e. node address, and a lifetime value. In this broadcast, the altruist uses an altruist announcement packet. The lifetime value indicates for how long the altruist node is willing to do more work in the soft-state approach. The other nodes in the altruists neighborhood can freely decide whether they use the service offered by the altruist or not.

Problem 4.12:

Small sensor devices are called motes. Motes are equipped with a radio, a processor, and a suite of sensors. An operating system makes it possible to deploy dedicated networks of sensors that can locate each other and route data without any advance knowledge of network topology.

5

Distributed Sensor Networks

5.1. INTRODUCTION

Ubiquitous computing envisages everyday objects as being augmented with computation and communication capabilities. While such artifacts retain their original use and appearance, their augmentation can seamlessly enhance and extend their usage, opening up novel interaction patterns and applications.

Bluetooth supports the paradigm of spontaneous networking, wherein nodes can engage in communications without advance knowledge of each other. A procedure-named inquiry can be used to discover which other Bluetooth units are within communication range and connections are then established, based on information exchanged during the inquiry. Once a unit has discovered another unit, connection is established very fast, since information exchanged in the inquiry procedure can be exploited.

Networks of wireless sensors are the result of rapid convergence of three key technologies: digital circuitry, wireless communications, and Micro-ElectroMechanical Systems (MEMS). Advances in hardware technology and engineering design have led to reductions in size, power consumption, and cost. This has enabled compact, autonomous nodes, each containing one or more sensors, computation and communication capabilities, and a power supply.

The millimeter-scale nodes, called smart dust, explore the limits on size and power consumption in autonomous sensor nodes. Size reduction is paramount in making the nodes inexpensive and easy to deploy. Smart dust incorporates the requisite sensing, communication, and computing

Wireless Sensor Network Designs A. Hać
 ISBN: 0-470-86736-1

hardware, along with a power supply, in a volume of no more than a few cubic millimeters, while still achieving good performance in terms of sensor functionality and communications capability.

5.2. BLUETOOTH IN THE DISTRIBUTED SENSOR NETWORK

Ubiquitous computing devices experience special communication with other smart devices by using minimal power and possibly without the help of a central infrastructure. Wireless communication technologies lack robustness, consume too much energy, or require an infrastructure to become viable candidates. To evaluate the suitability of the Bluetooth standard for such communication requirements, Kasten and Langheinrich (2001) integrated a Bluetooth module into the prototype of a distributed sensor network node, developed within the European Smart-Its research project. While Bluetooth offers robust and convenient dedicated communication, experiments suggest that the Bluetooth standard could benefit from improved support for symmetric communication establishment and slave-to-slave communication.

The goal of the Smart-Its project is to attach small, unobtrusive computing devices, so-called Smart-Its, to established real-world objects. While a single Smart-It is able to perceive context information from its integrated sensors, a number of special connected Smart-Its can gain collective awareness by sharing this information.

Sharing information requires a suitable communication technology, which preferably should be wireless in order to be in line with the unobtrusive nature of the devices. Since there is no central authority in a Smart-Its sensor network, nodes within the network must also be able to communicate without an advance knowledge of each other, and without the help of a background infrastructure, though they may utilize services when available. Moreover, the communication technology must be robust, scale well, and use the limited energy of the autonomous device efficiently. The communication technology employed should adhere to a broadly used standard to lever from existing communication services in the environment. These needs have prompted a search for a suitable communication technology for the Smart-Its sensor network.

Bluetooth is a communication standard that provides special configuration of master/slave piconets with the maximum of eight active units. It supports spontaneous connections between devices without their requiring detailed knowledge about each other. Bluetooth allows data transfers between units over distances of, nominally, up to 10 meters. The gross data rates of 1 Mbps is shared among all participants of a piconet. Bluetooth operates in the license-free 2.4 GHz ISM (Industry Scientific Medical) spectrum (2.400–2.484 GHz) and uses Frequency Hopping Spread Spectrum (FHSS) to minimize interference. The technology is geared toward low energy consumption, and targets the consumer mass market with world-wide availability and low price.

The Smart-Its embeds computation into real-world objects by attaching small, unobtrusive, and autonomous computing devices to them. These devices, the Smart-Its, integrate sensing, processing, and communication capabilities, which can be customized to the objects to which they are attached.

While a single Smart-It is able to perceive context information from the integrated sensors, a number of dedicated connected Smart-Its can gain collective awareness by sharing this information. A group of Smart-Its augmented objects can thus establish a common context that can be exploited by applications and services located in the environment.

Examples of application scenarios of collective awareness of Smart-Its are an anti-credit-card-theft mode, where a Smart-It-enabled credit card only functions if a sufficient number of Smart-It-enabled personal artifacts, such as clothes or car keys, are thus around, thus rendering the card useless when lost or stolen.

The device should be easy to use, program, and debug. At the same time, it should be small enough to serve as a demonstrator for a Smart-It sensor node, but large enough for easy handling.

To facilitate rapid prototyping, the Smart-It unit implements a limited functional core, which includes a processor, memory, and Bluetooth communications. The sensing capabilities are not included. A versatile external interface, with analog and digital I/O, is provided that allows for integration of single sensors or even a daughter board for sensing. These diverse interfaces, ranging from simple analog output over serial interfaces to bus systems, such as I^2C (Inter Integrated Circuit), allow for identifying appropriate sensors and sensing algorithms for use within a Smart-Its environment. The RS232 serial port is used mainly for debugging purposes.

Keeping in line with the unobtrusive nature of the ubiquitous computing paradigm, the Smart-Its devices should be able to operate autonomously for extended periods of time. This includes consciously choosing the components regarding their power consumption, as well as providing a suitable power source. For easy handling, the device is powered from an externally attached rechargeable battery pack, rather than having a battery housing mounted on the device. This way the small batteries are attached for normal operation, while still being able to operate the device using bulky but more powerful batteries for extended testing.

For easy layout of the circuit board, Kasten and Langheinrich (2001) chose a single voltage plane and an overall low component count. The system is in-circuit programmable to minimize turnaround times.

5.2.1. Bluetooth Components and Devices

Commercial Bluetooth solutions are available as self-contained transceiver modules. These are shielded subsystems designed to be used as add-on peripherals. They feature an embedded CPU (Central Processor Unit) and different types of memory, as well as baseband and radio circuits. The modules offer a generic Host Controller Interface (HCI) to the lower layers of the Bluetooth protocol stack, while the higher layers of the protocol, as well as applications, must be implemented on the host system. Since the in-system CPU and memory are not available for installing user specific implementations, even a minimal stand-alone Bluetooth node needs an additional host CPU in order to execute applications and the corresponding higher layers of the Bluetooth protocol. Transport layers for communication between the Bluetooth module and the host system are standardized for UART (Universal Asynchronous Receiver/Transmitter), RS232, and USB (Universal Serial Bus).

To run the higher Bluetooth protocol layers and applications, Kasten and Langheinrich chose the Atmel ATmega 103L microcontroller as the host CPU. The unit is in-system programmable, and features an 8-bit RISC (Reduced Instruction Set Computer) core with up to 4 MIPS (Millions of Instructions per Second) at 4 MHz, a serial UART as well as several power modes. The embedded memory consists of 128 kbytes flash memory and 4 kbytes of internal SRAM (Static RAM). The data memory can be extended up to 64 kbytes, requiring only two external components, the SRAM (Static RAM) and an address latch. The external memory is directly addressable by the 16-bit data-memory address bus, i.e. without paging. A more powerful system is used to allow more complex on-board preprocessing even though a less powerful processor with less memory could, potentially, have delivered sensor data to the Bluetooth module.

A Smart-Its prototype is placed on top of a battery pack (mostly hidden behind the board) used for testing and evaluation. All components are mounted onto a 4×6 cm two-layer printed circuit board. The unit has several external interfaces. A serial UART port is available for data transfer and debugging at speeds up to 57.6 kbps. There are two 8-bit general-purpose I/O ports, eight 10-bit analog-to-digital converters, and two edge or level triggered interrupt lines to interface external sensors or other components.

Four LEDs (Light Emitting Diode) can be used for debugging and status information. For example, one LED can be used to flash a heart beat signal when the unit is operational. A voltage regulator is used to supply the necessary operating voltage of 3.3 V from the battery pack.

Jumpers providing access to each of the main component's individual power supply lines allow for exact monitoring of power consumption and duty cycles. The connector is the in-circuit system programming interface (SPI) of the MCU (MicroController Unit). The Bluetooth module and an external 2.4 GHz antenna are mounted on top of a plane, to shield the system from RF interference.

The Bluetooth module is attached to the microcontroller unit by a UART implemented in software (shown in Figure 5.1). Kasten and Langheinrich (2001) do not use the hardware UART (universal asynchronous receiver/transmitter) provided by the microcontroller, as that would have required additional circuitry for multiplexing pins shared between the UART and the in-circuit programming ports. Therefore Kasten and Langheinrich implemented a second software UART in the C programming language. Timing constraints prohibit data transfers exceeding 9.6 kbps, effectively limiting the gross data rate of Bluetooth. Kasten and Langheinrich valued low component count and low circuit complexity over higher data rates.

The system software is implemented in the C programming language, providing low-level drivers, a simple scheduler (which supports event-driven scheduling of application tasks), and the host portion of the Bluetooth protocol stack. There are system dependent drivers for both UART ports, analog-to-digital converters, general purpose I/O, random number generator, system clock, and sensors.

An open source and several commercial implementations of the host portion of the Bluetooth stack are available. The commercially available

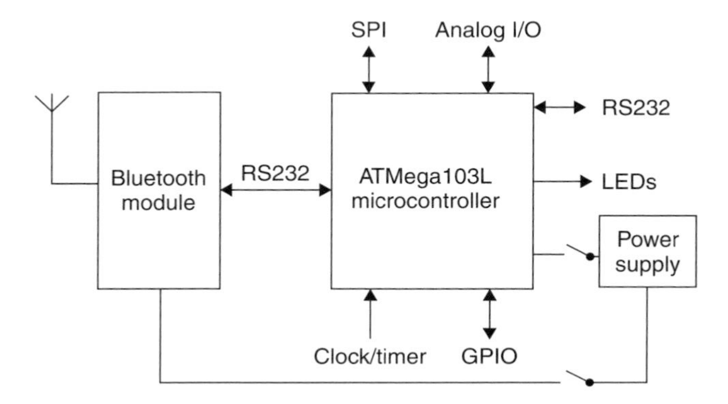

Figure 5.1 System overview.

software stacks make very high demands on the system, both in terms of required operating system features (particularly multithreading) as well as program and data memory provisions. Experiences with the software had shown that about 2 kbytes of data memory would be sufficient for a minimal implementation, most of which is used as buffer space. Kasten and Langheinrich (2001) used the open source implementation due to its immediate availability.

Kasten and Langheinrich ported the host portion of the Bluetooth protocol stack from the open source Linux implementation to the microcontroller environment. Supported layers are HCI and the Logical Link Control and Adaptation Protocol (L2CAP). The Linux version of the Bluetooth stack requires multithreading capabilities and access to the serial port. In this system, these functions are taken care of by the scheduler and the low-level drivers. The main obstacle in porting is the limited memory capacity of the microcontroller.

5.2.2. Bluetooth Communication and Networking

Bluetooth is the standard for *ad hoc* networking. It was originally conceived as a cable replacement technology and may serve well in that application domain. However, its particular design makes it less suited for other applications in the domain of dedicated networking.

Bluetooth has been optimized to support a large number of communications taking place in the same area at the same time. It organizes all communications into piconets, each piconet serving up to eight participants. Multiple piconets with overlapping coverage areas are referred to as scatternets. It is possible to interconnect piconets by using the units participating in different piconets on a time-division multiplex basis. However, since the radio can only tune to a single piconet carrier at any instant in time, a unit can only communicate in one piconet at a time.

Piconets are managed by a single master that implements centralized control over channel access. All other participants in a piconet are designated slaves. Communication is strictly slave-to-master (or vice versa), but can never be slave-to-slave. During the existence of a piconet, master and slave roles can be switched. This is desirable, for example, when a slave wants to take over an existing piconet fully. Likewise, a slave in an existing piconet may want to set up a new piconet, establishing itself as its master and the current piconet master as slave. The latter case implies a double role for the original piconet master; it becomes a slave in the new piconet while still maintaining the original piconet as master. A unit can be slave in two piconets or be master in one, and slave in another piconet.

The limit of eight nodes in a piconet is inadequate for setting up a densely connected sensor network. To communicate with more than eight nodes at the same time requires some sort of time multiplexing, where additional nodes have to be parked and unparked repeatedly. Setting up additional piconets requires gateway nodes to alternate between their respective piconets, since Bluetooth only supports units in one active piconet at a time.

Applications usually need slave-to-slave communication, which is not provided in the Bluetooth standard. To channel all slave-to-slave traffic through the master increases both traffic and energy consumption. Alternatively, a slave can switch roles with the master, or set up an additional piconet. These solutions incur substantial communication and configuration overhead. The communication protocols may alleviate some of this overhead, and lessen its impact on communication and power usage.

The default state of a Bluetooth unit is standby. In this state, the unit is in a low-power mode, with all components but the internal clock shut off. In standby there can be no connections open.

When there is an active connection to a Bluetooth unit, it is said to be in connect state. In connect state, Bluetooth knows four different power modes: active, sniff, hold, and park. In active mode, the Bluetooth unit actively participates on the channel. Data transmission can start almost instantaneously, but at the expense of increased power consumption (compared to the remaining three modes).

When low-power operation is favored over short response times, units can make use of one of the three power-saving modes: sniff, hold, and park. All low-power modes reduce the duty cycle of different units within a piconet. In sniff mode, slave units only listen in on the channel at specified times, agreed upon with the master. Hence, transmissions can only start at these times. The connections of a piconet can also be put on hold. In hold mode, every participant (including the master) can take some time off for sleeping. Prior to entering hold mode, master and slaves agree on a time when to return to active mode again. The time off can also be used for conducting other business, such as attending other piconets, or scanning for other units.

The park mode is a special mode for slaves that do not need to participate in a piconet, but nevertheless want to remain connected to avoid going through the connection establishment procedure again. Parked slaves do not count as regular, i.e. active, piconet members. In addition to the maximum of eight active members, there may be up to 255 parked slaves within a piconet.

Low-power modes are a trade-off between power consumption and response time. Increasing sleep time reduces power consumption but prolongs time before access can be made, and vice versa. Low-power modes are a powerful tool offering a range of options to applications when the transmission pattern is known in advance. When data traffic commences at a regular schedule, the sniff and park modes seem to be appropriate. For example, a Smart-Its node may want to dispense its sensor readings every 10-second to a node in the background infrastructure (implementing the master). That node would set itself up for sniff mode with a 10-second sleep cycle. Similarly, the hold mode serves applications communicating on a more irregular, yet predictable schedule.

If, however, time critical data transmissions start spontaneously, an application has no other option than to keep the Bluetooth module in an active mode. The example here may be a Smart-It node using a background service for processing audio clues. Transmission of the data would need to start immediately after observing the audio clue since buffering on the local Smart-Its device is not feasible.

The dominant component in a Smart-It, with respect to power consumption, is the Bluetooth module. Since the Bluetooth modules are engineering samples that do not implement low-power modes, no data is available on how these would alleviate the significant power usage of the modules in active mode. The improved Bluetooth products should reduce power consumption considerably since the system design allows for easy replacement of the Bluetooth transceiver module.

Bluetooth supports the paradigm of spontaneous networking, where nodes can engage in communications without advance knowledge of each other. A procedure-named inquiry can be used to discover which other Bluetooth units are within communication range. Connections are then established based on information exchanged during inquiry. Once a unit has discovered another unit, connection establishment is very fast, since information exchanged in the inquiry procedure can be exploited.

Inquiry is an asymmetric procedure, in which the inquiring unit and the inquired unit need to be in complementary modes, called inquiry and inquiry-scan. When a Bluetooth unit has been set to inquiry mode, it continuously sends out inquiry messages to probe for other units. Inquiry mode continues for a previously specified time, until a previously specified number of units have been discovered, or until stopped explicitly. Likewise, other Bluetooth units only listen (and reply) to inquiry messages when they have been explicitly set into inquiry-scan mode. In a unit, inquiry and inquiry-scan modes are mutually exclusive at any time.

When a unit in inquiry-scan mode recognizes an inquiry message, it replies to the inquirer. Thus, the complete inquiry procedure requires one broadcast message to be sent from the inquirer, and one message from every inquired unit back to the inquirer.

If an inquiry is initiated periodically, then the interval between two inquiry instances must be determined randomly, to avoid two Bluetooth units synchronizing their inquiry procedures in lock step. In a scenario where units are peers, i.e. when there is no dedicated inquirer, application software carries the burden of breaking the symmetry.

The power consumption increases considerably during inquiry. This is due to the asymmetric nature of the Bluetooth inquiry procedure, where the burden of expending power is mostly placed on the unit conducting the inquiry.

To save power, a unit in inquiry-scan mode does not continuously listen to inquiry messages. Instead, it only listens for a very short period of time (11.25 ms by default), which, under regular conditions, suffices for the inquiry message to get through with sufficiently high probability. Then the unit enters idle mode for a much longer interval (typically 1.28 s). However, the inquiring unit needs to send inquiry messages (and alternately listen for potential replies) during the entire interval, since it cannot know when the target unit is actually listening.

According to Salonidis *et al.* (2000), the expected delay for link formation (i.e., inquiry plus connection establishment) of peer units is 1 s when both units alternate between inquiry and inquiry-scan modes following uniform distribution. In the link-formation delay, device discovery is by far the dominating factor. However, Kasten and Langheinrich (2001) came to different, much higher time results for device discovery, both in theory as well as in experiments. The Salonidis *et al.* results do not include the fact that units in inquiry-scan mode only pay attention to inquiry messages for 11.25 ms out of 1.28 s, that is less than 1 % of the time. Therefore, we need to add (1.28 s − 11.25 ms)/2 to the expected delay. Also, the Salonidis *et al.* results assume an ideal, error-free environment, where messages are never lost.

Kasten and Langheinrich's experiments show that device discovery is much slower in real settings and often takes several seconds to complete. The experimental set-up consisted of two immobile Bluetooth evaluation boards, using the same Ericsson ROK 101 007 modules as the Smart-It prototypes, which were placed at a distance of about 1 meter. One unit was constantly set to inquiry-scan mode, while the other unit was dedicated to inquiry, both using Bluetooth default settings.

Instead of using two of the Smart-It units directly, Kasten and Langheinrich (2001) used the evaluation boards (where the host portion of the inquiring unit's stack was run on a Linux machine) since it offered a much finer timer granularity than what would have been possible using the Smart-Its prototype. In every test, the dedicated inquirer was conducting inquiry for exactly 12.8 seconds, even if the target device was discovered in less than

12.8 seconds. Prior to carrying out the next test the inquirer went back to standby mode for a time uniformly distributed between 0 and 12.8 seconds to avoid synchronization artifacts. The experiment was set in a typical office environment with little traffic from an IEEE 802.11 wireless LAN and no Bluetooth traffic.

The results of the experiments are as follows:

- The average inquiry delay is 2221 ms
- After 1910 ms, 4728 ms, and 5449 ms, the target unit had been found in 50, 95, and 99 % of all tests, respectively.

A possible reason for the high discovery delay may be that inquiry messages and replies to inquiry messages are lost or are not being recognized as such. Differences in the local clocks and the frequency hopping scheme may be reasons for the latter case.

The Bluetooth inquiry model in general seems to be geared toward settings where a dedicated unit is responsible for discovering a set of other units, e.g. a laptop computer periodically scanning for periphery. It seems less appropriate for truly symmetrical nodes. Also, in the laptop setting, a delay of several seconds for connection establishment would be tolerable. In distributed sensor networks such as the Smart-Its network, the nodes are mobile. An example is sports gear augmented with Smart-Its, e.g. bikes, skateboards, or a football. Based on the experienced mean discovery delay of 2221 ms, two Bluetooth devices traveling at a relative speed of 12.5 km/h (4.5 m/s) could set up a connection before moving out of communication range again. The lengthy connection establishment effectively prevents the use of Bluetooth in fast moving settings.

The inquiry message broadcast by an inquiring unit does not contain any information about the source. Instead, the inquired unit gives away information required for connection establishment, such as the unique device ID (identifier), in the inquiry response. Thus the inquired unit must reveal information about itself without knowing who is inquiring. This inquiry scheme may become a privacy concern, where personal belongings such as children's toys may be augmented with Smart-Its.

Because of power consumption, Bluetooth's inquiry is probably less suited for low-power nodes that will frequently have to scan their surroundings to discover new nodes or background services. On the other hand, this poses no problem for more powerful devices such as laptop computers: placing the power burden on the inquiring unit may be a desired feature in an asymmetrical communication setting, where it would relieve mobile low-power periphery.

Many commercial Bluetooth modules do not offer the full functionality of the specification, for example, Ericsson ROK 101 007 modules have several features missing.

Most importantly, the units do not implement point-to-multipoint connections. Therefore, piconets are limited to just two devices and interconnected piconets cannot be established. Consequently, broadcast and master–slave role switching has not been implemented either, since it only makes sense for piconets of three or more participants. When in connect state, the units cannot actively inquire other devices, nor can they be inquired, regardless of any traffic over that connection. In addition to the very high power consumption of the module, none of the low power modes (hold, sniff, and park) are supported.

Besides important features not being implemented, a number of other difficulties arose during the development of the first Smart-It prototypes. The Bluetooth modules were hard to obtain and came at a rather high price ($US 80 per piece). Secondly, product information, such as unimplemented features, and mechanical and electrical specifications were unavailable until the modules were shipped. For the assembly of the Smart-Its devices, a placement machine was required. Whereas all other components could be soldered manually, the ball grid array of the SMD (Surface Mount Device)-packaged Bluetooth unit could not. This meant that the assembly of only a few Smart-Its already required production-scale facilities solely due to the packaging of the Bluetooth units, thus greatly increasing cost.

Although this is not a particular problem in using Bluetooth in general, it indicates that experimentation and development for researchers without direct access to production scale facilities is made difficult.

5.2.3. Different Technologies

In the Smart Dust project at Berkeley, similar sized prototypes have already been built. However, instead of aiming for a sticker-sized form factor, Smart Dust is ultimately aiming at much smaller size of only a single cubic millimeter, i.e. dust-sized. Also, instead of radio communication, smart dust pursues active and passive optical communications for ultra-low-power consumption. While passive communication is very power conservative, it requires a central authority (Base Station Transceiver, BTS) that initiates communication with a modulated beam laser. Individual nodes reflect a constant laser beam from the BTS and use a deflectable mirror built in MEMS technology to modulate the reply onto the beam. Active communication scenarios using a built-in laser are also investigated.

Smart-Its are envisaged as providing a substantial amount of preprocessing, for example in the area of audio and video sensor data, not only on a single unit basis, but particularly as a collective, distributed processor made up of a Smart-Its. Consequently, Smart-Its aim at providing complex context information, while Smart Dust focuses on relaying direct sensor data to a more powerful central processor.

Within the Smart-Dust project, a range of prototypes has also been built that uses different communication technologies apart from optical communication. The weC Mote uses a custom protocol over a 916.5 MHz transceiver with a range of 20 meters and transmission rates of up to 5 kbps. It runs a custom micro-threaded operating system called Tiny OS, features various different on-board sensors and is used in the UCLA (University of California at Los Angeles) habitat monitoring project as part of a tiered environmental monitoring system.

Next to the Bluetooth technology, a number of comparable communication technologies exist that support some or all required communication aspects.

IEEE 802.11 for Wireless Local Area Networks (WLAN) and HiperLAN/2, offer dedicated (*ad hoc*) modes for peer-to-peer communication. Because 802.11 requires a dedicated access point (AP) for many features such as QoS (Quality of Service) or power saving, its *ad hoc* mode is very limited. In HiperLAN/2, mobile terminals take over the role of APs when being in *ad hoc* mode and thus can continue to support QoS (Quality of Service) and power saving. Since these technologies are mainly intended for scenarios where mobile clients communicate through base stations, their transmission power is considerably higher than that of Bluetooth (10–300 mW, compared to 1 mW in Bluetooth). WLAN devices that support transmit power control (TPC) might be a suitable alternative.

A standardization of IEEE 802.15 defines a Personal Area Network (PAN) standard. Its first incarnation (802.15.1) is based on Bluetooth and improves and extends the existing specification. IEEE 802.15.3 aims for high data rates of 20 Mbps or more, at low cost and low power consumption. IEEE 802.15.4 supplements a low data rate (10 kbps) standard, but using ultra-low power, complexity and cost.

The XI Spike communication platform, developed by Eleven Engineering in Canada, has stirred interest as a viable competitor to Bluetooth, operating at both the 915 MHz and 2.4 GHz ISM bands using frequency hopping and Direct Sequence Spread Spectrum (DSSS). Originally developed for the gaming industry (connecting game controllers to consoles), it: offers multiple data rates of up to 844 kbps using a transmission power of 0.75 mW; supports both peer-to-peer and broadcast communication; allows its embedded

RISC (Reduced Instruction Set Computer) processor to be used for user applications; and comes at a much lower price ($US 6.25) than any available Bluetooth module.

Several factors have contributed to the significant attention Bluetooth has received. One of the first international standards available, it greatly simplifies dedicated networking using the piconet communication paradigm. By using the freely available ISM band, Bluetooth devices can be used world wide without alterations. The frequency hopping technology makes transmissions robust against narrowband interferences (which might be frequent within the ISM band). Even though Bluetooth modules are rather expensive, prices are expected to drop to about $US 5 per unit once mass production is running full scale.

Originally intended as a cable replacement technology, Bluetooth modules (i.e. built fully to specification) are well suited for scenarios where a powerful master device (usually a laptop, PDA (Personal Digital Assistant) or mobile phone) connects seamlessly to a number of peripherals (e.g. a printer, keyboard, or mouse). With data rates of up to 1 Mbps, Bluetooth also offers more than enough bandwidth for ubiquitous computing applications such as simple sensor networks. However, scenarios involving a large number of identical low-power devices using dedicated networking in a true peer-to-peer fashion still experience the following obstacles when using Bluetooth as their communication technology:

- Asymmetrical communication set-up. Finding new communication partners requires one node to be in inquiry mode and the other in inquiry scan mode at the same time.
- Master–slave communication paradigm. Communication in piconets must always be conducted between master and slave, and two slaves must always involve the master node in order to communicate.
- Piconet concept. No more than seven slaves can be active in a piconet at any time; if more nodes need to be added, other active nodes must be put in the park mode. In the park mode, however, nodes cannot actively communicate.
- Scatternet concept. Even though nodes can be in more than one piconet at a time, they can only be active in one of them at a time; meanwhile communications within other piconets must be suspended.
- Power consumption. Even if the power consumption of current pre-series modules can be cut significantly, centralized control of the piconet as well as the asymmetric nature of inquiry and connection establishment, puts the burden of expending power onto a single device. Low-power modes may help but they do not apply to every situation.

Using a custom radio solution might render communication sensitive to interference unless spread spectrum solutions such as frequency hopping are used. Also, error correction, transmission power adaptation, and fundamental quality of service options that are already part of the Bluetooth standard would need to be reimplemented when using a custom solution.

5.3. MOBILE NETWORKING FOR SMART-DUST

Networking nodes must consume extremely low power, communicate at bit rates measured in kilobits per second, and potentially need to operate in high volumetric densities. These requirements need novel *ad hoc* routing and media access solutions. Smart dust enables range of applications, from sensor-rich smart spaces to self-identification and history tracking for virtually any kind of physical object.

5.3.1. Smart-Dust Technology

A smart dust mote is illustrated in Figure 5.2. Integrated into a single package are MEMS sensors, a semiconductor laser diode and MEMS beam-steering mirror for active optical transmission, a MEMS corner-cube retroreflector for passive optical transmission, an optical receiver, signal processing and control circuitry, and a power source based on thick-film batteries and solar cells. This self-powered package has the ability to sense and communicate.

These functions are incorporated while maintaining very low power consumption, thereby maximizing operating life given the limited volume available for energy storage. Within the design goal of 1 mm^3 volume, using the best available battery technology, the total stored energy is on the order of 1 joule. If this energy is consumed continuously over a day, the dust mote power consumption cannot exceed roughly 10 microwatts. The Smart-Dust functionality can be achieved only if the total power consumption of a dust mote is limited to microwatt levels, and if careful power management strategies are utilized, i.e. the various parts of the dust mote are powered on only when necessary. To enable dust motes to function over a span of days, solar cells could be employed to scavenge energy when the sun shines (roughly 1 joule per day) or when room lights are turned on (about 1 millijoule per day).

A communications architecture for ultra-low power can use communication technologies based on radio frequency (RF) or optical transmission techniques. Each technique has its advantages and disadvantages. RF presents

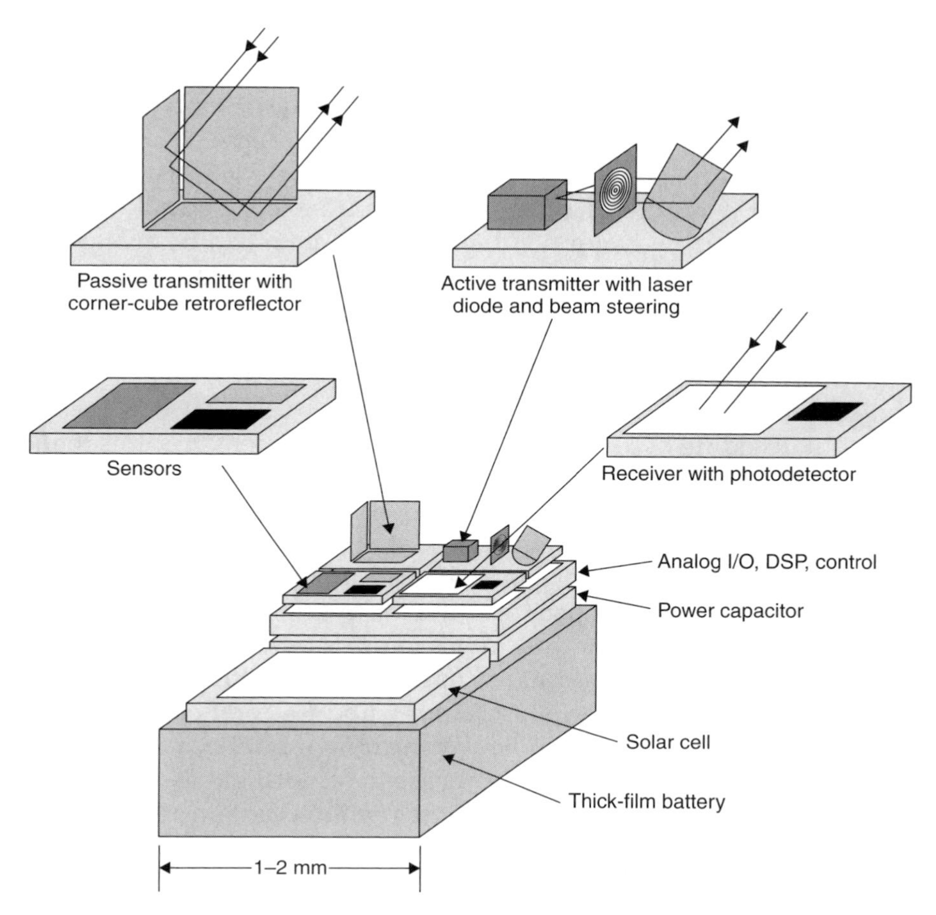

Figure 5.2 Smart-dust mote, containing microfabricated sensors, optical receiver, passive and active optical transmitters, signal processing and control circuitry, and power sources.

a problem because dust motes offer very limited space for antennas, and are demanding extremely short-wavelength (i.e. high-frequency) transmission. Communication in this regime may not be compatible with low-power operation. Furthermore, radio transceivers are relatively complex circuits, making it difficult to reduce their power consumption to the required microwatt levels. They require modulation, band-pass filtering and demodulation circuitry, and additional circuitry is required if the transmissions of a large number of dust motes are to be multiplexed using time-, frequency- or code-division multiple access.

An alternative is to employ free-space optical transmission. Kahn *et al.*'s (1999, 2000) studies have shown that when a line-of-sight path is available, well-designed free-space optical links require significantly lower energy per bit than their RF counterparts. There are several reasons for the power advantage of optical links. Optical transceivers require only simple baseband analog and digital circuitry; no modulators, active bandpass filters or demodulators are needed. The short wavelength of visible or near-infrared light (of the order of 1 μm) makes it possible for a millimeter-scale device to emit a narrow beam (i.e. high antenna gain can be achieved). As another consequence of this short wavelength, a Base-Station Transceiver (BTS) equipped with a compact imaging receiver can decode the simultaneous transmissions from a large number of dust motes at different locations within the receiver field of view, which is a form of space-division multiplexing.

Successful decoding of these simultaneous transmissions requires that dust motes do not block one another's line of sight to the BTS. Such blockage is unlikely, in view of the dust motes' small size. A second requirement for decoding of simultaneous transmission is that the images of different dust motes be formed on different pixels in the BTS imaging receiver.

Another advantage of free-space optical transmission is that a special MEMS structure makes it possible for dust motes to use passive optical transmission techniques, i.e. to transmit modulated optical signals without supplying any optical power. This structure is a Corner-Cube Retroreflector, or CCR. It comprises three mutually perpendicular mirrors of gold-coated polysilicon. The CCR has the property that any incident ray of light is reflected back to the source (provided that it is incident within a certain range of angles centered about the cube's body diagonal). If one of the mirrors is misaligned, this retroreflection property is spoiled. The microfabricated CCR includes an electrostatic actuator that can deflect one of the mirrors at kilohertz rates. A CCR illuminated by an external light source can transmit back a modulated signal at kilobits per second. Since the dust mote itself does not emit light, the passive transmitter consumes little power. Using a microfabricated CCR, Chu *et al.* (1997) have demonstrated data transmission at a bit rate of up to 1 kilobit per second, and over a range up to 150 meters, using a 5-milliwatt illuminating laser.

CCR-based passive optical links require an uninterrupted line-of-sight path. Moreover, a CCR-based passive transmitter is inherently directional; a CCR can transmit to the BTS only when the CCR body diagonal happens to point directly toward the BTS, within a few tenths of degrees. A passive transmitter can be made more omnidirectional by employing several CCRs oriented in different directions, at the expense of increased dust-mote size. If a dust mote employs only one or a few CCRs, the lack of omnidirectional transmission has important implications for feasible network routing strategies.

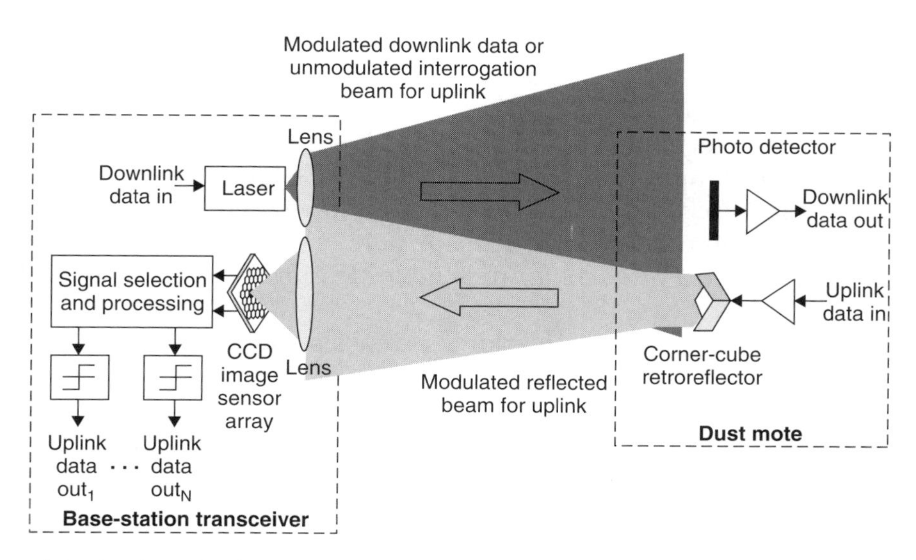

Figure 5.3 Design of a free-space optical network in which a base station transceiver communicates simultaneously with a collection of many dust motes (only one dust mote is shown). A single laser at the base station supplies optical power for the downlink and the uplink.

Figure 5.3 illustrates a free-space optical network utilizing the CCR-based passive uplink. The BTS contains a laser whose beam illuminates an area containing dust motes. This beam can be modulated with downlink data, including commands to wake up and query the dust motes. When the illuminating beam is not modulated, the dust motes can use their CCRs to transmit uplink data back to the base station. A high-frame-rate CCD video camera at the BTS sees these CCR signals as lights blinking on and off. It decodes these blinking images to yield the uplink data. Kahn *et al.*'s analysis shows that this uplink scheme achieves several kilobits per second over hundreds of meters in full sunlight. At night, in clear, still air, the range should extend to several kilometers. The camera uses an imaging process to separate the simultaneous transmissions from dust motes at different locations by using space-division multiplexing. The ability of a video camera to resolve these transmissions is a consequence of the short wavelength of visible or near-infrared light. This does not require any coordination among the dust motes, and thus does not complicate their design.

When the application requires dust motes to use active optical transmitters, MEMS technology can be used to assemble a semiconductor laser, a collimating lens and a beam-steering micromirror, as shown in Figure 5.2. Active transmitters make possible peer-to-peer communication between dust motes, provided there exists a line-of-sight path between them. Power consumption imposes a trade-off between bandwidth and range. The dust motes can communicate over longer ranges (tens of kilometers) at low data rates or higher bit rates (megabits per second) over shorter distances. The relatively high power consumption of semiconductor lasers (of the order of 1 milliwatt) means that these active transmitters should be used for short-duration, burst-mode communication only. Sensor networks using active dust-mote transmitters require a protocol for dust motes to aim their beams toward the receiving parties.

Development of mobile networking protocols for Smart Dust involves critical limitations as follows:

- the free-space optical links require uninterrupted line-of-sight paths;
- the passive and active dust mote transmitters have directional characteristics that must be considered in system design, and
- there are trade-offs between bit rate, energy per bit, distance and directionality in these energy-limited, free-space optical links.

An unbroken line-of-sight path is normally required for operation of free-space optical links for smart dust. These links cannot operate reliably using non-line-of-sight propagation, which would rely on reflections from one or more objects between the transmitter and receiver.

A fixed dust mote without a line-of-sight path to the BTS can communicate with the BTS via multi-hop routing, provided that a suitable multi-hop path exists. The existence of such a path is more likely when the dust mote density is higher. Multi-hop routing increases latency, and requires dust motes to be equipped with active optical transmitters. Constraints on size and power consumption of the dust mote digital circuitry dictate the need for low-complexity dedicated multi-hop routing algorithms.

In most smart dust systems, the BTS interrogating-beam angular spread should be matched to the field of view of the BTS imaging receiver. These two should be matched in all systems using passive dust mote transmitters, and in systems using active dust mote transmitters when the application involves frequent bidirectional transmission between the BTS and dust motes.

An active dust mote transmitter is based on a laser diode. It should employ a narrow beam width, typically of the order of a few degrees or less. This necessitates equipping the dust mote with an active beam-steering mechanism.

The dust mote's transmitter and receiver have different angular spreads. This leads to nonreciprocal link characteristics, wherein a dust mote may receive from another node, but be unable to transmit to it, or vice versa. As a consequence, a dust mote may receive queries from other nodes, and may attempt to answer them, unaware that its transmissions are in vain. When dust motes are fixed, in order to conserve dust mote power, the other nodes should acknowledge this dust mote's transmissions, and this dust mote should not answer further queries from nodes that do not acknowledge its transmissions.

It is known that in free-space optical networks, nonreciprocity can lead to hidden nodes, which can cause collisions during medium access. For example, this effect is observed in networks having a shared-bus physical topology, and using MAC protocols based on random time-division multiplexing, such as CSMA/CA (Carrier Sense Multiple Access/Collision Avoidance) with RTS/CTS (Request to Send/Clear to Send). In Smart-Dust networks, the uplink (dust mote to BTS) uses space-division multiplexing. Uplink collisions will not occur as long as the dust motes are sufficiently separated that their transmissions are detected by different pixels in the BTS imaging receiver. Collisions during active peer-to-peer communications are a potential problem in smart dust networks. A peer-to-peer collision avoidance scheme must cope with a dynamic network configuration, while not introducing excessive complexity or latency.

5.3.2. Communication and Networking

The application in sensor networks is primarily sensor read-out, the key protocol issues are to perform read-out from a large volume of sensors co-located within a potentially small area. Random access to the medium is both energy consuming and bandwidth inefficient. It is useful to exploit passive and broadcast-oriented techniques when possible. The free-space approach supports multiple simultaneous read-out of sensors, mixes active and passive approaches using demand access techniques, and provides efficient and low-latency response to areas of a sensor network that are undergoing frequent changes.

A single wide beam from the BTS can simultaneously probe many dust motes. The imaging receiver at the BTS receives multiple reflected beams from the motes, as long as they are sufficiently separated in space as to be resolved by the receiver's pixel array. The probe beam sweeps the three-dimensional space covered by the base station on a regular basis, most likely determined by the nature of the application and its need for moment-by-moment sensor readings.

To save transmit power, if the mote must use active communications, then it is best to use the active transmitter in a high-bit-rate, short-burst mode. Demand access methods can be used to combine the low latency advantages of active communications with the low-power advantages of the passive approach.

When the mote needs to transmit information, it actively transmits a short-duration burst signal to the BTS. The BTS, detecting this signal, then probes in the general geographical area from which the burst was detected. Assuming that the passive transmitter (i.e. CCR) is properly oriented toward the BTS, the mote can respond by modulating the reflected probe beam with the data it needs to transmit.

This communications structure has much in common with cellular and satellite networks. The paging channel is acquired using contention access techniques. The BTS grants a channel to the node requesting attention. In a cellular network, this is accomplished by assigning a frequency, time-slot, and/or code to the node. In the scheme described for dust motes, the channel is granted by the incident probe beam.

There are as many channels (paging or data) as there are resolvable pixels at the BTS. The BTS has no way of distinguishing between simultaneously communicating dust motes if they fall within the same pixel in the imaging array. One possible way to deal with this is to introduce the time-slotted techniques found in Time Division Multiple Access (TDMA) communications systems. A wide-aperture beam from the BTS can be modulated to offer a common time base by which to synchronize the motes. The BTS can then signal to an individual mote the particular time slot it has assigned to it for communication. The mote must await its time slot to communicate, whether it uses an active or a passive transmitter.

Probe beam revisit rates could be determined in an application-specific manner. The areas where changes are happening most rapidly should be revisited most frequently. If sensor readings are not changing much, then occasional samples are sufficient to obtain statistically significant results. It is better to spend probe dwell time on those sensors that are experiencing the most rapid reading changes, and for which infrequent visits lead to the greatest divergence from the current sensor values.

It is useful for sensors to operate in ensembles. Rather than implementing a broad range of sensors in a single integrated circuit, it is possible simply to deploy a mixture of different sensors in a given geographical area and allow them to self-organize.

Sensors are typically specialized for detecting certain signatures. One kind detects motion, another heat, and a third sound. When one sensor detects its critical event signature, it makes other nearby sensors aware of its detection. They then orient their sensing function in a particular, signature-specific way. For example, a simple motion-detecting sensor might cue more sophisticated sensors detecting thermal or other radiation properties.

A cuing system can be designed in a centralized scheme. The motion sensor communicates with the BTS, which in turn communicates with a nearby heat sensor. If passive communications techniques can be used, this may well be the most power-efficient way to propagate the detection information.

The centralized/passive schemes cannot be used if the line-of-sight path is blocked, or if the probe revisit rate is too infrequent to meet detection latency constraints. In these cases, the detecting mote must employ an active transmitter. If the line-of-sight path is blocked, then the mote will need to use special, multi-hop techniques to communicate with the BTS or nearby sensor nodes.

In an special scheme, a node transmits for a short burst and waits for an ACK (acknowledgement) response from any listening node to determine that its transmission has been received. Determining true reachability between pairs of motes requires a full four-phase handshake. This must be executed in the context of appropriate timeouts and made robust to dynamic changes in the positions of the communicating nodes, which may be floating in the air.

Routing tables can be constructed from such pairwise discovery of connectivity. However, standard routing algorithms, like RIP (Routing Information Protocol), OSPF (Open Shortest Path First), and DVMRP (Distance Vector Multicast Routing Protocol), assume bidirectional and symmetric links. This is not always the case for smart dust. It may be possible for mote A to communicate with mote B, but not vice versa. Even if the communication is bidirectional, it need not exhibit the same bandwidth or loss characteristics in both directions. Therefore, new routing algorithms must be developed to deal with the general case of links that are unidirectional and/or asymmetric in their performance.

The efforts of IETF (Internet Engineering Task Force) Unidirectional Link Routing Working Group focus on supporting high-bandwidth unidirectional links where all nodes have at least low-bandwidth bidirectional links (e.g. a high-bandwidth satellite link superimposed on nodes interconnected via slow-speed telephone links).

One possible improvement is to make use of emerging MEMS technology for on-board inertial navigation circuits to make sensors more aware of near neighbors even as they drift out of line-of-sight of the BTS. The BTS can determine the relative location of dust motes within its field of view. It could then disseminate this near-neighbor information to motes able to observe its probe beam. The on-board inertial navigation capability, combined with these periodic relative location snapshots, could assist motes in orienting their laser and detector optics to improve their ability to establish links with nearby motes.

Several projects have recently been initiated to investigate a variety of communications research aspects of distributed sensor networks. The Factoid Project at the Compaq Palo Alto Western Research Laboratory (WRL) is developing a portable device small enough to be attached to a key chain. The device collects announcements from broadcasting devices in the environment, and these can be uploaded to a user's home-base station. The prototype devices are much larger than smart dust motes, communication is accomplished via RF transmission, and the networking depends on short-range, point-to-point links.

The Wireless Integrated Network Sensors (WINS) Project at UCLA is developing low-power MEMS-based devices that, in addition to sensing and actuating, can also communicate. The essential difference is that WINS has chosen to concentrate on RF communications over short distances.

The Ultralow Power Wireless Sensor Project at MIT is another project that focuses on low-power sensing devices that also communicate. The primary thrust is extremely low-power operation. The prototype system transmits over a range of data rates, from 1 bit/s to 1 megabit/s, with transmission power levels that span from 10 microwatts to 10 milliwatts. The RF communications subsystem is developed for the project by Analog Devices. Optical technologies are not investigated. The design addresses the multi-hop wireless networking protocol issues.

5.4. SUMMARY

The Smart-Its embeds computation into real-world objects by attaching small, unobtrusive, and autonomous computing devices to them. These devices, the Smart-Its, integrate sensing, processing, and communication capabilities, which can be customized to the objects to which they are attached.

The dominant component in a Smart-It, with respect to power consumption, is the Bluetooth module. Since the Bluetooth modules are engineering samples that do not implement low-power modes, no data is available on how these would alleviate the significant power usage of the modules in active mode. The improved Bluetooth products should reduce power consumption considerably since the system design allows for easy replacement of the Bluetooth transceiver module.

Smart dust is an integrated approach to networks of millimeter-scale sensing/communicating nodes. Smart dust can transmit passively using novel optical reflector technology. This provides an inexpensive way to probe a sensor or to acknowledge that information was received. Active optical transmission is also possible, but consumes more power. It is used when passive techniques cannot be used, such as when the line-of-sight path between the dust mote and BTS is blocked.

PROBLEMS

Learning Objectives

After completing this chapter you should be able to:

- demonstrate an understanding of the distributed sensor networks;
- discuss what is meant by Bluetooth communication and networking;
- explain what mobile networking for smart dust is;
- demonstrate an understanding of smart-dust technology.

Practice Problems

Problem 5.1: How are the piconets managed?
Problem 5.2: What is the limit on the number of nodes in a piconet, and how does it affect communication in a sensor network?
Problem 5.3: How does a unit in inquiry-scan mode save the power?
Problem 5.4: What are the limitations in development of mobile networking protocols for smart dust?

Practice Problem Solutions

Problem 5.1:

Piconets are managed by a single master that implements centralized control over channel access. All other participants in a piconet are designated slaves. Communication is strictly slave-to-master (or vice versa), but can never be slave-to-slave. During the existence of a piconet, master and slave roles can be switched.

Problem 5.2:

The limit of eight nodes in a piconet is inadequate for setting up a densely connected sensor network. To communicate with more than eight nodes at the same time requires some sort of time multiplexing, where additional nodes have to be parked and unparked repeatedly. Setting up additional piconets requires gateway nodes to alternate between their respective piconets, since Bluetooth only supports units in one active piconet at a time.

Problem 5.3:

To save power, a unit in inquiry-scan mode does not continuously listen to inquiry messages. Instead, it only listens for a very short period of time (11.25 ms by default), which, under regular conditions, suffices for the inquiry message to get through with sufficiently high probability. Then the unit enters idle mode for a much longer interval (typically 1.28 s). However, the inquiring unit needs to send inquiry messages (and alternately listen for potential replies) during the entire interval, since it cannot know when the target unit is actually listening.

Problem 5.4:

Development of mobile networking protocols for Smart Dust involves critical limitations as follows:

- the free-space optical links require uninterrupted line-of-sight paths;
- the passive and active dust-mote transmitters have directional characteristics that must be considered in system design, and
- there are trade-offs between bit rate, energy per bit, distance and directionality in these energy-limited, free-space optical links.

6

Clustering Techniques in Wireless Sensor Networks

6.1. INTRODUCTION

Advances in MEMS technology have resulted in cheap and portable devices with formidable sensing, computing and wireless communication capabilities. A network of these devices is invaluable for automated information gathering and distributed microsensing in many civil, military and industrial applications. The use of wireless media for communication provides a flexible means of deploying these nodes without a fixed infrastructure, possibly in an inhospitable terrain. Once deployed, the nodes require minimal external support for their functioning.

Topology discovery algorithm for wireless sensor networks finds a set of distinguished nodes to construct the approximate topology of the network. The distinguished nodes reply to the topology discovery probes, thereby minimizing the communication overhead. The algorithm forms a tree of clusters, rooted at the monitoring node, which initiates the topology discovery process. This organization is used for efficient data dissemination and aggregation, duty cycle assignments and network state retrieval. The mechanisms are distributed, use only local information, and are highly scalable.

The vision of ubiquitous computing is based on the idea that future computers will merge with their environment until they become completely invisible to the user. Distributed wireless microsensor networks are an important component of this ubiquitous computing and small dimensions

Wireless Sensor Network Designs A. Hać
 ISBN: 0-470-86736-1

are a design goal for microsensors. The energy supply for the sensors is a main constraint in the intended miniaturization process. It can be reduced only to a limited degree since energy density of conventional energy sources increases slowly. In addition to improvements in energy density, energy consumption can be reduced. This approach includes the use of energy-conserving hardware. Moreover, a higher lifetime for sensor networks can be accomplished through optimized applications, operating systems, and communication protocols. Particular modules of the sensor hardware are turned off when they are not needed.

Wireless distributed microsensor systems enable fault-tolerant monitoring and control of a variety of applications. Due to the large number of microsensor nodes that may be deployed and the long required system lifetimes, replacing the battery is not an option. Sensor systems must utilize the least possible energy while operating over a wide range of scenarios. These include power-aware computation and communication component technology, low-energy signaling and networking, system partitioning considering computation and communication trade-offs, and a power-aware software infrastructure.

Many dedicated network protocols (e.g. routing, service discovery, etc.) use flooding as the basic mechanism for propagating control messages. In flooding, a node transmits a message to all of its neighbors which, in turn, transmit to their neighbors until the message has been propagated to the entire network. Typically, only a subset of the neighbors is required to forward the message in order to guarantee complete flooding of the entire network. If the node geographic density (i.e. the number of neighbors within a node's radio reach) is much higher than what is strictly required to maintain connectivity, the flooding becomes inefficient because of redundant, superfluous forwarding. This superfluous flooding increases link overhead and wireless medium congestion. In a large network, with a heavy load, this extra overhead can have a severe impact on performance.

6.2. TOPOLOGY DISCOVERY AND CLUSTERS IN SENSOR NETWORKS

Wireless sensor networks pose many challenges, primarily because the sensor nodes are resource constrained. Energy is constrained by the limited battery power in sensor nodes. The form factor is an important node design consideration for easy operability and specified deployment of these nodes, which limit the resources in a node. The protocols and applications designed for sensor networks should be highly efficient and optimize the resources used.

Sensor network architectures use massively distributed and highly complex network systems comprising hundreds of tiny sensor nodes. These nodes experience various modes of operation while maintaining local knowledge of the network for scalability. The nodes may also use networking functionalities such as routing cooperatively to maintain network connectivity. The behavior of the network is highly unpredictable because of randomness in individual node state and network structure.

Topology discovery algorithm for sensor networks uses data dissemination and aggregation, duty-cycle assignments and network-state retrieval. Network topology provides information about the active nodes, their connectivity, and the reachability map of the system.

The topology discovery algorithm uses the wireless broadcast medium of communication. The nodes know about the existence of other nodes in their communication range by listening to the communication channel. The algorithm finds a set of distinguished nodes, and by using their neighborhood information constructs the approximate topology of the network. Only distinguished nodes reply to the topology discovery probes, thereby reducing the communication overhead of the process. These distinguished nodes form clusters comprising nodes in their neighborhood. These clusters are arranged in a tree structure, rooted at the monitoring or the initiating node.

The tree of clusters represents a logical organization of the nodes and provides a framework for managing sensor networks. Only local information between adjacent clusters flows from nodes in one cluster to nodes in a cluster at a different level in the tree of clusters. The clustering also provides a mechanism for assigning node duty cycles so that a minimal set of nodes is active in maintaining the network connectivity. The cluster heads incur only minimal overhead to set up the structure and maintain local information about its neighborhood.

Sensor networks have fundamentally different architecture than wired data networks. Nodes are designed with a low cost and small form factor for easy deployment in large numbers. Hence limited memory, processor and battery power is provided. Energy constraints also limit the communication range of these devices. These nodes have various modes of operation with different levels of active and passive states for energy management. They maintain only local knowledge of the network as global information storage is not scalable, and may provide networking functionalities like routing, to maintain cooperatively the network connectivity.

The behavior of the network can be highly unpredictable because of the operating characteristics of the nodes and the randomness in which the network is set up. Hence the algorithms consider failure of a network as a rule rather than as an exception, and can handle this more efficiently.

A sensor network model incorporates the specific features as follows:

- Network topology describes the current connectivity and reachability of the network nodes and assists routing operations and future node deployment.
- The energy map provides the energy levels of the nodes in different parts of the network. The spatial and temporal energy gradient of the network nodes coupled with network topology can be used to identify the low energy areas of the network.
- The usage pattern describes node activity, data transmitted per unit of time, and emergency tracking in the network.
- The cost model provides equipment cost, energy cost, and human cost for maintaining the network at desired performance level.
- Network models take into account that sensor networks are highly unpredictable and unreliable.

The above models form the Management Information Base (MIB) for sensor networks. To update the MIB with the current state of the network, a monitoring node measures various network parameters. Measurements have spatial and temporal error, and the measurement probes have to operate at a finer granularity. A probe uses energy from the system, and in this way can change the state of the network.

The models are used for different network management functions as follows:

- Sensors are deployed randomly with little or no prior knowledge of the terrain. Future deployment of sensors depends upon the network state.
- Setting network operating parameters involves routing tables, node duty cycles, timeout values of various events, position estimation, etc.
- Monitoring network states using network models involves periodic measurements to obtain various states like network connectivity, energy maps, etc.
- Reactive network maintenance is served by monitoring the network when the regions of low network performance are traced to identify the reasons for poor performance. Corrective measures like deployment of new sensors or directing network traffic around those regions are useful.
- Proactive network maintenance allows predicting of future network states from periodic measurement of network states to determine the dynamic behavior of the network, and to predict the future state. This is useful for predicting network failures and for taking a preventive action.
- Design of sensor-network models with cost factor and usage patterns is used for design of sensor network architectures.

6.2.1. Topology Discovery Algorithm

The topology discovery algorithm used in sensor networks constructs the topology of the entire network from the perspective of a single node. The algorithm has three stages of execution as follows:

- A monitoring node requires the topology of the network to initiate a topology discovery request.
- This request diverges throughout the network reaching all active nodes.
- A response action is set up that converges back to the initiating node with the topology information.

The request divergence is through controlled flooding so that each node forwards exactly one topology discovery request. Note that each node should send out at least one packet for other nodes to know its existence. This also ensures that all nodes receive a packet if they are connected. Various methods may be employed for the response action.

When topology discovery request diverges, every node receives the information about neighboring nodes. In the response action, each node can reply with its neighborhood list. To illustrate the response action of these methods, the network in Figure 6.1 is presented with node A as the initiating node. The topology discovery request reaches node B from node A, and nodes C and D from node B. Requests are forwarded only once so that no action takes place even though node C and D may hear requests from each other.

In the direct response approach we flood the entire network with the topology discovery request. When a node receives a topology discovery request it forwards this message and sends back a response to the node from which the request was received. The response action for the nodes in Figure 6.1 is as follows:

- node B replies to node A;
- node C replies to node B; node B forwards the reply to node A;

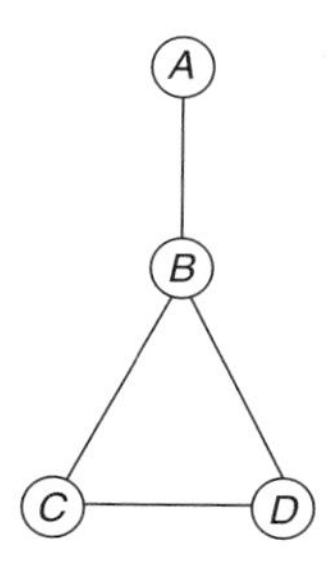

Figure 6.1 An example illustrating topology discovery.

- node D replies to node B; node B forwards the reply to node A;
- node A gets the complete topology.

Note that even though parent nodes can hear the children while they forward a request (for example, node A knows about node B when node B forwards), this is not useful because its neighborhood information is incomplete. Hence an exclusive response packet is needed for sending the neighborhood information.

In an aggregated response, all active nodes send a topology discovery request but wait for the children nodes to respond before sending their own responses. After forwarding a topology discovery request, a node gets to know its neighborhood list and children nodes by listening to the communication channel. Once this is set up, the node waits for responses from its children nodes. Upon receiving the responses, the node aggregates the data and sends it to its own parent. The response action for the nodes in Figure 6.1 is as follows:

- nodes C and D forward request; node B listens to these nodes and deduces them to be its children;
- node C replies to node B; Node D replies to node B;
- node B aggregates information from nodes C, D and itself; node B forwards the reply to node A;
- node A gets the complete topology.

In a clustered-response approach, the network is divided into set of clusters. Each cluster is represented by one node (called the cluster head) and each node is part of at least one cluster. Thus each node is in the range of at least one cluster head. The response action is generated only by the cluster heads, which send the information about the nodes in its cluster. Similarly to the aggregated response method, the cluster heads can aggregate information from other cluster heads before sending the response. The response action for the nodes in Figure 6.1 is as follows:

- assume that node B is a cluster head and nodes C and D are in its cluster;
- nodes C and D do not reply;
- only node B replies to node A;
- node A does not receive the information about the link $C \longleftrightarrow D$.

The information may be incomplete in using the clustered response approach. Direct and aggregated response methods provide an accurate view of the network topology. The clustered response creates a reachability map in which all reachable cluster heads allow all other nodes to be reachable from at least one cluster head.

The overhead incurred in topology discovery by the clustered-response approach is significantly lower than the direct or aggregated-response approaches.

6.2.2. Clusters in Sensor Networks

The communication overhead for the clustered response approach depends on the number of clusters that are formed and the length of the path connecting the clusters. Thus, to achieve the minimum communication overhead, the following problems need to solved:

- *Set cover problem*: to find a minimum cardinality set of cluster heads, which have to reply.
- *The Steiner tree problem*: to form a minimal tree with the set of the cluster heads.

These are the combinatorial optimization problems. Moreover, for an optimal solution we need to have global information about the network whereas the nodes only have local information. Thus, a heuristics approach is used, which provides an approximate solution to the problems. The algorithm is simple and completely distributed, and can thus be applied to sensor networks.

The topology discovery algorithm for finding the cluster heads is based on the simple greedy $\log(n)$-approximation algorithm for finding the set cover. At each stage a node is chosen from the discovered nodes that cover the maximum remaining undiscovered nodes. In the case of topology discovery, the neighborhood sets and vertices in the graph are not known at runtime, thus the implementation of the algorithm is not straightforward. Instead the neighborhood sets have to be generated as the topology discovery request propagates through the network. Two different node-coloring approaches are used to find the set of cluster heads during request propagation: the first approach uses three colors and the second approach uses four colors. The response generation mechanism is the same in both cases.

In the request propagation with three colors, all nodes, which receive a topology discovery request packet and are alive, are considered to be discovered nodes. The node coloring describes the node state as follows:

- White is an undiscovered node, or node that has not received a topology discovery packet.

- Black is a cluster head node, which replies to topology discovery request with its neighborhood set.
- Grey is a node that is covered by at least one black node, i.e. it is a neighbor of a black node.

Initially all nodes are white. When the topology discovery request propagates, each node is colored black or gray according to its state in the network. At the end of the initial phase of the algorithm, each node in the network is either a black node or the neighbor of a black node (i.e. grey node). All nodes broadcast a topology discovery request packet exactly once in the initial phase of the algorithm. Thus all nodes have the neighborhood information by listening to these transmissions. The nodes have the neighborhood lists available before the topology acknowledgment is returned.

Two heuristics are used to find the next neighborhood set determined by a new black node, which covers the maximum number of uncovered nodes. The first heuristic uses a node coloring mechanism to find the required set of nodes. The second heuristic applies a forwarding delay that is inversely proportional to the distance between the receiving and sending nodes. These heuristics provide a solution quite near to the centralized greedy set cover solution. The process is as follows:

- The node that initiates the topology discovery request is assigned the color black and broadcasts a topology discovery request packet.
- All white nodes become grey nodes when they receive a packet from a black node. Each grey node broadcasts the request to all its neighbors with a random delay inversely proportional to its distance from the black node from which it received the packet.
- When a white node receives a packet from a grey node, it becomes a black node with some random delay. If, in the meantime, that white node receives a packet from a black node, it becomes a grey node. The random delay is inversely proportional to the distance from that grey node from which the request was received.
- Once nodes are grey or black, they ignore other topology discovery request packets.

A new black node is chosen to cover the maximum number of as-yet uncovered elements. This is achieved by having a forwarding delay inversely proportional to the distance between the sending and receiving nodes. The heuristic behind having a forwarding delay inversely proportional to distance from the sending node is explained as follows.

The coverage region of each node is the circular area centered at the node with radius equal to its communication range. The number of nodes covered by a single node is proportional to its coverage area times the local node density. The number of new nodes covered by a forwarding node is proportional to its coverage area minus the already-covered area. This is illustrated in Figure 6.2 where node A makes nodes B and C grey. Node B forwards a packet before node C does, so that more new nodes can receive the request. The delay makes node D more likely to be black than is node E. The intermediate node between two black nodes (node B in Figure 6.2) is always within the range of both the black nodes since three colors were used for their formation.

In the request propagation with four colors, all nodes that receive a topology discovery request packet and are alive, are considered to be discovered nodes. The node coloring describes the node state as follows:

- White is an undiscovered node, or nodes, that has not received a topology discovery packet.
- Black is a cluster head node that replies to a topology discovery request with its neighborhood set.
- Grey is a node that is covered by at least one black node, i.e. it is a neighbor of a black node.
- Dark grey is a discovered node that is not currently covered by any neighboring black node and is hence two hops away from a black node. The white node changes to dark grey on receiving a request from a grey node.

This method propagates in similar fashion to the three-color method. Initially all nodes are white. When the topology discovery request propagates,

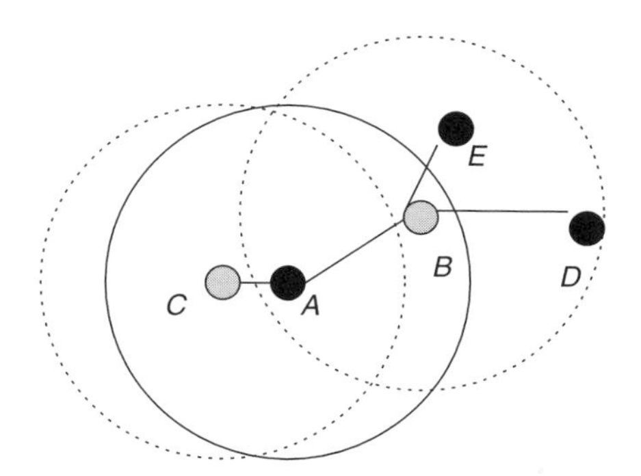

Figure 6.2 Illustration of the delay heuristic for three colors.

each node is colored black, gray or dark grey according to their state in the network. Thus at the end all nodes in the network are either black nodes or neighbors of black nodes (i.e. gray nodes) as follows:

- The node that initiates the topology discovery request is assigned color black and broadcasts a topology discovery request packet.
- All white nodes become grey nodes when they receive a packet from a black node. These grey nodes broadcast the request to all their neighbors with a delay inversely proportional to its distance to the black node from which they received the request.
- When a white node receives a packet from grey node it becomes dark grey. It broadcasts this request to all its neighbors and starts a timer to become a black node. The forwarding delay is inversely proportional to its distance from the grey node from which it received this request.
- When a white node receives a packet from dark grey node, it becomes a black node with some random delay. If, in the meantime, that white node receives a packet from a black node, it becomes a grey node.
- A dark grey node waits for a limited time for one of its neighbors to become black. When the timer expires, the dark grey node becomes a black node because there is no black node to cover it.
- Once nodes are grey or black they ignore other topology discovery request packets.

The heuristic behind having four colors for the algorithm is explained by using Figure 6.3. A new black node should be chosen so that it covers the maximum number of as-yet uncovered elements. The black nodes are separated from each other by two hops so that nodes belong to only one black node neighborhood (for instance, nodes *A* and *D*). This may not be possible in all cases and some black nodes are formed just one hop away from another

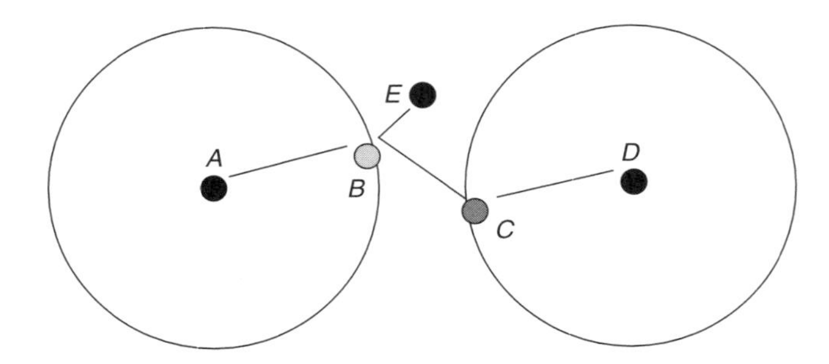

Figure 6.3 Illustration of the delay heuristic for four colors.

(for instance, node E). The heuristic behind the forwarding delay principle is similar to the three color heuristic.

The number of clusters formed by the four color heuristic is slightly lower than by the three color heuristic. In four color heuristic the clusters are formed with lesser overlap. There are some solitary black nodes, created from dark grey nodes that timed out to become black, which do not need to cover any of their neighbors. Thus, even though the number of black nodes is similar to the three color heuristic, the number of bytes transmitted is lower. However, the three-coloring approach generates a tree of clusters, which is more amenable to the network management applications.

In the topology discovery algorithm response mechanism, the first phase of the algorithm is to set up the node colors. The initiating node becomes the root of the black node tree where the parent black nodes are at most two hops away (using four colors) and one hop away (using three colors) from its children black node. Each node has the following information:

- A cluster is identified by the black node, which heads the cluster.
- A grey node knows its cluster ID (identifier).
- Each node knows its parent black node, which is the last black node from which the topology discovery was forwarded to reach this node.
- Each black node knows the default node to which it forwards packets in order to reach the parent black node. This node is essentially the node from which the black node received the topology discovery request.
- All nodes have their neighborhood information.

Using the above information, the steps for topology discovery algorithm response are described as follows:

- When a node becomes black, it sets up a timer to reply to the discovery request. Each black node waits for this time period during which it receives responses from its children black nodes.
- The node aggregates all neighborhood lists from its children and itself, and when its time period for acknowledgment expires, it forwards the aggregated neighborhood list to the default node the next hop to its parent.
- All forwarding nodes in between black nodes may also add their adjacency lists to the list to black nodes.

For the algorithm to work properly, timeouts of acknowledgments should be properly set. For example, the timeouts of children black nodes should always expire before a parent black node. Thus, timeout value is set inversely proportional to the number of hops a black node is away from the monitoring node. An upper bound on the number of hops between extreme nodes is required. If the extent of deployment region and communication range of nodes is known initially, the maximum number of hops can be easily calculated. However, if that information is not available to the nodes, the topology discovery runs in stages where it discovers only a certain extent of the area at each stage. A typical tree of clusters obtained by the topology discovery algorithm is shown in Figure 6.4. The example shows a 100×100 square meters area with 200 nodes and communication range of 20 meters. The arrow represents the initiating node. The characteristics of the clusters are as follows:

- The total surface area and the communication range of nodes bound the maximum number of black nodes formed.
- The number of nodes in each cluster depends on the local density of network.
- The depth of the tree is bounded.
- Routing paths are near optimal for data flow between source (sensor nodes) and sink (monitoring node).

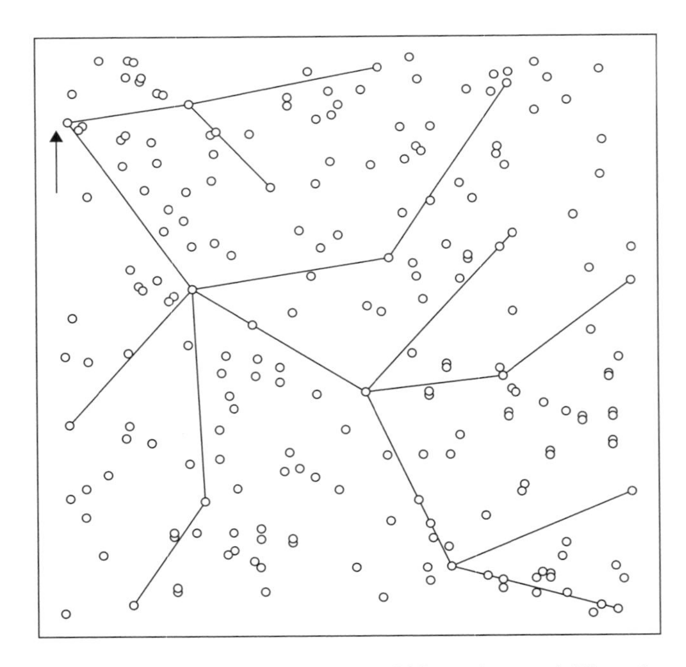

Figure 6.4 Illustration of a tree of clusters with 200 nodes and 20 meters range.

These algorithms assume a zero error rate for channels. However, minor adjustments to the protocols are needed to account for dropped packets due to channel errors.

The topology discovery initially floods the entire network. Hence channel error is not a problem as long as topology discovery requests reach a node from any path. Since the sensor networks under consideration here are dense, with many paths existing between source and destination, channel error does not create a significant impact. The number of black nodes formed may be increased due to packet losses.

However, topology acknowledgment packets are returned through single prescribed paths and hence packets may be lost. Also, the algorithm decreases the redundancy of topology information propagated among different packets, and the loss of a packet may be significant. As the packets are aggregated while moving up the cluster tree, the magnitude of loss may increase.

This problem has a simple solution if all links are symmetrical. If node A can listen to node B when node B is transmitting, then node B can listen to node A when node A transmits. When a topology discovery response has to be sent from node A, it forwards the packet to its default node (say node B). Node B, upon receiving this packet, will again forward it to its own default node, the next hop to the parent. Now node A can listen to any packet forwarded by node B and hence node A would know whether node B forwarded the same packet. If node A does not hear such a packet, it retransmits the packet assuming that node B never received the packet due to channel error.

Eavesdropping can be used as an indirect acknowledgment mechanism for reliable transmission. The only added overhead for this simple method is that every forwarded packet has to be stored at a node until the packet is reliably transmitted. The node uses energy while listening to transmissions and cannot switch itself off immediately after forwarding a packet.

6.2.3. Applications of Topology Discovery

The main purpose of the topology discovery process is to provide the network administrator with the network topology as follows:

- *Connectivity map.* The direct response and the aggregated response mechanisms provide the entire connectivity map of the region. Note that clustered response methods cannot provide this information.
- *Reachability map.* The topology discovery algorithm mechanism provides a reachability map of the region. The connectivity map is a superset of the reachability map.

- *Energy model*. When a node forwards the topology discovery request, it can include its available energy in the packet. Each node can cache the energy information of all its neighbors. If a node does not become black, it can discard the cached value. Thus all black nodes have energy information about all their neighbors, which can be sent as part of the reply. A black node can also estimate the energy consumption of nodes in its cluster by listening to the transmitted packets.
- *Usage model*. As in the previous case, each node can transmit the number of bytes received and transmitted by this node during the last several minutes. A black node will have this information cached at the time it sends its response.

This way the topology discovery algorithm provides different views of the network to the user.

We assume that in a sensor network the information flows from a sensor to the monitoring node with some control information transmitted from monitoring node to the sensors. The topology discovery process sets up a tree of clusters rooted at the initiating node. Thus, any data flow from a sensor to the monitoring node has to flow up the tree of clusters.

Each cluster has a minimal number of nodes, which are active to transfer packets between a parent and child cluster pair. Whenever a sensor needs to send some data to the monitor, it can just wake up and broadcast. The duty cycle assignment mechanism ensures that at least one node is active and responsible for forwarding the data to the next cluster. There is also at least one node in the next cluster active to receive this packet.

Each black node covers a region given by its communication range. The parent black node, logically, also covers the area covered by its children black nodes. Thus the area covered propagates up the tree and the monitor covers the entire area. The area covered by each black node may be cached during the topology response phase. The parent black node receives such areas from its children and, in turn, makes the larger area to approximate to its logical coverage region. Region based queries from the monitor node can be channeled to the appropriate region by the black nodes using their coverage information. On the return path the data may be aggregated at the black nodes.

The duty cycle of nodes for data forwarding is set up as follows. Each node in a cluster has at least the following information: the cluster ID (identifier) and the parent black node, which is the last black node from which the topology discovery request was forwarded. In each cluster, by using this information, the sets of nodes between two clusters are chosen to forward packets between clusters. At least one node in each set is active at a given time to maintain a link between a parent and child cluster pair.

In the assignment with location information, the nodes have knowledge about their geographical location. After a black node has sent a topology acknowledgment, it has knowledge of both its parent black node and the children black nodes. By using this information, the sets of nodes for each parent and child pair, need to be set up, so that in any set only one node needs to be active to transfer or receive packets from the clusters.

Figure 6.5 shows a general case in which a cluster (with black node B) may be formed as child of another cluster (with black node A). Since three colors are used to set up the tree of clusters, there is an intermediate node between the clusters (node C).

The communication range of nodes is equal to R. In a circular area with radius $R/2$, shown by the dotted circle, nodes always form a completely connected graph, as each node is within communication range of other nodes. This region is centered at the midpoint (point P) of a parent and child cluster pair. If there is at least one node active in both clusters inside this region, then a packet can be forwarded from one cluster to the other cluster. The algorithm to set up the sets of nodes is as follows.

- Black nodes send a packet with information about its parent cluster and children clusters to all its neighbors. This packet also contains the location information about the black nodes, which are the heads of the respective clusters.
- Nodes decide to be a part of the packet forwarding set by considering a circular region of radius $R/2$ centered at midpoint of the particular pair of black nodes.
- If a node is inside such region for a particular packet pair, this node becomes an active forwarding node for that cluster pair with some random delay.
- When the node becomes a forwarding node it sends a packet to signal this event. All other nodes go to the sleep mode for this pair of clusters. However, they may be in an active mode for the other pairs of clusters.

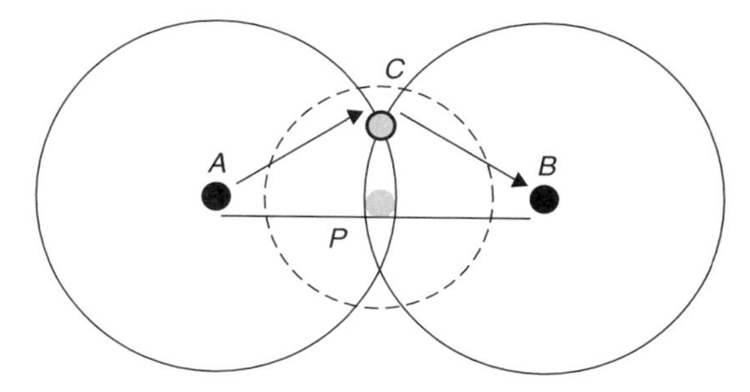

Figure 6.5 Assigning up the duty cycle with location information.

- A node may give up its active state for a cluster pair after this node has spent a certain amount of energy. The node sends a signal so that one of the other sleeping nodes can become active. When the active node receives a response signal from another node it goes to sleep mode for that cluster pair.
- Although the circular region of radius $R/2$ overlaps two clusters, there may not be other nodes in both clusters. Since all nodes can receive transmission from each other in this region, when an active node in a cluster receives a signal about activation of another node in the other cluster, it signals the black node that there exists at least one node in each cluster and the overlap regions may be used for packet forwarding.
- The intermediate node between two black nodes is used for forwarding if the overlap region does not have the other nodes in both clusters.
- During forwarding, the black node listens to all packets and forwards only the packets from the sending node which is out of range for the active forwarding node.

Figure 6.6 illustrates how a packet is forwarded between clusters. There are two clusters with black nodes A and B. The respective forwarding nodes in the overlap regions are node C and node D. When node P sends a packet, node A determines if node P is within the range of node C. If it is not, then node A forwards the packet to node C. Otherwise, node C can listen to the packet from node P. Node C forwards the packet in the overlapping region where node D receives it. Note that since node C is in the range of node P, the black node A does not need to forward this packet.

In the assignment without location information, the nodes do not have information about their location. The three-color cluster tree has the property that any parent and child pair is, at the most, one hop away from each other. This means that there is, at most, one intermediate node between any two black nodes.

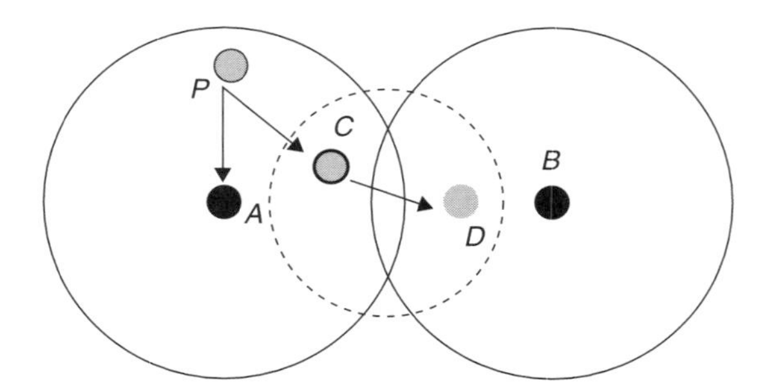

Figure 6.6 Node forwarding between clusters.

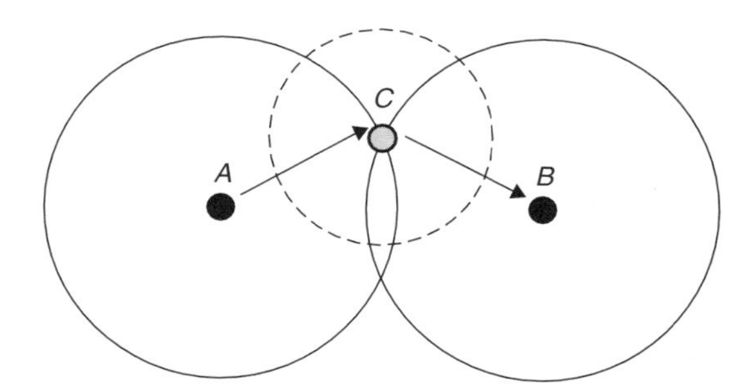

Figure 6.7 Assigning up the duty cycle without location information.

In the algorithm discussed, the locations of black nodes were known, and the actual mid point was calculated. The nodes inside a circular region of radius $R/2$ centered at this point, were considered for forwarding. This is only possible with location information.

In the approach without location information, a circle of radius $R/2$ is centered at the intermediate node between two black nodes. Figure 6.7 illustrates the mechanism.

The intermediate node C sends out a message to set up the forwarding nodes. Nodes within a distance of $R/2$ (shown by dotted circle) from this intermediate node consider themselves for forwarding between a particular pair of clusters. The remaining procedure is exactly the same as the approach with location information.

Due to lack of location information, a black node cannot decide the reachability of packets between forwarding nodes. The black node, instead of forwarding a packet immediately, waits for a random amount of time before forwarding the packet. In the meantime, if the black node hears that the active forwarding node forwarded the same packet, it does not forward this packet.

6.3. ADAPTIVE CLUSTERING WITH DETERMINISTIC CLUSTER-HEAD SELECTION

Reducing the power consumption of wireless microsensor networks increases the network lifetime. A communication protocol LEACH (Low-Energy Adaptive Clustering Hierarchy) can be extended from stochastic cluster-head selection algorithm to include a deterministic component. This way, depending on the network configuration an increase of network lifetime can be accomplished. Lifetime of microsensor networks is defined by using three metrics FND (First Node Dies), HNA (Half of the Nodes Alive), and LND (Last Node Dies).

LEACH is a communication protocol for microsensor networks. It is used to collect data from distributed microsensors and transmit it to a base station. LEACH uses the following clustering-model: some of the nodes elect themselves as cluster-heads, which collect sensor data from other nodes in the vicinity and transfer the aggregated data to the base station. Since data transfers to the base station dissipate much energy, the nodes take turns with the transmission by rotating the cluster-heads, which leads to balanced energy consumption of all nodes and hence to a longer lifetime of the network.

Modification of LEACH's cluster-head selection algorithm reduces energy consumption. For a microsensor network the following assumptions are made:

- the base station (BS) is located far from the sensors and is immobile;
- all nodes in the network are homogeneous and energy constrained;
- all nodes are able to reach the BS;
- nodes have no location information;
- the propagation channel is symmetric;
- cluster-heads perform data compression.

The energy needed for the transmission of one bit of data from node A to node B, is the same as to transmit one bit from node B to node A because of the symmetric propagation channel. Cluster-heads collect nk-bit messages from n adjacent nodes and compress the data to $(c \times n)$ k-bit messages, which are transmitted to the BS, with $c \leqslant 1$ as the compression coefficient.

The operation of LEACH is divided into rounds, each of which consists of a set-up and a steady-state phase. During the set-up phase, cluster-heads are determined and the clusters are organized. During the steady-state phase, data transfers to the base station occur.

LEACH cluster-heads are stochastically selected by each node n determining a random number between 0 and 1. If the number is less than a threshold $T(n)$, the node becomes a cluster-head for the current round.

Considering a single round of LEACH, a stochastic cluster-head selection will not automatically lead to minimum energy consumption during data transfer for a given set of nodes. All cluster-heads can be located near the edges of the network or adjacent nodes can become cluster-heads. In these cases some nodes have to bridge long distances in order to reach a cluster-head. However, considering two or more rounds, a selection of favorable cluster-heads results in an unfavorable cluster-head selection in later rounds, since LEACH tries to distribute energy consumption among all nodes. An example case is shown in Figure 6.8. In the bad-case scenario, cluster-heads

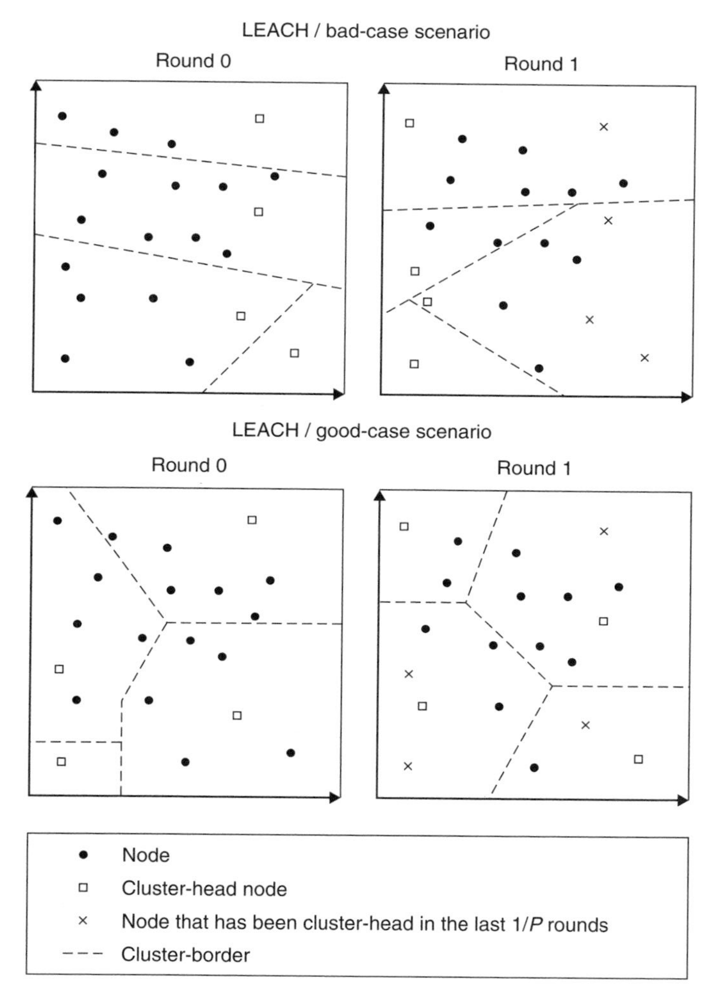

Figure 6.8 LEACH network with $P = 0.2$, $n = 20$, and network dimension of 100×100 meters. Above, cluster-heads are placed in proximity to each other and near the edges which leads to high energy consumption since nodes have to transmit over long distances. Below, energy is saved by uniformly distributing cluster-heads over the network. The set of nodes that have not been cluster-heads in round 0 and 1 is equal for both cases.

are selected unfavorably near the edges, in round 0 on the right-hand side and in round 1 on the left-hand side of the network. In the good-case scenario cluster-heads are not distributed optimally across the network, but better than in the bad-case scenario. A selection of favorable cluster-heads will not automatically lead to a higher energy consumption in later rounds.

A selection of favorable cluster-heads in earlier rounds does not result in an unfavorable cluster-head selection in later rounds. Therefore, energy savings from earlier rounds will not be consumed by higher energy dissipation in later rounds. Regarding energy consumption, a deterministic cluster-head selection algorithm can out perform a stochastic algorithm.

The definition of the lifetime of a microsensor network is determined by the kind of service it provides. Hence, three approaches to defining lifetime are considered. In some cases it is necessary for all nodes to stay alive as long as possible, since network quality decreases considerably as soon as one node dies. Scenarios for this case include intrusion or fire detection. In these scenarios it is important to know when the first node dies. The metric First Node Dies (FND) denotes an estimated value for this event in a specific network configuration. Furthermore, sensors can be placed in proximity to each other. Thus, adjacent sensors could record related or identical data. Hence, the loss of a single or a few nodes does not automatically diminish the quality of service in the network. In this case, the metric Half of the Nodes Alive (HNA) denotes an estimated value for the half-life period of a microsensor network. Finally, the metric Last Node Dies (LND) gives an estimated value for the overall lifetime of a microsensor network.

For a cluster-based algorithm like LEACH, the metric LND is not interesting since more than one node is necessary to perform the clustering algorithm. The discussion of algorithms includes the metrics FND and HNA.

An approach to increasing the lifetime of a LEACH network is to include the remaining energy level available in each node. This can be achieved by reducing the threshold relative to the node's remaining energy. This modification of the cluster-head threshold can increase the lifetime of a LEACH microsensor network by 30 % for FND and by more than 20 % for HNA.

A modification of the threshold equation by the remaining energy has a crucial disadvantage, in that after a certain number of rounds the sensor network cannot perform, although there are still nodes available with enough energy to transmit data to the base station. The reason for this is a cluster-head threshold that is too low, because the remaining nodes have a very low energy level.

A possible solution of this problem is a further modification of the threshold equation, which is expanded by a factor that increases the threshold for any node that has not been cluster-head for a certain number of rounds. The chance of this node becoming a cluster-head increases because of a higher threshold. A possible blockade of the network is solved. This way the data is transmitted to the base station as long as the nodes are alive.

6.4. SENSOR CLUSTERS' PERFORMANCE

The goals of a wireless sensor network are to detect events of interest and estimate parameters that characterize these events. The resulting information is transmitted to one or more locations outside the network. For example, a typical scenario might include a number of sensors spread over an outdoor area for the purpose of determining vehicle traffic. The first step is to determine if there is a vehicle present, and the second step is to classify the type of vehicle. Parameters such as speed, direction, and cargo are of interest. Figure 6.9 shows a conceptual diagram of the three layers in the physical system. The cluster layer is where the collaborative signal processing occurs, while the wireless Mobile *Ad hoc* Network (MANET) is responsible for routing and dissemination of the information. Note that conceptually, the wireless network is larger than the sensor network, because it includes additional nodes.

The issues in designing a sensor network include:

- selection of the collaborative signal processing algorithms run at each sensor node;
- selection of multi-hop networking algorithms, and
- optimal matching of sensor requirements with communications performance.

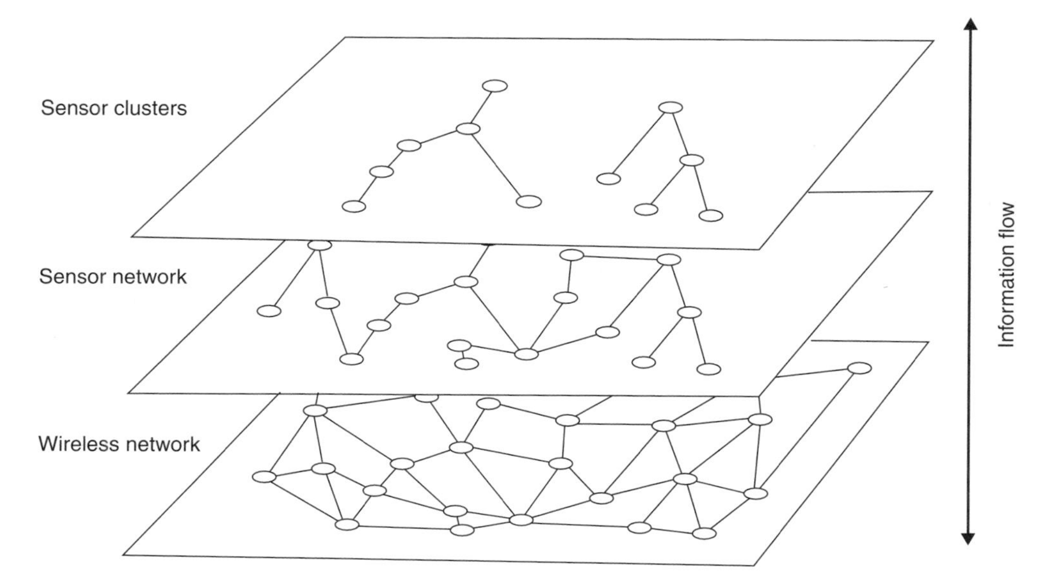

Figure 6.9 The conceptual layers in wireless sensor network.

For military networks, additional issues are:

- low probability of detection and exploitation;
- resistance to jamming, reliability of data;
- latency;
- survivability of the system.

To make the design and optimization efforts tractable, the problem should be decomposed as much as possible. This is done by clearly defining interfaces between the different layers containing the various sensor, networking, and communication processes. Moreover, the wireless sensor network must be coupled with the environment and the target(s); there are two or more transmission media, one for the radio propagation and the other for the propagation of the sensor input (acoustic, seismic, etc.).

To reduce the amount of power spent on long distance radio transmissions, the sensor nodes are aggregated into clusters. This concept is especially useful when the ranges of the sensors are relatively short. During the process of distributed detection, estimation and data fusion, the radio transmissions are among nodes within a cluster, under the control of a cluster-head or master node. While it is quite possible that all the nodes in a cluster are identical, it may be more desirable to provide the cluster-head with more functionality. Location awareness using GPS (Global Positioning System) and a longer-range radio are two useful additions.

Figure 6.10 shows the processing occurring at different layers in the protocol stack for such a cluster-based system. A short range radio (Radio 1) is used to communicate among the sensors in a cluster. The sensor layer is responsible for the collaborative signal processing, which processing can include beamforming, as well as distributed detection, estimation and data fusion. The system operates by using an emitter which generates observations at one or more sensors. In the figure, only node A receives a particular observation. The sensor layer processes the observation and makes a tentative decision, thereby performing data reduction down to a few bits. (For beamforming, either the raw data or a finely quantized version thereof is transmitted instead, requiring significantly more bandwidth.) This information is placed in a very short data packet that is to be sent to all other nodes in the cluster (Nodes B_i and Node C), assumed to be within one hop. Therefore, the packet can bypass the transport and network layers and go directly to the MAC layer for transmission at the appropriate time.

Upon reception of the packet, the other nodes update their tentative decisions. These decisions may then be re-broadcast to all nodes in the cluster. The number of iterations depends on the distributed algorithm, and

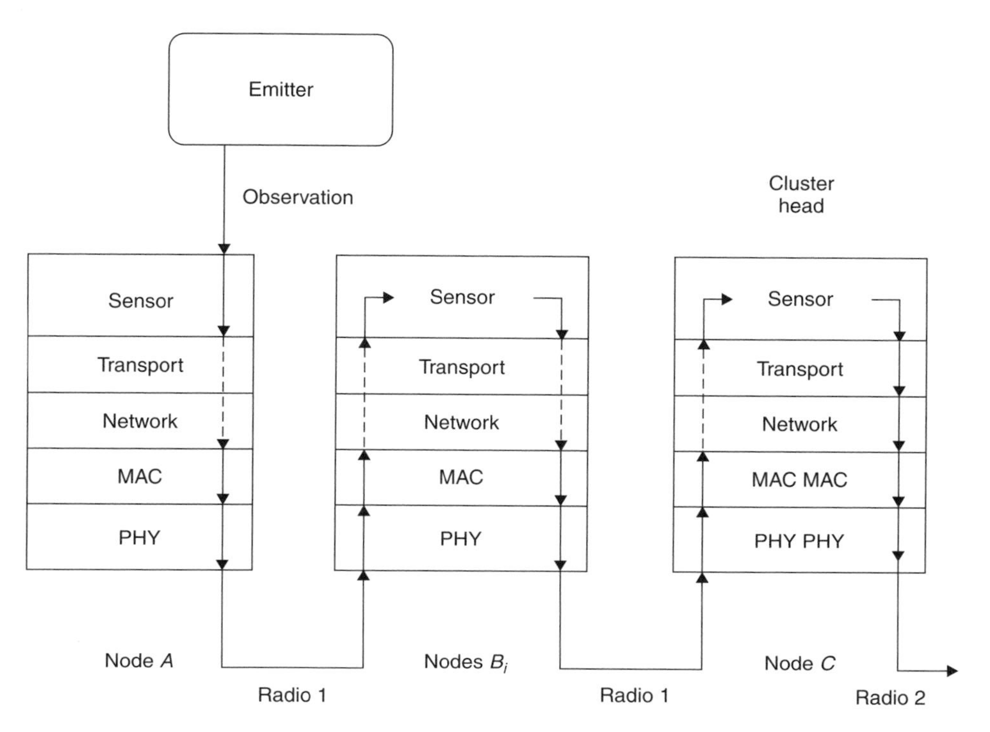

Figure 6.10 The movement of data through the different protocol layers in a cluster-based wireless sensor network.

eventually convergence is achieved. A number of parameters, such as the decision and a confidence measure, now need to be transmitted from the cluster to a remote location using the larger mobile *ad-hoc* network.

A summary packet is generated and sent down to the network layer, as shown in the figure by the solid lines in the right side of node *C*. The network layer uses its routing protocol to select the next hop in the MANET. The network packet is encapsulated by the MAC and transmitted. The actual transmission may use the same radio system as used for cluster-based processing, albeit with increased power and changes in other radio parameters. For example, the virtual subnet approach uses different channels for intracluster and intercluster communications. However, it is also possible to use a completely different radio, as shown in Figure 6.10.

6.4.1. Distributed Sensor Processing

In distributed detection, a number of independent sensors each make a local decision, often a binary one, and then these decisions are combined at a fusion center to generate a global decision. In the Neyman–Pearson formulation, a bound on the global probability of false alarm is assumed, and the goal is to determine the optimum decision rules at the sensors and at the fusion center that maximize the global probability of detection. Alternatively, the Bayesian formulation can be used, and the global probability of error be computed. If the communication network is able to handle the increased load, performance can be improved through the use of decision feedback.

Swaszek and Willet (1995) proposed an extensive feedback approach that they call 'parleying'. The basic idea is that each sensor makes an initial binary decision that is then distributed to all the other sensors. The goal is to achieve a consensus on the given hypothesis through multiple iterations. The algorithm constrains the consensus to be optimum in that it matches the performance of a centralized processor having access to all the data. The main algorithmic performance issue is the number of parleys (iterations) required to reach this consensus. An extension to the parley algorithm uses soft decisions in order to reduce both the number of channel accesses required and the total number of bits transmitted. The parley algorithm leads to the same global decision being made at each node in the cluster.

When classifying a target, the true misclassification probability is the main metric of interest. For any parameter, the maximum likelihood (or maximum *a posteriori*, if possible) estimate of these parameters is desired, along with the variance of the estimate. Additionally, the total energy expended in making the detection decision and doing any parameter estimation and classification is important.

While the use of decision feedback in a sensor cluster can certainly improve performance, there is an additional cost in the complexity of the sensor nodes and a possible increase in transmission energy requirements. Advances in integrated circuitry mitigate the first problem, and the use of short range transmissions helps with the second one. In general, the trend is to put more signal processing in the node in order to reduce the number of transmissions. Cluster-based collaborative signal processing provides a good trade-off between improved performance and low energy use. Within a node, multispectral or multimode sets of colocated sensors, combined in a kind of local data fusion, may be used to improve the performance. This type of data fusion is generally different from the data fusion that may occur at a fusion center (cluster-head).

The overall utility of the sensor network may be improved if each sensor is context-aware, that is, it has some knowledge of its environment. Schmidt *et al.* (1999) studied the use of context awareness for adapting the operating parameters of a GSM (Global Standard for Mobile) cellular phone and a personal digital assistant, and proposed a four-layer architecture. The lowest layer is the sensor layer, which consists of the actual hardware sensors. For each sensor, a number of cues is created. Cues are abstractions of a sensor, and they allow calibration and post-processing; when a sensor is replaced by one of a different type, only the cues must be modified.

Typical cues include:

- the average of the sensor data over a given interval;
- the standard deviation over the same interval;
- distance between the first and third quartiles;
- first derivative of the sensor data.

Multiple sets of contexts can be defined from the cues. For example, a single context is the terrain surrounding the sensor node, such as forest, urban area, open field, etc. Here, the choices are mutually exclusive, but this is not a requirement. Another context is the number of other sensor nodes with direct (single-hop) radio connectivity. A third context is the required level of transmission security or stealth. Determining the cue to context mapping is, in general, a difficult challenge. Once the sensor's context is known, parameters such as transmit power, waveform, distributed detection algorithm, etc., can be set.

Regardless of the application, there are certain critical features that can determine the efficiency and effectiveness of a dedicated network. These features can be categorized into quantitative features and qualitative features. Quantitative features include:

- *Network settling time*: the time required for a collection of mobile wireless nodes to organize itself automatically and transmit the first message reliably.
- *Network join time*: the time required for an entering node or group of nodes to become integrated into the special network.
- *Network depart time*: the time required for the network to recognize the loss of one or more nodes, and reorganize itself to route around the departed nodes.
- *Network recovery time*: the time required for a collapsed portion of the network, due to traffic overload or node failures, to become functional again once the load is reduced or the nodes become operational.
- *Frequency of updates (overhead)*: the number of control packets required in a given period to maintain proper network operation.
- *Memory requirement*: the storage space requirements in bytes, including routing tables and other management tables.
- *Network scalability*: the number of nodes that the dedicated network can scale to and reliably preserve communication.

Qualitative critical features include:

- *Knowledge of nodal locations*: does the routing algorithm require local or global knowledge of the network?
- *Effect of topology changes*: does the routing algorithm need complete restructuring or only incremental updates?
- *Adaptation to radio communication environment*: do nodes use estimated knowledge of fading, shadowing, or multi-user interference on links in their routing decisions?
- *Power consciousness*: does the network employ routing mechanisms that consider the remaining battery life of a node?
- *Single or multichannel*: does the routing algorithm utilize a separate control channel? In some applications, multichannel execution may make the network vulnerable to counter measures.
- *Bidirectional or unidirectional links*: does the routing algorithm perform efficiently on unidirectional links, e.g. if bidirectional links become unidirectional?
- *Preservation of network security*: do routing and MAC layer policies support the survivability of the network, in terms of low probability of detection, low probability of intercept, and security?
- *QoS routing and handling of priority messages*: does the routing algorithm support priority messaging and reduction of latency for delay sensitive real-time traffic? Can the network send priority messages and voice even when it is overloaded with routine traffic levels?
- *Real-time voice and video services*: can the network support simultaneous real-time multicast voice or video while supporting traffic loads associated with situation awareness, and other routine services?

Thread-task level metrics include average power expended in a given time period to complete a thread (task), including power expended in transmitting control messages and information packets, and task completion time. Diagnostic metrics, which characterize network behavior at the packet level, include end-to-end throughput (average successful transmission rate) and delay, average link utilization, and packet loss rate.

The performance of the sensor network depends on the routing of the underlying dedicated network. MANET routing algorithms include the dynamic source routing protocol (DSR) and the *ad-hoc* on-demand distance vector routing protocol (AODV), either of which can be used as basis for the underlying wireless network. Perhaps of more relevance is the zone routing protocol

(ZRP), which is a hybrid of proactive and reactive routing protocols. This means that the network is partitioned into zones, and the routes from a node to all other nodes in its zone are determined. Routes to nodes in other zones are found as needed. ZRP may allow the sensor network to implement decision feedback among all nodes in a zone in a straightforward manner.

As an example, consider the sensor network shown in Figure 6.11. The sensors have been placed along the roads, with the greatest concentration at the fork. The Linked Cluster Algorithm (LCA) was used to self-organize the network, leading to the creation of four clusters. Clusters 1 and 2 overlap, as

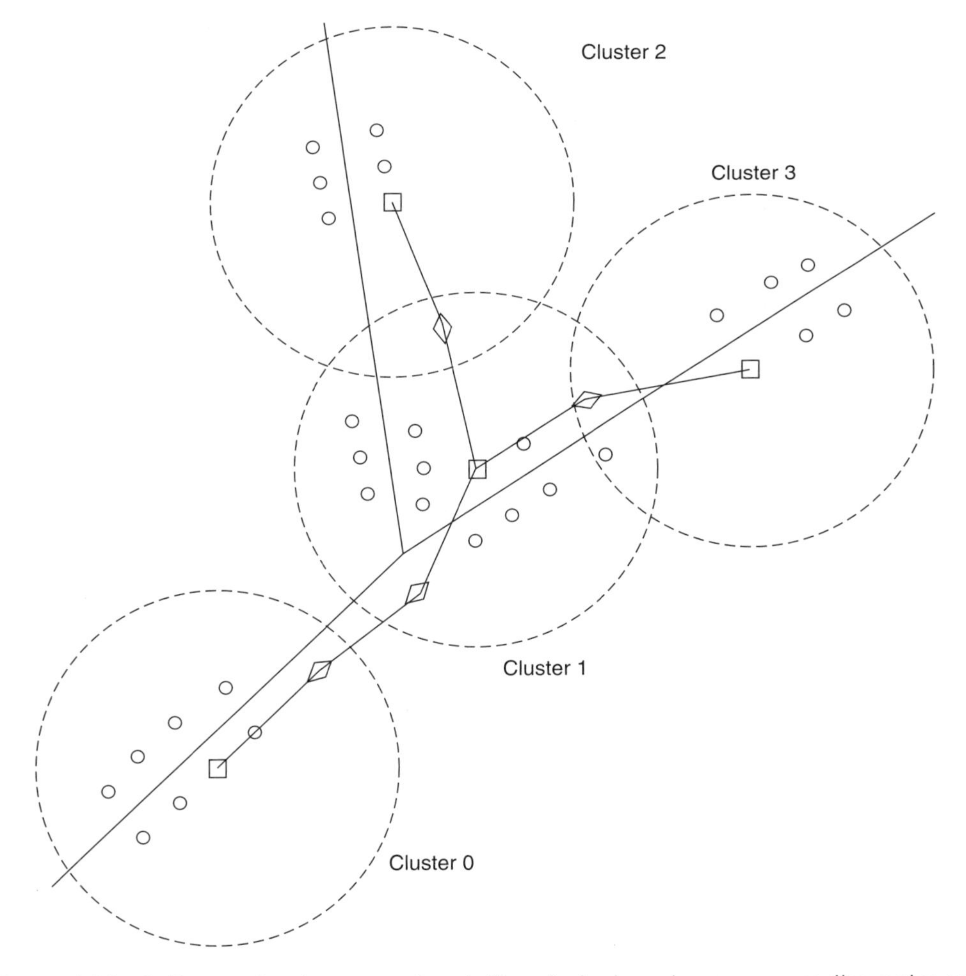

Figure 6.11 Self-organized sensor network. The cluster heads are squares, the gateway nodes are diamonds, and the ordinary sensor nodes are circles. The transmission areas of the four cluster heads are indicated by the four large circles.

do clusters 1 and 3, so only a single gateway node is used to connect each pair. Since clusters 0 and 1 do not overlap, a pair of gateways is created; the resulting backbone network that connects the cluster heads is shown in the illustration.

The numbering of the nodes in the LCA determines which nodes become cluster-heads and gateways. Since the initial topology is known, four specific nodes are assigned the highest node numbers, thereby ensuring that they would become cluster-heads. Essentially, by choosing the cluster-heads in advance, the clusters have shapes that are well suited to collaborative signal processing. To decompose a cluster further into subclusters, for example, cluster 1 could easily be divided into two or three sensor groups for the purpose of distributed detection. Once a decision is reached in a subcluster, it would be sent to the cluster-head for dissemination.

6.5. POWER-AWARE FUNCTIONS IN WIRELESS SENSOR NETWORKS

The design of micropower wireless sensor systems has gained increasing importance for a variety of civil and military applications. With advances in MEMS technology and its associated interfaces, signal processing, and RF circuitry, the focus has shifted away from limited macrosensors communicating with base stations, to creating wireless networks of communicating microsensors that aggregate complex data to provide rich, multidimensional pictures of the environment. While individual microsensor nodes are not as accurate as their macrosensor counterparts, the networking of a large number of nodes enables high quality sensing networks with the additional advantages of easy deployment and fault tolerance. These are the characteristics that make microsensors ideal for deployment in otherwise inaccessible environments, where maintenance would be inconvenient or impossible.

The unique operating environment and performance requirements of distributed microsensor networks require fundamentally new approaches to system design. As an example, consider the expected performance versus longevity of the microsensor node, compared with battery-powered portable devices. The node, complete with sensors, DSP (Digital Signal Processing), and radio, is capable of a tremendous diversity of functionality. Throughout its lifetime, a node may be called upon to be a data gatherer, a signal processor, and a relay station. Its lifetime, however, must be of the order of months to years, since battery replacement for thousands of nodes is not an option. In contrast, much less capable devices, such as cellular telephones, are only expected to run for days on a single battery charge. High diversity also exists within the environment and user demands upon the sensor network. Ambient noise in the environment, the rate of event arrival, and the user's quality requirements of the data may vary considerably over time.

A long node lifetime under diverse operating conditions demands power-aware system design. In a power-aware design, the node's energy consumption displays a graceful scalability in energy consumption at all levels of the system hierarchy, including the signal processing algorithms, operating system, network protocols, and even the integrated circuits themselves. Computation and communication are partitioned and balanced for minimum energy consumption. Software that understands the energy–quality tradeoff collaborates with hardware that scales its own energy consumption accordingly.

Once the power-aware microsensor nodes are incorporated into the framework of a larger network, additional power-aware methodologies emerge at the network level. Decisions about local computation versus radio communication, the partitioning of computation across nodes, and error correction on the link layer offer a diversity of operational points for the network.

A network protocol layer for wireless sensors allows for sensor collaboration. Sensor collaboration is important for two reasons. First, data collected from multiple sensors can offer valuable inferences about the environment. For example, large sensor arrays have been used for target detection, classification and tracking. Second, sensor collaboration can provide trade-offs in communication versus computation energy. Since it is likely that the data acquired from one sensor are highly correlated with data from its neighbors, data aggregation can reduce the redundant information transmitted within the network. When the distance to the base station is large, there is a large advantage in using local data aggregation (e.g. beamforming) rather than direct communication. Since wireless sensors are energy constrained, it is important to exploit such trade-offs to increase system lifetimes and improve energy efficiency.

The energy-efficient network protocol LEACH (Low Energy Adaptive Clustering Hierarchy) utilizes clustering techniques that greatly reduce the energy dissipated by a sensor system. In LEACH, sensor nodes are organized into local clusters. Within the cluster is a rotating cluster-head. The cluster-head receives data from all other sensors in the cluster, performs data aggregation, and transmits the aggregate data to the end-user. This greatly reduces the amount of data that is sent to the end-user for increased energy efficiency. LEACH can achieve reduction in energy of up to a factor of 8 over conventional routing protocols such as multi-hop routing. However, the effectiveness of a clustering network protocol is highly dependent on the performance of the algorithms used for data aggregation and communication.

It is important to design and implement energy-efficient sensor algorithms for data aggregation and link-level protocols for the wireless sensors.

Beamforming algorithms is one class of algorithms that can be used to combine data. Beamforming can enhance the source signal and remove uncorrelated noise or interference. Since many types of beamforming algorithms exist, it is important to make a careful selection based upon their computation energy and beamforming quality.

Algorithm implementations for a sensor network can take advantage of the network's inherent capability for parallel processing to reduce energy further. Partitioning a computation among multiple sensor nodes and performing the computation in parallel permits a greater allowable latency per computation, allowing energy savings through frequency and voltage scaling.

As an example, consider a target tracking application that requires sensor data to be transformed into the frequency domain through 1024-point FFT (Fast Fourier Transform). The FFT (Fast Fourier Transform) results are phase shifted and summed in a frequency-domain beamformer to calculate signal energies in 12 uniform directions, and the Line-Of-Bearing (LOB) is estimated as the direction with the most signal energy. By intersecting multiple LOBs at the base station, the source's location can be determined. Figure 6.12 demonstrates the tracking application performed with traditional clustering techniques for a seven sensor cluster. The sensors (S1–S6) collect data and transmit the data directly to the cluster-head (S7), where the FFT, beamforming and LOB estimation are performed. Measurements on the SA-1100 at an operating voltage of 1.5 V and frequency of 206 MHz show that the tracking application dissipates 27.27 mJ of energy.

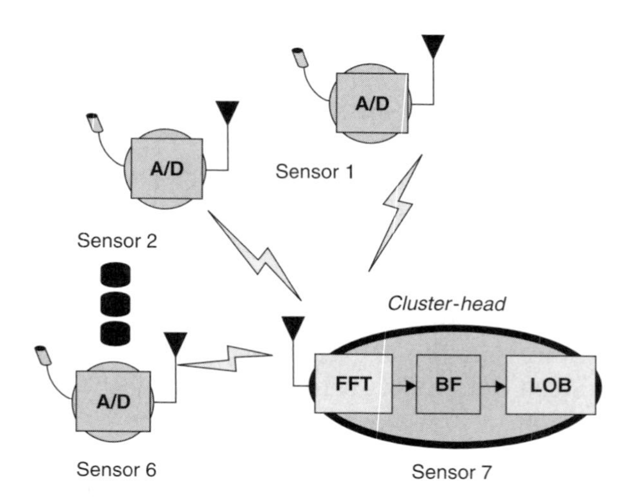

Figure 6.12 Approach 1: All computation is done at the cluster-head.

Distributing the FFT computation among the sensors reduces energy dissipation. In the distributed processing scenario of Figure 6.13, the sensors collect data and perform the FFTs before transmitting the FFT results to the cluster-head. At the cluster-head, the FFT results are beamformed and the LOB estimate is found. Since the seven FFTs are done in parallel, the supply voltage and frequency can be reduced without sacrificing latency. When the FFTs are performed at 0.9 V, and the beamforming and LOB estimation at the cluster-head are performed at 1.3 V, then the tracking application dissipates 15.16 mJ, a 44 % improvement in energy dissipation.

Energy–quality trade-offs appear at the link layer as well. One of the primary functions of the link layer is to ensure that data is transmitted reliably. Thus, the link layer is responsible for some basic form of error detection and correction. Most wireless systems utilize a fixed error correction scheme to minimize errors and may add more error protection than necessary to the transmitted data. In a energy-constrained system, the extra computation becomes an important concern. Thus, by adapting the error correction scheme used at the link layer, energy consumption can be scaled while maintaining the Bit Error Rate (BER) requirements of the user.

Error control can be provided by various algorithms and techniques, such as convolutional coding, BCH (Bose–Chaudhuri–Hocquenghem) coding, and turbo coding. The encoding and decoding energy consumed by the various algorithms can differ considerably. As the code rate increases, the algorithm's energy also increases. Hence, given bit error rate and latency requirements, the lowest power FEC (Forward Error Control) algorithm that satisfies these needs should continuously be chosen. Power consumption can

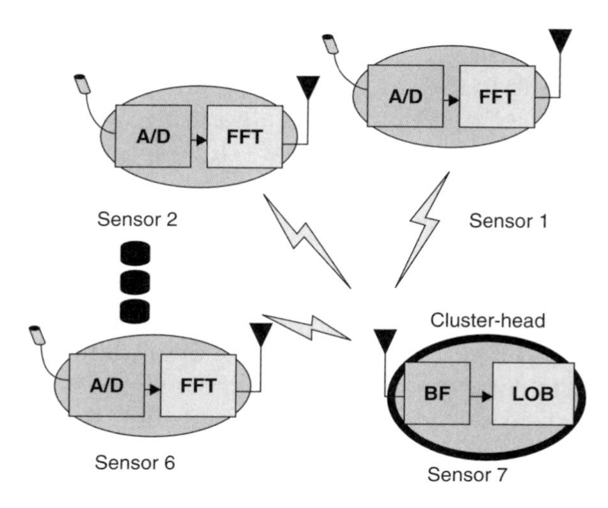

Figure 6.13 Approach 2: Distribute the FFT computation among all sensors.

be further reduced by controlling the transmit power of the physical radio. For a given bit error rate, FEC lowers the transmit power required to send a given message. However, FEC also requires additional processing at the transmitter and receiver, increasing both the latency and processing energy. This is another computation versus communication trade-off that divides available energy between the transmit power and coding processing to best minimize total system power.

6.5.1. Power Aware Software

The overall energy efficiency of wireless sensor networks crucially depends on the software that runs on them. Although dedicated circuits can be substantially more energy efficient, the flexibility offered by general purpose processors and DSPs have engineered a shift towards programmable solutions. Power consumption can be substantially reduced by improving the control software and the application software.

The embedded operating system can dynamically reduce system power consumption by controlling shutdown, the powering down of all or parts of the node when no interesting events occur, and dynamic voltage scaling. Dynamic power management using node shutdown, in general, is a nontrivial problem. The sensor node consists of different blocks, each characterized by various low power modes and overheads to transition to them. The node sleep states are a combination of various block shutdown modes. If the overheads in transitioning to sleep states were negligible, then a simple greedy algorithm could make the system go into the deepest sleep state as soon as it is idle. However, in reality, transitioning to a sleep state and waking up has a latency and energy overhead. Therefore, implementing the right policy for transitioning to the available sleep states is critical.

It is highly desirable to structure the algorithms and software such that computational accuracy can be traded off with energy consumption. Transforming software such that most significant computations are accomplished first improves the energy–quality scalability can be improved. Consider an example of a sensor node performing an FIR filtering operation. If the energy availability to the node were reduced, the algorithm may be terminated early to reduce the computational energy. In an unscalable software implementation, this would result in severe quality degradation. By accumulating the partial products corresponding to the most significant coefficients first (by sorting them in decreasing order of magnitude), the scalable algorithm produces far more accurate results at lower energies.

An application programming interface is an abstraction that hides the underlying complexity of the system from the end-user. Hence, a wireless sensor network API is a key enabler in allowing end-users to manage the tremendous operational complexity of such networks. While end-users are experts in their respective application domains (say, remote climate monitoring), they are not necessarily experts in distributed wireless networking and do not wish to be bothered with the internal network operation. By defining high-level objects, a functional interface and the associated semantics, APIs make the task of application development significantly easier.

An API consists of a functional interface, object abstractions, and detailed behavioral semantics. Together, these elements of an API define the ways in which an application developer can use the system. Key abstractions in a wireless sensor network API are the nodes, base station, links, messages, etc. The functional interface itself is divided into the following:

- functions that gather the state (of the nodes, part of a network, a link between two nodes, etc.);
- functions that set the state (of the nodes, of a cluster or the behavior of a protocol);
- functions that allow data exchange between nodes and the base station;
- functions that capture the desired operating point from the user at the base station;
- functions that help visualize the current network state;
- functions that allow users to incorporate their own models (for energy, delay, etc.).

An API is much more than the sum of its functional interface and object abstractions. This is because of the (often implicit) application development paradigm associated with it. The API is especially crafted to promote application development based on certain philosophies which the designers of the network consider to be optimal in the sense of correctness, robustness and performance. For example, a good overall application framework for wireless sensor networks is the Get-Optimize-Set paradigm. This paradigm basically implies collecting the network state, using this state information along with the knowledge of the desired operating point to compute the new optimal state, and then setting the network to this state. The entire application code is based on this template.

Power aware computation and communication is the key to achieving long network lifetimes due to the energy constrained nature of the nodes. An important responsibility of the API is not only to allow the end-user to construct the system in a power-aware manner but also to encourage such an approach. For starters, functions in a high quality network API have explicit energy, quality, latency and operating point annotations. Hence, instead of demanding a certain function from the network, one can demand a certain function subject to constraints (energy, delay, quality, etc.). Next, the API has basic energy modeling allowing the end-user to calibrate the energy efficiency of the various parts of the application. For users requiring models beyond the level of sophistication that the API offers, there are modeling interfaces that allow users to register arbitrarily complex models. Next, a good wireless sensor API allows what have come to be known in the software community as thick and thin clients. These adjectives refer to the complexity and overhead of typical application layers. Finally, the Get-Optimize-Set paradigm promulgated by the API allows the network to beat the optimal operating point thus enhancing energy efficiency.

6.6. EFFICIENT FLOODING WITH PASSIVE CLUSTERING

Clustering and route aggregation techniques have been proposed to reduce the flooding overhead. These techniques operate in a proactive, background mode. They use explicit control packets to elect a small set of nodes (cluster-heads, gateways or flood-forwarding nodes), and restrict to such a set the flood forwarding function. These proactive schemes cause traffic overhead in the network.

A flooding mechanism based on passive, on-demand clustering reduces flooding overhead without loss of network performance. Passive clustering is an on-demand protocol which dynamically partitions the network into clusters interconnected by gateways. Passive clustering exploits data packets for cluster formation, and is executed only when there is user data traffic. Passive clustering has the following advantages compared with active clustering and route aggregation techniques.

(1) Passive clustering eliminates cluster set-up latency and extra control overhead (by exploiting on-going packets).
(2) Passive clustering uses an efficient gateway selection heuristic to elect the minimum number of forwarding nodes (thus reducing superfluous flooding).
(3) Passive clustering reduces node power consumption by eliminating the periodic, background control packet exchange.

Multi-hop *ad hoc* networks (MANETs) are self-creating, self-organizing and self-administrating without deploying any kind of infrastructure. They offer special benefits and versatility for wide applications in military (e.g. battlefields, sensor networks, etc.), commercial (e.g. distributed mobile computing, disaster discovery systems, etc.), and educational (e.g. conferences, conventions, etc.) environments, where fixed infrastructure is not easily acquired. With the absence of pre-established infrastructure (e.g. no router, no access point, etc.), two nodes communicate with one another in a peer-to-peer fashion. Two nodes communicate directly if they are within transmission range of each other. Otherwise, nodes can communicate via a multi-hop route with the cooperation of other nodes. To find such a multi-hop path to another node, each MANET node widely use flooding or broadcast (e.g. hello messages). Many *ad hoc* routing protocols, multicast schemes, or service discovery programs depend on massive flooding.

In flooding, a node transmits a message to all neighbors. The neighbors in turn relay the message to their neighbors until the message has been propagated to the entire network. This is blind flooding with performance related to the average number of neighbors (neighbor degree) in the CSMA/CA network. As the neighbor degree becomes higher, the blind flooding suffers from the increases in:

- redundant and superfluous packets;
- the probability of collision, and
- congestion in wireless medium.

When topology or neighborhood information is available, only a subset of neighbors is required to participate in flooding to guarantee the complete flooding of the network. This is efficient flooding. The characteristics of MANETs (e.g. node mobility, the limited bandwidth and resource), however, make collecting topological information very difficult. It generally needs extra overhead due to the periodic message exchanges or event driven updates with optional deployment of GPS (Global Positioning System). For this reason, many on-demand dedicated routing schemes and service discovery protocols use blind flooding. With periodic route table exchanges, proactive *ad-hoc* routing schemes, unlike on-demand routing methods, can gather topological information without a significant overhead (through piggybacking topology information or learning neighbors). Thus, a few proactive *ad hoc* routing mechanisms use route aggregation methods so that the route information is propagated by only a subset of nodes in the network.

Passive clustering is an efficient flooding suitable for on-demand protocols, and does not require the deployment of GPS or explicit periodic control messages. This scheme has several contributions compared with other efficient flooding mechanisms (such as multipoint relay, neighbor coverage, etc.) as follows:

(1) Passive clustering does not need any periodic messages, instead, it exploits existing traffic to piggyback its small control messages. Based on passive clustering technique, it is very resource efficient regardless of the degree of neighbor nodes or the size of the network. Passive clustering provides scalability and practicality for choosing the minimal number of forwarding nodes in the presence of dynamic topology changes. Therefore, it can be easily applied to on-demand routing schemes to improve the performance and scalability.
(2) Passive clustering does not have any set-up latency, and it saves energy with no traffic.
(3) Passive clustering maintenance is well adaptive to dynamic topology and resource availability changes.

The problem of finding a subset of dominant forwarding nodes in MANETs is NP-complete. Thus, the work on efficient flooding focuses on developing efficient heuristics that select a suboptimal dominant set with low forwarding overhead.

There are several heuristics to reduce rebroadcasts. Upon receiving a flooding packet, a node decides whether it relays the packet to its neighbor or not, by using one of following heuristics:

- probabilistic scheme where this node rebroadcasts the packet with the randomly chosen probability;
- counter-based scheme where this node rebroadcasts if the number of received duplicate packets is less than a threshold;
- distance-based scheme that uses the relative distance between hosts to make the decision;
- location-based scheme based on pre-acquired location information of neighbors;
- cluster-based scheme where only cluster-heads and gateways forward the packet.

The passive clustering is different from those ideas in that it provides a platform of efficient flooding based on locally collected information (e.g. neighbor information, cluster states, etc.). Each node participates in flooding based on the role or state in the cluster structure.

Another approach to efficient flooding is to exploit topological information. With the node mobility and the absence of pre-existing infrastructure in the *ad-hoc* network, all works use the periodic hello message exchange method to collect topological information. Passive clustering does not require periodic control messages to collect topological information. Instead, it exploits on-going data packets to exchange cluster-related information.

The two schemes are called self pruning and dominant pruning. Self pruning is similar to the neighbor-coverage scheme. With self-pruning schemes, each forwarding node piggybacks the list of its own neighbors on the outgoing packet. A node rebroadcasts (becomes a forwarding node) only when this node has neighbors not covered by forwarding nodes. While the self-pruning heuristic utilizes information of directly connected neighbors only, the dominant-pruning scheme extends the range of neighbor information to two-hop-away neighbors. The dominant-pruning scheme is similar to Multipoint Relay (MPR) scheme in which a node periodically exchanges a list of adjacent nodes with its neighbors so that each node can collect the information of two-hop-away neighbors. Each node, based on the gathered information, selects the minimal subset of forwarding neighbors, which covers all neighbors within two hops. Each sender piggybacks its chosen forwarding nodes (MPRNs) onto the outgoing broadcast packet. Moreover, based on topological information, many schemes choose a dominant set. They still depend on the periodic hello messages to collect topological information. The extra hello messages, however, consume resources and drop the network throughput in MANETs. The extra traffic brings about congestion and collision as geographic density increases. The collision probability of hello messages in a single-hop and a two-hop network as the number of neighbors increases shows that the neighbor degree increases the collision probability of broadcast (the collision probability is more than 0.1 with more than 15 neighbors), and hidden terminals aggravate the collision in the multi-hop network. We assume there is no other traffic except for hello messages in the network. With user-data packets, the collision probability of hello messages increases. Thus, it is very difficult to collect the complete neighbor topology using hello messages.

These schemes (e.g. neighbor-coverage, MPR, etc.) are not scalable to offered load and the number of neighbors.

Clustering selects forwarding nodes, and groups nodes into clusters. A representative of each group (cluster) is named a cluster-head and a node belonging to more than two clusters at the same time is a gateway. Other members are the ordinary nodes. The transmission area of the cluster-head defines a cluster. Two-hop clustering is used where any node in a cluster can reach any other node in the same cluster with, at most, two hops. With clustering, nonordinary nodes can be the dominant forwarding nodes as shown in Figure 6.14.

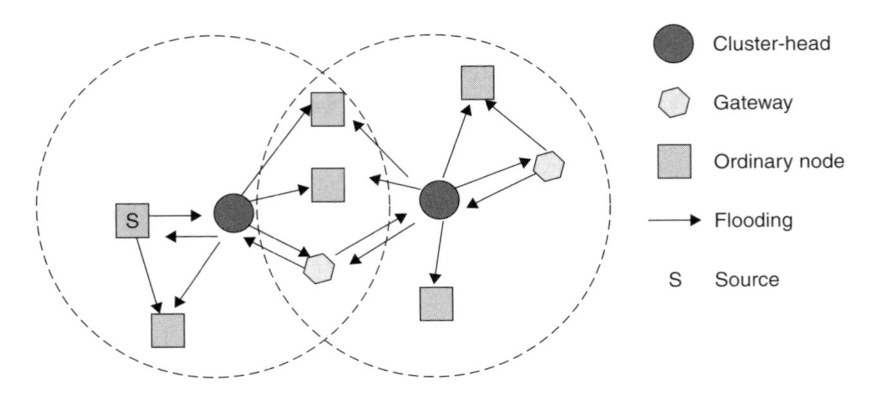

Figure 6.14 An example of efficient flooding with clustering. Only cluster-heads and gateways rebroadcast and ordinary nodes stop forwarding.

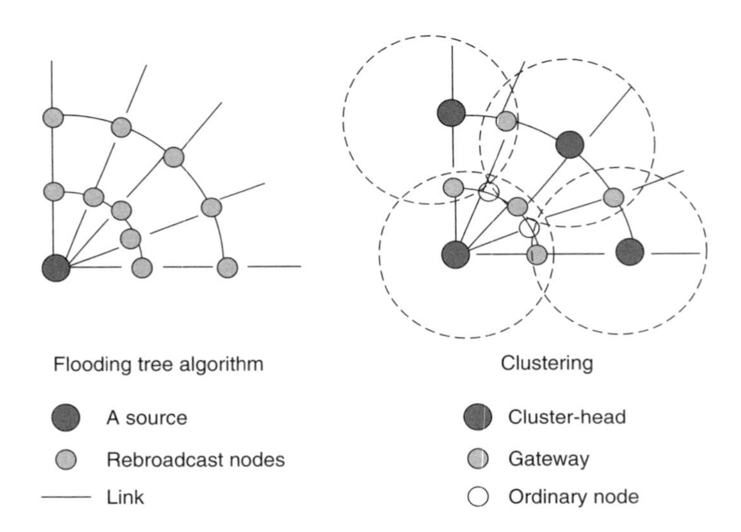

Figure 6.15 In flooding tree algorithms, every neighbor of a source has to rebroadcast since each neighbor is, at most, one adjacent node of some node. In clustering, however, ordinary nodes are not forwarding nodes.

Figure 6.15 illustrates the difference between clustering and the MPR scheme. Clustering partitions the network into several groups based on the radio range of a cluster head. The network topology, therefore, does not have a serious impact on the clustering performance. MPR, on the other hand, chooses the dominant set using topological information so that the performance of MPR is closely related to the network topology.

Clustering in *ad-hoc* networks includes hierarchical routing schemes, the master election algorithms, power control, reliable broadcast, and efficient broadcast. The cluster architecture has also been used for efficient flooding. Some clustering schemes are based on the complete knowledge of the neighbors. However, the complete knowledge of neighbor information in such networks is difficult to collect and requires a control overhead caused by periodic exchanges of hello messages. The clustering algorithms use a large number of gateways in the dense network and do not use a gateway reduction mechanism to select a minimal number of gateways. The clustering incurs a maintenance cost in case of a high mobility.

The three important observations are as follows:

(1) The selection mechanism to choose the dominant set should be efficient and dynamic. Otherwise, the scheme cannot be used effectively and practically.
(2) In a MANET, collecting accurate topological information is very difficult and carries an overhead.
(3) Clustering scheme is independent of the network topology unlike the route aggregation protocols (e.g. MPR).

6.6.1. Passive Clustering

Passive clustering is an on-demand protocol. It constructs and maintains the cluster architecture only when there are on-going data packets that piggyback cluster-related information (e.g. the state of a node in a cluster, the IP (Internet Protocol) address of the node). Each node collects neighbor information through packet receptions. Passive clustering, therefore, eliminates set-up latency and major control overhead of clustering protocols.

Passive clustering has the following mechanisms for the cluster formation:

- First Declaration Wins rule, and
- Gateway Selection Heuristic.

With the First Declaration Wins rule, a node that first claims to be a cluster-head rules the remaining nodes in its clustered area (radio coverage). There is no waiting period (to make sure all the neighbors have been checked) unlike that in all the weight-driven clustering mechanisms. The Gateway Selection Heuristic provides a procedure to elect the minimal number of gateways (including distributed gateways) required to maintain the connectivity in a distributed manner.

Passive clustering maintains clusters using implicit timeout. A node assumes that some nodes are out of locality if they have not sent any data for longer than timeout duration. With reasonable offered load, a node can respond to dynamic topology changes.

When a node joins the network, it sets the cluster state to *initial*. Moreover, the state of a floating node (a node that does not belong to a cluster yet) also sets to *initial*. Because passive clustering exploits on-going packets, the implementation of passive clustering resides between layers 3 and 4.

The IP option field for cluster information is as follows:

- Node ID (identifier) is the IP (Internet Protocol) address of the sender node. This is different from the source address of the IP packet;
- state of the cluster is the cluster state of the sender node;
- if a sender node is a gateway, then it tags two IP addresses of cluster heads which are reachable from the gateway;

The passive clustering algorithm is as follows:

- Cluster states. There are six possible states; *initial, cluster-head, ordinary node, gateway, cluster-head gateway, gateway ready*, and *distributed gateway*.
- The packet handling. Upon sending a packet, each node piggybacks cluster-related information. Upon a packet reception, each node extracts cluster-related information of neighbors and updates the neighbor information table.
- A cluster-head declaration is done by a node in *initial* state which changes its state to *cluster-head ready* (a candidate cluster-head) when a packet arrives from another node that is not a cluster-head. With outgoing packet, a *cluster-head ready* node can declare as a cluster-head. This helps the connectivity because it reduces isolated clusters.
- A node becomes a member of a cluster once it has heard or overheard a message from any cluster head. A member node can serve as a gateway or an ordinary node depending on the collected neighbor information. A member node can settle as an ordinary node only after it has learned enough neighbor gateways. In passive clustering, however, the existence of a gateway can be found only by overhearing a packet from that gateway. Thus, another internal state is *gateway ready*, for a candidate gateway node that has not yet discovered enough neighbor gateways. A gateway selection mechanism is developed to reduce the total gateways in the network. A candidate gateway finalizes its role as a gateway upon sending a packet (announcing the gateway's role). A candidate gateway node can become an ordinary node any time with the detection of enough gateways.

A gateway is a bridge node that connects two adjacent clusters. Thus, a node that belongs to more than two clusters at the same time is eligible to be a gateway. Only one gateway is needed for each pair of two adjacent clusters. The gateway selection mechanism allows only one gateway for each pair of two neighboring cluster-heads. However, it is possible that there is no potential gateway between two adjacent clusters, that is, two cluster-heads are not mutually reachable via a two-hop route. If there is a three-hop route between two nodes, then the clustering scheme selects those intermediate nodes as distributed gateways. Without the knowledge of complete two-hop neighbors' information, choosing a minimal number of distributed gateways is difficult. Topological knowledge carries an overhead and works inefficiently, thus, a counter-based distributed gateway selection mechanism is considered.

The gateway selection mechanism can be summarized as follows:

- Gateway means that a node belonging to more than two clusters at the same time becomes a candidate gateway. Upon sending a packet, a potential gateway chooses two cluster-heads from among known cluster-heads. This node will serve as an intermediate node between those cluster-heads. This node cannot be an intermediate node of two cluster-heads that were announced by another neighbor gateway node. If the node finds two cluster-heads, then it finalizes its role as a gateway and announces two cluster-heads to neighbors. If a gateway has received a packet from another gateway that has announced the same pair of cluster-heads, then this node compares the node ID of itself with that of the sender. If this node has the lower ID, it keeps its role as the gateway. Otherwise, it chooses a different pair of cluster-heads or changes its state. If this node can find another pair of neighbor cluster-heads that is not announced by any other gateway, then it keeps its state as *gateway* for the new pair of cluster-heads, otherwise it changes its state to *ordinary node*.
- Passive clustering allows one distributed gateway for each cluster-head and each node. A node that belongs to only one cluster can be an ordinary node when at least two (distributed) gateways are known to this node. Otherwise, it keeps the candidate gateway state. A candidate gateway node can be a distributed gateway if there is no neighbor-distributed gateway that also belongs to the same cluster. If an ordinary node has received a packet from a distributed gateway and no gateway is a neighbor node of that node, then this node changes to a distributed gateway.

Figure 6.16 shows an example of cluster architecture developed by passive clustering. With moderate on-going traffic, passive clustering allows only one gateway for each pair of clusters and enough distributed gateway nodes.

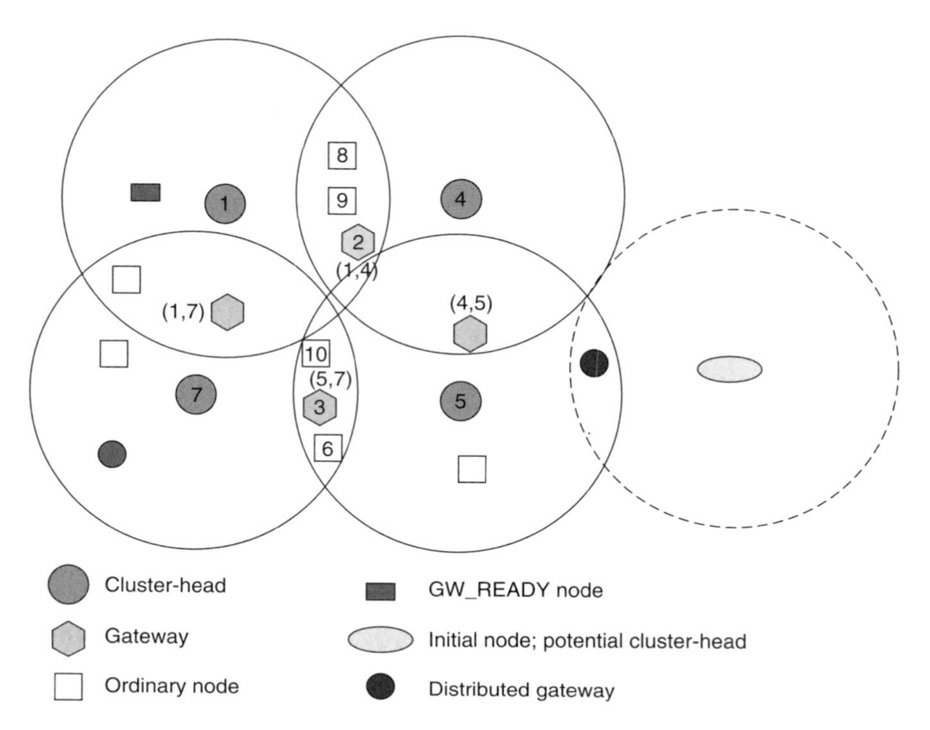

Figure 6.16 An example of a gateway selection heuristic. There is at most one gateway between any pair of two cluster-heads. A gateway can survive only when this node is the only gateway for an announced pair of cluster-heads or this node has the lowest ID among contention gateways (who announced the same pair of cluster heads).

The overhead and flooding efficiency of passive clustering needs to be analyzed. For the message overhead, passive clustering adds 8 bytes or 16 bytes to each outgoing packet. In analysis control, message overhead is considered, as the number of messages is more important than the size of each packet in dedicated networks using IEEE 802.11 DCF protocol.

Passive clustering mechanisms are more efficient than distributed tree algorithms in respect of processing overhead. The computational overhead of passive clustering is $O(Avg\_Neighbor)$ where *Avg_Neighbor* denotes the number of active neighbors. Upon receiving a packet, each node updates its neighbor table and changes its state if necessary. A cluster-head only updates its neighbor table. A member node, in addition, adjusts its state based on gateway selection heuristic. Each node computes with $O(Avg\_Neighbor)$ computational complexity upon receiving a packet. With an outgoing packet, each node simply piggybacks cluster-related information. The complexity is $O(1)$.

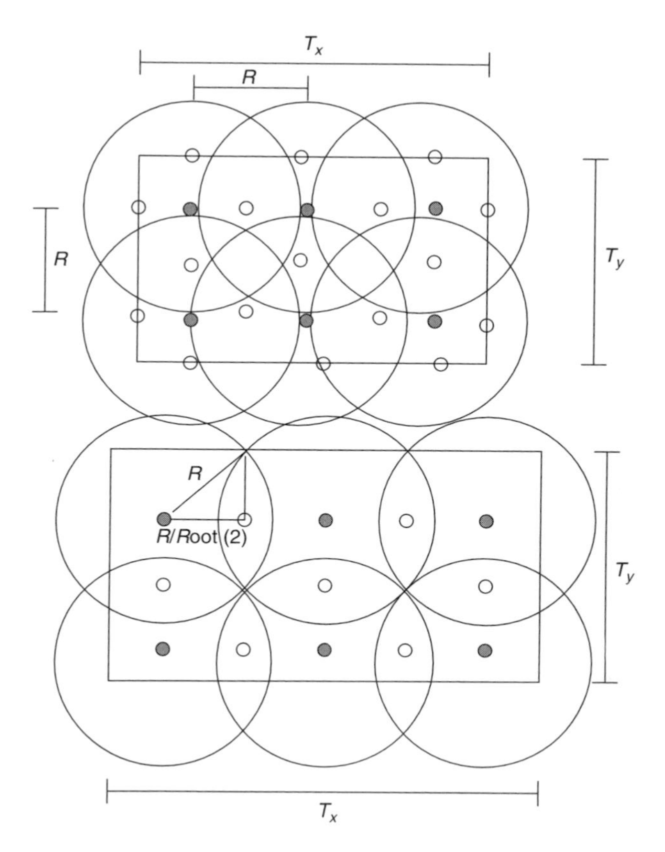

Figure 6.17 The average and most dense case of cluster architecture.

Passive clustering divides nodes into several groups based on the transmission range of the representative node (cluster-head). Thus, the number of forwarding nodes is stable regardless of the geographical density of the network. The reduction rate improves in proportion to the geographical density.

Figure 6.17 illustrates the most dense and average case of cluster construction with the assumption that there are infinite number of nodes placed randomly, and the network size is $(T_x \times T_y)$ where T_x is the horizontal size and T_y is the vertical size of the network area.

6.7. SUMMARY

Routing and data dissemination in sensor networks requires a simple and scalable solution.

The topology discovery algorithm for wireless-sensor networks selects a set of distinguished nodes, and constructs a reachability map based on their information. The topology discovery algorithm logically organizes the network in the form of clusters and forms a tree of clusters rooted at the monitoring node. We discussed the applications of tree of clusters for efficient data dissemination and aggregation, duty-cycle assignments and network-state retrieval. The topology discovery algorithm is completely distributed, uses only local information, and is highly scalable.

To achieve optimal performance in a wireless sensor network, it is important to consider the interactions among the algorithms operating at the different layers of the protocol stack. While there has been much research on partitioning a MANET into clusters, most of this work has focused on doing so for routing and resource allocation purposes. For sensor networks, a key addition is how the self-organization of the network into clusters affects the sensing performance.

Distributed microsensor networks hold great promise in applications ranging from medical monitoring and diagnosis to target detection, home automation, hazard detection, and automotive and industrial control. However, even within a single application, the tremendous operational and environmental diversity inherent to the microsensor network demand an ability to make trade-offs between quality and energy dissipation. Hooks for energy–quality scalability are necessary not only at the component level, but also throughout the node's algorithms and the network's communication protocols. Distributed sensor networks designed with built-in power awareness and scalable energy consumption will achieve maximal system lifetime in the most challenging and diverse environments.

Passive clustering can reduce redundant flooding with negligible extra protocol overhead. Moreover, passive clustering can be applied to reactive, on-demand routing protocols with substantial performance gains.

Performance of blind flooding is severely impaired especially in large and dense networks.

PROBLEMS

Learning Objectives

After completing this chapter you should be able to:

- demonstrate understanding of the clustering techniques in wireless sensor networks;

- discuss what is meant by topology discovery in wireless sensor networks;
- explain what the clusters in sensor networks are;
- demonstrate understanding of adaptive clustering with deterministic cluster-head selection;
- explain what the power-aware functions in wireless sensor networks are;
- explain what efficient flooding with passive clustering is;
- demonstrate understanding of passive clustering.

Practice Problems

Problem 6.1: What are the functions of clustering techniques?
Problem 6.2: What are the stages of execution in the topology discovery algorithm?
Problem 6.3: What heuristics are used in a node coloring mechanism?
Problem 6.4: What are the design issues in sensor network?
Problem 6.5: What are cues?
Problem 6.6: What is passive clustering?
Problem 6.7: What are the advantages of passive clustering?

Practice Problem Solutions

Problem 6.1:

A tree of clusters represents a logical organization of the nodes and provides a framework for managing sensor networks. Only local information between adjacent clusters flows from nodes in one cluster to nodes in a cluster at a different level in the tree of clusters. The clustering also provides a mechanism for assigning node duty cycles so that a minimal set of nodes is active in maintaining the network connectivity. The cluster-heads incur only minimal overhead to set up the structure and maintain local information about its neighborhood.

Problem 6.2:

The topology discovery algorithm used in sensor networks constructs the topology of the entire network from the perspective of a single node. The algorithm has three stages of execution as follows:

- a monitoring node requires the topology of the network to initiate a topology discovery request;

- this request diverges throughout the network reaching all active nodes;
- a response action is set up which converges back to the initiating node with the topology information.

Problem 6.3:

Two heuristics are used to find the next neighborhood set determined by a new black node, which covers the maximum number of uncovered nodes. The first heuristic uses a node-coloring mechanism to find the required set of nodes. The second heuristic applies a forwarding delay inversely proportional to the distance between receiving and sending node. These heuristics provide a solution quite near to the centralized greedy set cover solution.

Problem 6.4:

The issues in designing a sensor network include:

(1) selection of the collaborative signal processing algorithms run at each sensor node;
(2) selection of the multi-hop networking algorithms, and
(3) optimal matching of sensor requirements with communications performance.

For military networks, additional issues are: low probability of detection and exploitation, resistance to jamming, reliability of data, latency, and survivability of the system.

Problem 6.5:

Cues are abstractions of a sensor, and they allow calibration and post-processing; when a sensor is replaced by one of a different type, only the cues must be modified. Typical cues include:

- the average of the sensor data over a given interval;
- the standard deviation over the same interval;
- distance between the first and third quartiles;
- first derivative of the sensor data.

Problem 6.6:

Passive clustering is an on-demand protocol that dynamically partitions the network into clusters interconnected by gateways. Passive clustering exploits data packets for cluster formation, and is executed only when there is user data traffic.

Problem 6.7:

Passive clustering has the following advantages:

(1) it eliminates cluster set-up latency and extra control overhead (by exploiting on-going packets);
(2) it uses an efficient gateway selection heuristic to elect the minimum number of forwarding nodes (thus reducing superfluous flooding);
(3) it reduces node power consumption by eliminating the periodic, background control-packet exchange.

7

Security Protocols for Wireless Sensor Networks

7.1. INTRODUCTION

Thousands to millions of small sensors form self-organizing wireless networks. Security for these sensor networks is not easy since these sensors have limited processing power, storage, bandwidth, and energy.

A set of Security Protocols for Sensor Networks, SPINS, explores the challenges for security in sensor networks. SPINS include μTESLA (the micro version of the Timed, Efficient, Streaming, Loss-tolerant Authentication protocol), providing authenticated streaming broadcast, and SNEP (Secure Network Encryption Protocol), which provides data confidentiality, two-party data authentication, and data freshness, with low overhead. An authenticated routing protocol uses SPINS building blocks.

A sensor network should not leak sensor readings to neighboring networks. In many applications (e.g. key distribution), nodes communicate highly sensitive data. The standard approach for keeping sensitive data secret is to encrypt the data with a secret key that only intended receivers possess, hence achieving confidentiality. Given the observed communication patterns, secure channels are set up between nodes and base stations, and later bootstrap other secure channels as necessary.

Authenticated broadcast requires an asymmetric mechanism, otherwise any compromised receiver could forge messages from the sender. Asymmetric cryptographic mechanisms have high computation, communication, and

Wireless Sensor Network Designs A. Hać
 ISBN: 0-470-86736-1

storage overhead, which makes their usage on resource-constrained devices impractical. μTESLA overcomes this problem by introducing asymmetry through a delayed disclosure of symmetric keys, which results in an efficient broadcast authentication scheme.

Wireless networks, in general, are more vulnerable to security attacks than wired networks, due to the broadcast nature of the transmission medium. Furthermore, wireless sensor networks have an additional vulnerability because nodes are often placed in a hostile or dangerous environment where they are not physically protected.

In a target tracking application, nodes that detect a target in an area exchange messages containing a timestamp, the location of the sending node, and other application-specific information. When one of the nodes acquires a certain number of messages such that the location of the target can be approximately determined, the node sends the location of the target to the user.

7.2. SECURITY PROTOCOLS IN SENSOR NETWORKS

Small sensor devices are inexpensive, low-power devices. They have limited computational and communication resources. The sensors form a self-organizing wireless network in a multi-hop routing topology. Typical applications may periodically transmit sensor readings for processing.

The network consists of nodes (small battery-powered devices) that communicate with a more powerful base station, which in turn is connected to an outside network. The energy source on the devices is a small battery. Communication over radio is the most energy-consuming function performed by these devices, so that the communications overhead needs to be minimized. The limited energy supplies create limits for security, hence security needs to limit consumption of processor power. However, limited power supply limits the lifetime of keys. Base stations differ from nodes in having longer-lived energy supplies and having additional communications connections to outside networks.

These constraints make it impractical to use secure algorithms designed for powerful workstations. For example, the working memory of a sensor node is insufficient even to hold the variables (of sufficient length to ensure security) that are required in asymmetric cryptographic algorithms, let alone perform operations with them.

Asymmetric digital signatures for authentication are impractical for sensor networks for a number of reasons, such as, long signatures with a high communication overhead of 50–1000 bytes per packet, a very high overhead to create and verify the signature. Also, symmetric solutions for broadcast authentication are impractical: an improved k-time signature scheme requires over 300 bytes per packet. TESLA protocol adapted for sensor networks to become practical for broadcast authentication is called μTESLA.

Adding security to a highly resource-constrained sensor network is feasible. The security building blocks facilitate the implementation of a security solution for a sensor network by using an authenticated routing protocol and a two-party key agreement protocol. The choice of cryptographic primitives and the security protocols in the sensor networks is affected by the severe hardware and energy constraints.

A general security infrastructure that is applicable to a variety of sensor networks needs to define the system architecture and the trust requirements.

Generally, the sensor nodes communicate using RF (Radio Frequency), thus broadcast is the fundamental communication primitive. The baseline protocols account for this property, which affects the trust assumptions, and is exploited to minimize energy usage.

The sensor network forms around one or more base stations, which interface the sensor network to the outside network. The sensor nodes establish a routing forest, with a base station at the root of every tree. Periodic transmission of beacons allows nodes to create a routing topology. Each node can forward a message towards a base station, recognize packets addressed to it, and handle message broadcasts. The base station accesses individual nodes using source routing. The base station has capabilities similar to the network nodes, except that it has enough battery power to surpass the lifetime of all sensor nodes, sufficient memory to store cryptographic keys, and means for communicating with outside networks.

In the sensor applications there is limited local exchange and data processing. The communication patterns within the network fall into three categories:

- node to base station communication, e.g. sensor readings;
- base station to node communication, e.g. specific requests;
- base station to all nodes, e.g. routing beacons, queries or reprogramming of the entire network.

The security goal is primarily to address these communication patterns, and to adapt the baseline protocols to other communication patterns, i.e. node to node or node broadcast.

The sensor networks may be deployed in untrusted locations. While it may be possible to guarantee the integrity of each node through dedicated secure microcontrollers, such an architecture may be too restrictive and does not generalize to the majority of sensor networks. Perrig *et al.* (2001b) assume that individual sensors are untrusted. The SPINS key setup prevents compromising of one node spreading to other nodes.

Basic wireless communication is not secure. Because it is broadcast, any adversary can eavesdrop on the traffic, and inject new messages or replay and change old messages. Hence, SPINS does not place any trust assumptions on the communication infrastructure, except that messages are delivered to the destination with nonzero probability.

Since the base station is the gateway for the nodes to communicate with the outside world, compromising the base station can render the entire sensor network useless. Thus the base stations are a necessary part of the trusted computing base. All sensor nodes trust the base station: at creation time, each node is given a master key which is shared with the base station. All other keys are derived from this key.

Each node trusts itself and, in particular, the local clock is trusted to be accurate, i.e. to have a small drift. This is necessary for the authenticated broadcast protocol.

7.2.1. Sensor Network Security Requirements

Message authentication is important for many applications in sensor networks. Within the building sensor network, authentication is necessary for many administrative tasks (e.g. network reprogramming or controlling sensor node duty cycle). At the same time, an adversary can easily inject messages, so the receiver needs to make sure that the data used in any decision-making process originates from the correct source. Informally, data authentication allows a receiver to verify that the data really was sent by the claimed sender.

In the case of two-party communication, data authentication can be achieved through a purely symmetric mechanism: the sender and the receiver share a secret key to compute a Message Authentication Code (MAC) for all communicated data. When a message with a correct MAC arrives, the receiver knows that it must have been originated by the sender.

This style of authentication cannot be applied to a broadcast setting without placing much stronger trust assumptions on the network nodes. If one sender wants to send authentic data to mutually untrusted receivers, using a symmetric MAC is insecure. A receiver knows the MAC key, and hence could impersonate the sender and forge messages to other receivers. Hence, an asymmetric mechanism is needed to achieve authenticated broadcast. Authenticated broadcast can also be constructed from symmetric primitives, and asymmetry be introduced with delayed key disclosure and one-way function key chains.

In communication, data integrity ensures the receiver that the received data is not altered in transit by an adversary. In SPINS, data integrity is achieved through data authentication, which is a stronger property.

Since all the sensor networks stream some forms of time-varying measurement, and they are guaranteed confidentiality and authentication, we also must ensure that each message is fresh. Informally, data freshness implies that the data is recent, and it ensures that no adversary has replayed old messages. Two types of freshness are defined: weak freshness, which provides partial message ordering, but carries no delay information, and strong freshness, which provides a total order on a request–response pair, and allows for delay estimation. Weak freshness is required by sensor measurements, while strong freshness is useful for time synchronization within the network.

The following notation is used to describe security protocols and cryptographic operations:

- A, B are principals, such as communicating nodes;
- N_A is a nonce generated by A (a nonce is an unpredictable bit string, usually used to achieve freshness);
- $M_1|M_2$ denotes the concatenation of messages M_1 and M_2;
- K_{AB} denotes the secret (symmetric) key which is shared between A and B;
- $\{M\}_{K_{AB}}$ is the encryption of message M with the symmetric key shared by A and B;
- $\{M\}_{(K_{AB},IV)}$, denotes the encryption of message M, with key K_{AB}, and the initialization vector IV which is used in encryption modes such as cipher-block chaining (CBC), output feedback mode (OFB), or counter mode (CTR);
- secure channel is a channel that offers confidentiality, data authentication, integrity, and freshness.

Security requirements are achieved by using two security building blocks: SNEP and μTESLA. SNEP provides data confidentiality, two-party data authentication, integrity, and freshness. μTESLA provides authentication for data broadcast. The security for both mechanisms is bootstrapped with a shared secret key between each node and the base station. The trust to node-to-node interactions can be extended from the node-to-base-station trust.

SNEP has low communication overhead since it only adds bytes per message. SNEP, like many cryptographic protocols, uses a counter, but transmitting the counter value is avoided by keeping state at both end points. SNEP achieves semantic security, a strong security property that prevents eavesdroppers from inferring the message content from the encrypted message.

The same simple and efficient protocol also gives us data authentication, replay protection, and weak message freshness.

Data confidentiality is one of the most basic security primitives and it is used in almost every security protocol. A simple form of confidentiality can be achieved through encryption, but pure encryption is not sufficient. Another important security property is semantic security, which ensures that an eavesdropper has no information about the plain text, even if it sees multiple encryptions of the same plain text. For example, even if an attacker has an encryption of a 0 bit and an encryption of a 1 bit, it will not help it distinguish whether a new encryption is an encryption of 0 or 1. The basic technique for achieving this is randomization: before encrypting the message with a chaining encryption function (i.e. DES-CBC (Data Encryption Standard – Cipher Block Chaining)), the sender precedes the message with a random bit string. This prevents the attacker from inferring the plain text of encrypted messages if it knows plain text-cipher text pairs encrypted with the same key.

However, sending the randomized data over the RF channel requires more energy. A cryptographic mechanism achieves semantic security with no additional transmission overhead. A shared counter is used between the sender and the receiver for the block cipher in counter mode (CTR). Since the communicating parties share the counter and increment it after each block, the counter does not need to be sent with the message. To achieve two-party authentication and data integrity, a message authentication code (MAC) is used.

The combination of these mechanisms forms the Secure Network Encryption Protocol (SNEP). The encrypted data has the following format: $E = \{D\}_{(K_{encr},C)}$, where D is the data, the encryption key is K_{encr}, and the counter is C. The MAC is $M = MAC(K_{mac}, C|E)$. The keys K_{encr} and K_{mac} are derived from the master secret key K. The complete message that A sends to B is:

$$A \rightarrow B : \{D\}_{(K_{encr},C)}, MAC(K_{mac}, C|\{D\}_{(K_{encr},C)}).$$

SNEP offers the following properties:

- *Semantic security*: Since the counter value is incremented after each message, the same message is encrypted differently each time. The counter value is long enough never to repeat within the lifetime of the node.
- *Data authentication*: If the MAC verifies correctly, a receiver can be assured that the message originated from the claimed sender.
- *Replay protection*: The counter value in the MAC prevents replaying of old messages. Note that if the counter were not present in the MAC, an adversary could easily replay messages.

- *Weak freshness*: If the message verified correctly, a receiver knows that the message must have been sent after the previous message it received correctly (that had a lower counter value). This enforces a message ordering and yields weak freshness.
- *Low communication overhead*: The counter state is kept at each end point and does not need to be sent in each message. (In case the MAC does not match, the receiver can try out a fixed, small number of counter increments to recover from message loss. In case the optimistic resynchronization fails, the two parties engage in a counter exchange protocol, which uses the strong freshness protocol).

Plain SNEP only provides weak data freshness, because it only enforces a sending order on the messages within node B, but no absolute assurance to node A that a message was created by B in response to an event in node A.

Node A achieves strong data freshness for a response from node B through a nonce N_A (which is a random number sufficiently long such that it is unpredictable). Node A generates N_A randomly and sends it along with a request message R_A to node B. The simplest way to achieve strong freshness is for B to return the nonce with the response message R_B in an authenticated protocol. However, instead of returning the nonce to the sender, the process can be optimized by using the nonce implicitly in the MAC computation. The entire SNEP protocol providing strong freshness for B's response is:

$$A \rightarrow B : N_A, R_A$$

$$B \rightarrow A : \{R_B\}_{(K_{encr},C)}, MAC(K_{mac}, N_A|C|\{R_B\}_{(K_{encr},C)}).$$

If the MAC verifies correctly, node A knows that node B generated the response after it sent the request. The first message can also use plain SNEP if confidentiality and data authentication are needed.

7.2.2. Authenticated Broadcast

Asymmetric digital signatures for authentication are impractical for multiple reasons. They require long signatures with a high communication overhead of 50–1000 bytes per packet, and a very high overhead to create and verify the signature. One-time signature schemes that are based on symmetric cryptography (one-way functions without trap doors) have a high overhead: Gennaro and Rohatgi's broadcast signature based on Lamport's one-time signature requires over 1 kbyte of authentication information per packet, and Rohatgi's improved k-time signature scheme requires over 300 bytes per packet.

TESLA protocol provides efficient authenticated broadcast. However, TESLA is not designed for such limited computing environments as are encountered in sensor networks. TESLA authenticates the initial packet with a digital signature, which is too expensive to compute on sensor nodes since even fitting the code into the memory is a major challenge. For the same reason, one-time signatures are a challenge for use on sensor nodes.

Standard TESLA has an overhead of approximately 24 bytes per packet. For networks connecting workstations this is usually not significant. Sensor nodes, however, send very small messages that are around 30 bytes long. It is simply impractical to disclose the TESLA key for the previous intervals with every packet: with 64-bit keys and MACs, the TESLA-related part of the packet would constitute over 50 % of the packet.

The one-way key chain does not fit into the memory of a sensor node, so pure TESLA is not practical for a node to broadcast authenticated data.

The μTESLA solves the following inadequacies of TESLA in sensor networks:

- TESLA authenticates the initial packet with a digital signature, which is too expensive for sensor nodes. μTESLA uses only symmetric mechanisms.
- Disclosing a key in each packet requires too much energy for sending and receiving. μTESLA discloses the key once per epoch.
- It is expensive to store a one-way key chain in a sensor node. μTESLA restricts the number of authenticated senders.

The μTESLA is discussed for the case where the base station broadcasts authenticated information to the nodes, and the case where a node is the sender.

μTESLA requires that the base station and nodes are loosely time synchronized, and each node knows an upper bound on the maximum synchronization error. To send an authenticated packet, the base station simply computes a MAC on the packet with a key that is secret at that point in time. When a node gets a packet, it can verify that the corresponding MAC key has not yet been disclosed by the base station (based on its loosely synchronized clock, its maximum synchronization error, and the time schedule at which keys are disclosed). Since a receiving node is assured that the MAC key is known only by the base station, the receiving node is assured that no adversary could have altered the packet in transit. The node stores the packet in a buffer, and at the time of key disclosure, the base station broadcasts the verification key to all receivers. When a node receives the disclosed key, it can easily verify the correctness of the key. If the key is correct, the node can now use it to authenticate the packet stored in its buffer.

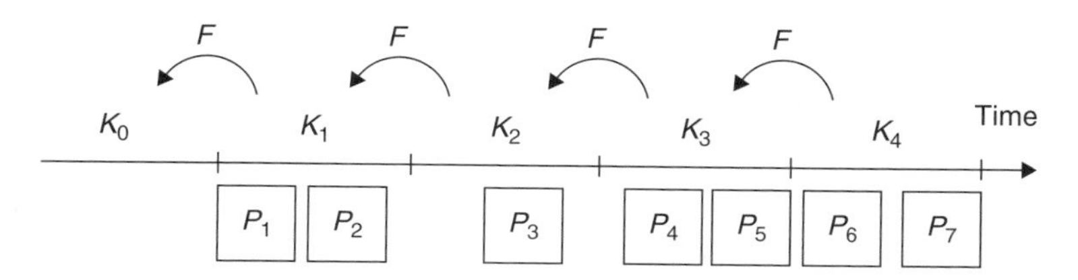

Figure 7.1 Using a time-released key chain for source authentication.

Each MAC key is one in a key chain, generated by a public one-way function F. To generate the one-way key chain, the sender chooses the last key K_n of the chain randomly, and repeatedly applies F to compute all other keys: $K_i = F(K_{i+1})$. Each node can easily perform time synchronization and retrieve an authenticated key of the key chain for the commitment in a secure and authenticated manner, using the SNEP building block.

Figure 7.1 shows an example of μTESLA. Each key of the key chain corresponds to a time interval and all packets sent within one time interval are authenticated with the same key. The time until keys of a particular interval are disclosed is two time intervals in this example. The receiver node is assumed to be loosely time synchronized and knows K_0 (a commitment to the key chain) in an authenticated way. Packets P_1 and P_2 sent in interval one, contain a MAC with key K_1. Packet P_3 has a MAC using key K_2. So far, the receiver cannot authenticate any packets yet. Let us assume that packets P_4, P_5, and P_6 are all lost, as well as the packet that discloses key K_1, so the receiver can still not authenticate P_1, P_2, or P_3. In interval four, the base station broadcasts key K_2, which the node authenticates by verifying $K_0 = F(F(K_2))$, and hence knows also $K_1 = F(K_2)$, so it can authenticate packets P_1, P_2 with K_1, and P_3 with K_2.

Instead of adding a disclosed key to each data packet, the key disclosure is independent of the packets broadcast, and is tied to time intervals. Within the context of μTESLA, the sender broadcasts the current key periodically in a special packet.

μTESLA has multiple phases: sender set-up, sending authenticated packets, bootstrapping new receivers, and authenticating packets. For simplicity, μTESLA is explained for the case where the base station broadcasts authenticated information, and the case where nodes send authenticated broadcasts.

During the sender set-up, the sender first generates a sequence of secret keys (or a key chain). To generate a one-way key chain of length, the sender chooses the last key K_n randomly, and generates the remaining values by successively applying a one-way function F [e.g. a cryptographic hash function such as MD5 (Message Digest 5)]: $K_j = F(K_{j+1})$. Because F is a one-way function, anybody can compute forward, e.g. compute $K_0, \ldots, K_j$

given K_{j+1}, but nobody can compute backward, e.g. compute K_{j+1} given only $K_0, \ldots, K_j$, due to the one-way generator function. This is similar to the S/Key (Secret Key) one-time password system.

During broadcasting of authenticated packets, the time is divided into intervals and the sender associates each key of the one-way key chain with one time interval. In time interval t, the sender uses the key of the current interval, K_t, to compute the MAC of packets in that interval. The sender will then reveal key K_t after a delay of δ intervals after the end of the time interval t. The key disclosure time delay δ is of the order of a few time intervals, as long as it is greater than any reasonable round-trip time between the sender and the receivers.

During bootstrapping of a new receiver, the important property of the one-way key chain is that once the receiver has an authenticated key of the chain, subsequent keys of the chain are self-authenticating, which means that the receiver can easily and efficiently authenticate subsequent keys of the one-way key chain using the one authenticated key. For example, if a receiver has an authenticated value K_i of the key chain, it can easily authenticate K_{i+1}, by verifying $K_i = F(K_{i+1})$. Therefore to bootstrap μTESLA, each receiver needs to have one authentic key of the one-way key chain as a commitment to the entire chain. Another requirement of μTESLA is that the sender and receiver are loosely time synchronized, and that the receiver knows the key disclosure schedule of the keys of the one-way key chain. Both the loose time synchronization, as well as the authenticated key chain commitment, can be established with a mechanism that provides strong freshness and point-to-point authentication. A receiver sends a nonce in the request message to the sender. The sender replies with a message containing its current time T_S (for time synchronization), a key K_i of the one-way key chain used in a past interval i (the commitment to the key chain), and the starting time T_i of interval i, the duration T_{int} of a time interval, and the disclosure delay δ (the last three values describe the key disclosure schedule).

$$M \rightarrow S : N_M$$

$$S \rightarrow M : T_S|K_i|T_i|T_{int}|\delta MAC(K_{MS}, N_M|T_S|K_i|T_i|T_{int}|\delta)$$

Since the confidentiality is not needed, the sender does not need to encrypt the data. The MAC uses the secret key shared by the node and base station to authenticate the data, the nonce N_M allows the node to verify freshness. Instead of using a digital signature scheme as in μTESLA, the node-to-base-station authenticated channel is used to bootstrap the authenticated broadcast.

During authenticating of the broadcast packets, when a receiver receives the packets with the MAC, it needs to ensure that the packet could not have been spoofed by an adversary. The threat is that the adversary already knows the disclosed key of a time interval and so it could forge the packet since it knows the key used to compute the MAC. Hence the receiver needs to be sure that the sender has not yet disclosed the key that corresponds to an incoming packet, implying that no adversary could have forged the contents. This is called the security condition, which receivers check for all incoming packets. Therefore, the sender and receivers need to be loosely time synchronized and the receivers need to know the key disclosure schedule. If the incoming packet satisfies the security condition, the receiver stores the packet (it can only verify it once the corresponding key is disclosed). If the security condition is violated (the packet had an unusually long delay), the receiver needs to drop the packet, since an adversary might have altered it.

As soon as the node receives a key K_j of a previous time interval, it authenticates the key by checking that it matches the last authentic key for which it knows K_i, using a small number of applications of the one-way function $F : K_i = F^{j-i}(K_j)$. If the check is successful, the new key K_j is authentic and the receiver can authenticate all packets that were sent within the time intervals i to j. The receiver also replaces the stored K_i with K_j.

When the nodes broadcast authenticated data, there are additional new problems. Since the node is severely memory limited, it cannot store the keys of a one-way key chain. Moreover, recomputing each key from the initial generating key K_n is computationally expensive. Another issue is that the node might not share a key with each receiver, hence sending out the authenticated commitment to the key chain would involve an expensive node-to-node key agreement. Broadcasting the disclosed keys to all receivers can also be expensive for the node and drain precious battery energy.

The two viable approaches for addressing this problem are as follows:

- The node broadcasts the data through the base station. It uses SNEP to send the data in an authenticated way to the base station, which subsequently broadcasts it.
- The node broadcasts the data. However, the base station keeps the one-way key chain and sends keys to the broadcasting node as needed. To conserve energy for the broadcasting node, the base station can also broadcast the disclosed keys, and/or perform the initial bootstrapping procedure for new receivers.

7.2.3. Applications

Secure protocols can be built out of the SPINS secure building blocks with an authenticated routing application, and a two-party key agreement protocol.

Using the μTESLA protocol, a lightweight, authenticated, dedicated (*ad hoc*) routing protocol builds an authenticated routing topology. Ad-hoc routing does not offer authenticated routing messages, hence, it is potentially easy for a malicious user to take over the network by injecting erroneous, replaying old, or advertising incorrect routing information.

The authenticated routing scheme assumes bidirectional communication channels, i.e. if node *A* hears node *B*, then node *B* hears node *A*. The route discovery depends on periodic broadcast of beacons. Every node, upon reception of a beacon packet, checks whether or not it has already received a beacon (which is a normal packet with a globally unique sender ID (identifier) and current time at base station, protected by a MAC to ensure integrity and that the data is authentic) in the current epoch. (Epoch means the interval of a routing updates.) If a node hears the beacon within the epoch, it does not take any further action. Otherwise, the node accepts the sender of the beacon as its parent to route towards the base station. Additionally, the node would repeat the beacon with the sender ID changed to itself. This route discovery resembles a distributed, breadth first search algorithm, and produces a routing topology.

However, in the above algorithm, the route discovery depends only on the receipt of a route packet, not on its contents. It is easy for any node to claim to be a valid base station. The μTESLA key disclosure packets can easily function as routing beacons. Only the sources of authenticated beacons are accepted as valid parents. Reception of a μTESLA packet guarantees that that packet originated at the base station, and that it is fresh. For each time interval, the parent is accepted as the first node sending a packet that is later successfully authenticated. Combining μTESLA key disclosure with the distribution of routing beacons allows us to charge the costs of the to transmission of the keys to network maintenance, rather than to the encryption system.

This scheme leads to a lightweight authenticated routing protocol. Since each node accepts only the first authenticated packet as the one to use in routing, it is impossible for an attacker to re-route arbitrary links within the sensor network. Furthermore, each node can easily verify that the parent forwarded the message: by our assumption of bidirectional connectivity, if the parent of a node forwarded the message, the node must have heard that.

The authenticated routing scheme above is just one way to build authenticated ad-hoc routing protocol using μTESLA. In protocols where base stations are not involved in route construction, μTESLA can still be used for security. In these cases, the initiating node will temporarily act as base station and beacons authenticated route updates. However, the node here will need to have significantly more memory resource than the sensor nodes explored here in order to store the key chain.

A convenient method to bootstrap secure connections is public-key cryptography protocols for symmetric-key set-up. Unfortunately, resource-constrained sensor nodes prevent us from using computationally expensive public-key cryptography. Therefore, the protocols are used solely from symmetric-key algorithms. Hence symmetric protocol that uses the base station is applied as a trusted agent for key set-up.

Assume that node A wants to establish a shared secret session key SK_{AB} with node B. Since A and B do not share any secrets, they need to use a trusted third party S, which is the base station in our case. In our trust set-up, both A and B share a secret key with the base station, K_{AS} and K_{BS}, respectively. The following protocol achieves secure key agreement as well as strong key freshness:

$$A \rightarrow B : N_A, A$$

$$B \rightarrow S : N_A, N_B, A, B, MAC(K_{BS}, N_A|N_B|A|B)$$

$$S \rightarrow A : \{SK_{AB}\}_{K_{AS}}, MAC(K'_{AS}, N_A|B|\{SK_{AB}\}_{K_{AS}})$$

$$S \rightarrow B : \{SK_{AB}\}_{K_{BS}}, MAC(K'_{BS}, N_B|A|\{SK_{AB}\}_{K_{BS}})$$

This protocol uses SNEP protocol with strong freshness. The nonces N_A and N_B ensure strong key freshness to both A and B. The SNEP protocol is responsible for ensuring confidentiality (through encryption with the keys K_{AS} and K_{BS}) of the established session key SK_{AB}, as well as message authentication (through the MAC using keys K'_{AS} and K'_{BS}) to make sure that the key was really generated by the base station. Note that the MAC in the second protocol message helps to defend the base station from denial-of-service attacks, so the base station only sends two messages to A and B if it received a legitimate request from one of the nodes.

A nice feature of the above protocol is that the base station performs most of the transmission work. Other protocols usually involve a ticket that the server sends to one of the parties, who forwards it to the other node, which requires more energy for the nodes to forward the message. The Kerberos key agreement protocol achieves similar properties, except that it does not provide strong key freshness. However, it would be straightforward to implement it with strong key freshness by using SNEP with strong freshness.

7.3. COMMUNICATION SECURITY IN SENSOR NETWORKS

Application messages are exchanged through the network, and the mobile code is sent from node to node. Because the security of mobile code greatly affects the security of the network, protection of the messages containing mobile code is an important part of communication security scheme.

The possible threats to a network if communication security is compromised are as follows:

(1) Insertion of malicious code is the most dangerous attack that can occur. Malicious code injected into the network could spread to all nodes, potentially destroying the whole network or, even worse, taking over the network on behalf of an adversary. A seized sensor network can either send false observations about the environment to a legitimate user or send observations about the monitored area to a malicious user.
(2) Interception of the messages containing the physical locations of sensor nodes allows an attacker to locate the nodes and destroy them. The significance of hiding the location information from an attacker lies in the fact that the sensor nodes have small dimensions and their location cannot be trivially traced. Thus, it is important to hide the locations of the nodes. In the case of static nodes, the location information does not age and must be protected through the lifetime of the network.
(3) Besides the locations of sensor nodes, an adversary can observe the application specific content of messages including message IDs, time stamps and other fields. Confidentiality of those fields in our example application is less important than confidentiality of location information, because the application-specific data does not contain sensitive information, and the lifetime of such data is significantly shorter.
(4) An adversary can inject false messages that give incorrect information about the environment to the user. Such messages also consume the scarce energy resources of the nodes. This type of attack is called sleep deprivation torture.

In the security scheme, the security levels are based on private key cryptography utilizing group keys. Applications and system software access the security API as a part of the middleware defined by the sensor network architecture. Since data contain some confidential information, the content of all messages in the network is encrypted.

The sensor nodes in the network are assumed to be allowed to access the content of any message.

The deployment of security mechanisms in a sensor network creates additional overhead. The latency increases due to the execution of the security related procedures, and the consumed energy directly decreases the lifetime of the network. To minimize the security related costs, the security overhead, and consequently the energy consumption, should relate to the sensitivity of the encrypted information. Following the taxonomy of the types of data in the network, three security levels are defined:

- security level I is reserved for mobile code, the most sensitive information sent through the network;
- security level II is dedicated to the location information conveyed in messages;
- the security level III mechanism is applied to the application specific information.

The strength of the encryption for each of the security levels corresponds to the sensitivity of the encrypted information. Therefore, the encryption applied at level I is stronger than the encryption applied at level II, while the encryption on level II is stronger than the one applied at level III.

Different security levels are implemented either by using various algorithms or by using the same algorithm with adjustable parameters that change its strength and corresponding computational overhead. Using one algorithm with adjustable parameters has the advantage of occupying less memory space.

RC6 (symmetric block cipher) is suitable for modification of its security strength because it has an adjustable parameter (number of rounds) that directly affects its strength. The overhead for the RC6 encryption algorithm increases with the strength of the encryption measured by the number of rounds.

The multicast model of communication inherent for the sensor network architecture suggests deployment of group keys. Otherwise, if each pair of nodes would require a key or a pair of keys, communication between the nodes would have to be unicast based. This would significantly increase the number of messages. Since the addition of security in a sensor network must not require the change of the whole sensor network architecture, group keys are utilized.

All nodes in the network share an initial set of master keys, and the number of keys depends on the estimated lifetime of the network. The longer the lifetime, the more keys are needed in order to expose less material for a known cipher text attack. The alternative approach, where the keys are established dynamically and propagated through the network, is not acceptable. A protocol that guarantees that all nodes received a key is required. Such a requirement is not feasible in a network where the nodes do not keep track of their neighbors.

One of the keys from the list of master keys is active at any moment. The algorithm for the selection of a particular key is based on a pseudo-random generator running at each node with the same seed. Periodically and synchronously on each node, a new random number is generated and used to provide and index an entry in the table of available master keys. This entry contains the active master key. The keys for three levels of security corresponding to the three types of data are then derived from the active master key.

In security level I, the messages containing mobile code are less frequent than messages that the application instances on different nodes exchange. This allows us to use strong encryption in spite of the resulting overhead. For information protected at this security level, nodes use the current master key. The set of master keys, the corresponding pseudo-random number generator, and a seed are credentials that a potential user must have in order to access the network. Once the user obtains those credentials, he/she can insert any code into the network. If a malicious user breaks the encryption on this level using a brute force attack, he/she can insert harmful code into the network.

In security level II (data that contains locations of sensor nodes) a security mechanism is provided that isolates parts of the network, so that breach of security in one part of the network does not affect the rest of the network.

According to the assumptions about the applications expected to run in sensor networks, the locations of sensor nodes are likely to be included in the majority of messages. Thus, the overhead that corresponds to the encryption of the location information significantly influences the overall security overhead in the network. This must be taken into account when the strength of the encryption at this level is determined. Since the protection level is lower for the location information than for mobile code, the probability that the key for level II can be broken is higher. Having the key, an adversary could potentially locate all nodes in the network. To constrain the damage to only one part of the network, the following security mechanism is proposed. Sensor nodes use location-based keys for level II encryption. The location-based keys enable separation between the regions where the location of nodes are compromised and those areas where nodes continue to operate safely.

The area covered by a sensor network is divided into cells. Nodes within one cell share a common location-based key, which is a function of a fixed location in the cell and the current master key. Between the cells, there is a bordering region whose width is equal to the transmission range. Nodes belonging to those regions have the keys for all adjacent cells. This ensures that two nodes within a transmission range from each other have a common key. The dimensions of the cells must be big enough for the localized nature of the algorithms in the network to ensure that the traffic among the cells is relatively low, compared with overall traffic. The areas can be of an arbitrary shape with the only requirement that the whole sensor terrain is covered. A division of the area in uniformly sized cells is the most appropriate solution, because it allows a fast and easy way for a node to determine its cell membership. The network is divided into hexagonal cells, since it ensures that the gateway nodes have at most three keys.

Part of the bootstrapping mechanism for sensor nodes is the process of determining their cell membership. In this process, the notion of extended cell is used. An extended cell is a hexagonal cell having the same center as the original cell, and the distance between its sides and the sides of the original cell is equal to the transmission range of the sensor nodes. The extended cell contains the original cell and corresponding bordering regions. Figure 7.2 shows three neighboring cells and their corresponding extended cells. Each node compares its location against each extended cell and determines whether it is in an extended cell or not. If a node is within the extended cell of C_x, it will have the key of C_x, K_{Cx}. The nodes within the bordering regions (shaded areas) have multiple keys. For example, the nodes that are adjacent to cells C_1 and C_2 have two keys: K_{C1} and K_{C2}, respectively.

In security level III, the application specific data is encrypted using a weaker encryption than the one used for the two other types of data. The weaker encryption requires a lower computational overhead for application specific data. Additionally, the high frequency of messages with application specific data prevents the use of stronger and resource consuming encryption. Therefore, an encryption algorithm demanding fewer computational resources is applied with a corresponding decrease in the strength of security.

The key used for the encryption of level III information is derived from the current master key. The MD5 (Message Digest 5) hash function accepts the master key and generates a key for level III. Since the master key is periodically changed, the corresponding key at this level follows those changes.

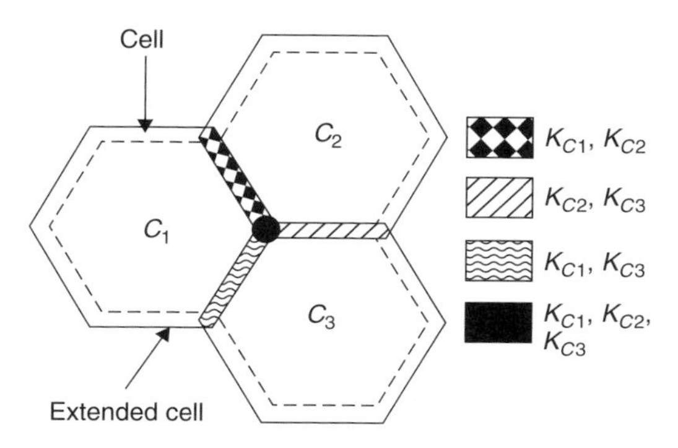

Figure 7.2 Cells, extended cells, and areas with multiple keys.

The major assumptions of security schemes are that the sensor nodes are perfectly time synchronized and have exact knowledge of their location. It is realistic for the nodes to be synchronized up to microseconds.

7.4. SUMMARY

As sensor networks deployment becomes widespread, security issues become a central concern. A suite of security building blocks is optimized for resource-constrained environments and wireless communication. SPINS (Security Protocol for Sensor Networks) has two secure building blocks: SNEP (Secure Network Encryption Protocol) and μTESLA (the micro version of the Timed, Efficient, Streaming, Loss-tolerant Authentication protocol). SNEP provides the following important baseline security primitives: data confidentiality, two-party data authentication, and data freshness. Efficient broadcast authentication is an important mechanism for sensor networks. μTESLA is a protocol that provides authenticated broadcast for severely resource-constrained environments. These protocols are practical even on minimal hardware: the performance of the protocol suite easily matches the data rate of the network. The suite of security building blocks can be used for building higher level protocols.

In the security scheme, the security levels are based on private key cryptography utilizing group keys. Applications and system software access the security API as a part of the middleware defined by the sensor network architecture. Since data contain some confidential information, the content of all messages in the network is encrypted.

PROBLEMS

Learning Objectives

After completing this chapter you should be able to:

- demonstrate understanding of the security protocols in sensor networks;
- discuss what is meant by design integration;
- explain what sensor network security requirements are;
- explain what an authenticated broadcast is;
- discuss communication security in sensor networks.

Practice Problems

Problem 7.1: How feasible is adding security to a sensor network?
Problem 7.2: What communication patterns should be considered by security?
Problem 7.3: What are the SPINS assumptions regarding security in wireless communication?
Problem 7.4: How is the base station considered in security of the network?
Problem 7.5: What is data freshness?
Problem 7.6: What is a secure channel?
Problem 7.7: What are the properties of SNEP?
Problem 7.8: What are the phases in μTESLA?

Practice Problem Solutions

Problem 7.1:

Adding security to a highly resource-constrained sensor network is feasible. The security building blocks facilitate the implementation of a security solution for a sensor network by using an authenticated routing protocol and a two-party key agreement protocol. The choice of cryptographic primitives and the security protocols in the sensor networks is affected by the severe hardware and energy constraints.

Problem 7.2:

The security goal is to adapt the baseline protocols to communication patterns, i.e. node to node or node broadcast, and to address primarily the following communication patterns:

- node-to-base-station communication, e.g. sensor readings;
- base-station-to-node communication, e.g. specific requests;
- Base station to all nodes, e.g. routing beacons, queries or reprogramming of the entire network.

Problem 7.3:

Basic wireless communication is not secure. Because it is broadcast, any adversary can eavesdrop on the traffic, and inject new messages or replay and change old messages. Hence, SPINS does not place any trust assumptions on the communication infrastructure, except that messages are delivered to the destination with nonzero probability.

Problem 7.4:

The base station is the gateway for the nodes to communicate with the outside world, hence, compromising the base station can render the entire sensor network useless. Thus, the base stations are a necessary part of the trusted computing base. All sensor nodes trust the base station: at creation time, each node is given a master key which is shared with the base station. All other keys are derived from this key.

Problem 7.5:

Informally, data freshness implies that the data is recent, and it ensures that no adversary has replayed old messages. Weak freshness provides partial message ordering, but carries no delay information. Strong freshness provides a total order on a request–response pair, and allows for delay estimation. Weak freshness is required by sensor measurements, while strong freshness is useful for time synchronization within the network.

Problem 7.6:

A secure channel is a channel that offers confidentiality, data authentication, integrity, and freshness.

Problem 7.7:

SNEP offers the following properties:

- *Semantic security*: Since the counter value is incremented after each message, the same message is encrypted differently each time. The counter value is long enough never to repeat within the lifetime of the node.
- *Data authentication*: If the MAC verifies correctly, a receiver can be assured that the message originated from the claimed sender.
- *Replay protection*: The counter value in the MAC prevents replaying of old messages. Note that if the counter were not present in the MAC, an adversary could easily replay messages.
- *Weak freshness*: If the message is verified correctly, a receiver knows that the message must have been sent after the previous correctly received message (that had a lower counter value). This enforces message ordering and yields weak freshness.
- *Low communication overhead*: The counter state is kept at each end point and does not need to be sent in each message. (In case the MAC does not match, the receiver can try out a fixed, small number of counter increments to recover from message loss. In case the optimistic resynchronization fails,

the two parties engage in a counter-exchange protocol, which uses the strong freshness protocol).

Problem 7.8:

μTESLA has multiple phases: Sender set-up, sending authenticated packets, bootstrapping new receivers, and authenticating packets.

8

Operating Systems for Embedded Applications

8.1. INTRODUCTION

Inferno is a well-designed, economical operating system that accommodates various providers of content and services from the equally varied transport and presentation platforms.

The goals of another Operating System (OS), Pebble, are flexibility, safety, and performance. Pebble's architecture includes:

- a minimal privileged mode nucleus, responsible for switching between protection domains;
- implementation of all system services by replaceable user-level components with minimal privileges (including the scheduler and all device drivers) that run in separate protection domains enforced by hardware memory protection;
- generation of code specialized for each possible cross-domain transfer.

The combination of these techniques results in a system with inexpensive cross-domain calls that makes it well-suited for efficiently specializing the operating system on a per-application basis and for supporting modern component-based applications.

The Pebble operating system supports complex embedded applications. This is accomplished through two key features:

Wireless Sensor Network Designs A. Hać
 ISBN: 0-470-86736-1

- safe extensibility, so that the system can be constructed from untrusted components and reconfigured while running;
- low interrupt latency, which ensures that the system can react quickly to external events.

Embedded systems are subject to tight power and energy constraints. The operating system has a significant impact on the energy efficiency of the embedded system. Conventional approaches to energy analysis of the OS and embedded software, in general, require the application software to be completely developed and integrated with the system software, and either that measurement on a hardware prototype or detailed simulation of the entire system be performed. This process requires significant design effort, and high-level or architectural optimizations on the embedded software.

8.2. THE INFERNO OPERATING SYSTEM

Inferno is an operating system for creating and supporting distributed services, developed at Lucent Technologies by Dorward *et al.* (1997). Inferno is used in a variety of network environments, for example those supporting advanced telephones, hand-held devices, TV set-top boxes attached to cable or satellite systems, and inexpensive Internet computers, as well as in conjunction with traditional computing systems.

Inferno's definitive strength lies in its portability and versatility across several dimensions:

- *Portability across processors*: it runs on Intel, Sparc, MIPS (Millions of Instructions Per Second), ARM, HP-PA (Hewlett Packard Precision Architecture), and PowerPC architectures, and is readily portable to others.
- *Portability across environments*: it runs as a stand-alone operating system on small terminals, and also as a user application under Windows NT, Windows 95, Unix (IRIX (a UNIX-based operating system from Silicon Graphics), Solaris, FreeBSD, Linux, AIX (a trademark of IBM, UNIX compatible operating system), HP-UX (Hewlett Packard UNIX operating system)) and Plan 9. In all of these environments, Inferno applications see the identical interface.
- *Distributed design*: the identical environment is established at the user's terminal and at the server, and each may import the resources (for example, the attached I/O devices or networks) of the other. Aided by

the communications facilities of the run-time system, applications may be split easily (and even dynamically) between client and server.

- *Minimal hardware requirements*: it runs useful applications stand-alone on machines with as little as 1 Mb of memory, and does not require memory-mapping hardware.
- *Portable applications*: Inferno applications are written in the type-safe language Limbo, whose binary representation is identical over all platforms.
- *Dynamic adaptability*: applications may, depending on the hardware or other resources available, load different program modules to perform a specific function. For example, a video player application might use any of several different decoder modules.

The role of the Inferno system is to create several standard interfaces for its applications:

- Applications use various resources internal to the system, such as a consistent virtual machine that runs the application programs, together with library modules that perform services as simple as string manipulation through more sophisticated graphics services for dealing with text, pictures, higher-level toolkits, and video.
- Applications exist in an external environment containing resources, such as data files that can be read and manipulated, together with objects that are named and manipulated, like files but more active. Devices (for example, a hand-held remote control, an MPEG decoder or a network interface) present themselves to the application as files.
- Standard protocols exist for communication within and between separate machines running Inferno, so that applications can cooperate.

Inferno uses interfaces supplied by an existing environment, either bare hardware or standard operating systems and protocols.

An Inferno-based service consists of many inexpensive terminals running Inferno as a native system, and a smaller number of large machines running Inferno as a hosted system. On these server machines Inferno interfaces to databases, transaction systems, existing OA&M (Operation, Administration, and Maintenance) facilities, and other resources provided under the native operating system. The Inferno applications themselves run either on the client or server machines, or both.

The purpose of most Inferno applications is to present information to the user, thus applications must locate the information sources in the network and construct a local representation of them. The user's terminal is also an information source and its devices represent resources to applications. Inferno uses the design of the Plan 9 operating system to present resources to these applications.

This design has the three following principles:

- All resources are named and accessed like files in a forest of hierarchical file systems.
- The disjoint resource hierarchies provided by different services are joined together into a single private hierarchical name space.
- A communication protocol, called Styx, is applied uniformly to access these resources, whether local or remote.

In practice, most applications see a fixed set of files organized as a directory tree. Some of the files contain ordinary data, but others represent more active resources. Devices are represented as files, and device drivers attached to a particular hardware box present themselves as small directories. These directories typically contain two files, data and control, which respectively perform actual device input/output and control operations. System services also use file names.

The device drivers and other internal resources respond to the procedural version of Styx protocol. The Inferno kernel implements a mount driver that transforms file system operations into remote procedure calls for transport over a network. On the other side of the connection, a server unwraps the Styx messages and implements them using resources local to it. Thus, it is possible to import parts of the name space (and thus resources) from other machines.

The Styx protocol lies above, and is independent of, the communications transport layer and is readily carried over various modem transport protocols.

Inferno applications are written in the programming language Limbo, which was designed specifically for the Inferno environment. Its syntax is influenced by the programming languages C and Pascal, and it supports the standard data types common to them, together with several higher-level data types such as lists, tuples, strings, dynamic arrays, and simple abstract data types.

In addition, Limbo supplies several advanced constructs carefully integrated into the Inferno virtual machine. In particular, a communication mechanism, called a channel, is used to connect different Limbo tasks on the same machine or across the network. A channel transports typed data in a machine-independent fashion, so that complex data structures (including channels themselves) may be passed between Limbo tasks or attached to files in the name space for language-level communication between machines.

Multi-tasking is supported directly by the Limbo language: independently scheduled threads of control may be spawned, and an alt statement is used to coordinate the channel communication between tasks (that is, alt is used to select one of several channels that are ready to communicate). By building channels and tasks into the language and its virtual machine, Inferno encourages a communication style that is easy to use and safe.

Limbo programs are built of modules, which are self-contained units with a well-defined interface containing functions (methods), abstract data types, and constants defined by the module and visible outside it. Modules are accessed dynamically, that is, when one module wishes to make use of another, it dynamically executes a load statement naming the desired module and uses a returned handle to access the new module. When the module is no longer in use, its storage and code will be released. The flexibility of the modular structure contributes to the small size of typical Inferno applications, and also to their adaptability.

Limbo is fully type checked at compile- and run-time; for example, pointers, besides being more restricted than in the C programming language, are checked before being dereferenced, and the type-consistency of a dynamically loaded module is checked when it is loaded. Limbo programs run safely on a machine without memory-protection hardware. Moreover, all Limbo data and program objects are subject to a garbage collector, built deeply into the Limbo run-time system. All system data objects are tracked by the virtual machine and freed as soon as they become unused.

Limbo programs are compiled into byte-codes representing instructions for a virtual machine called Dis. The architecture of the arithmetic part of Dis is a simple three-address machine, supplemented with a few specialized operations for handling some of the higher-level data types like arrays and strings. Garbage collection is handled below the level of the machine language, and the scheduling of tasks is similarly hidden. When loaded into memory for execution, the byte-codes are expanded into a format more efficient for execution; there is also an optional on-the-fly compiler that turns a Dis instruction stream into native machine instructions for the appropriate real hardware. This can be done efficiently because Dis instructions match well with the instruction-set architecture of today's machines. The resulting code executes at a speed approaching that of compiled C.

Underlying Dis is the Inferno kernel, which contains the interpreter and on-the-fly compiler, memory management, scheduling, device drivers, and protocol stacks. The kernel also contains the core of the file system (the name evaluator and the code that turns file system operations into remote procedure calls over communications links) as well as the small file systems implemented internally.

Inferno creates a standard environment for applications. Identical application programs can run under any instance of this environment, even in distributed fashion, and see the same resources. Depending on the environment in which Inferno itself is implemented, there are several versions of the Inferno kernel, Dis/Limbo interpreter, and device driver set.

When running as the native operating system, the kernel includes the interrupt handlers, graphics and other device drivers, needed to implement the abstractions presented to applications. For a hosted system, for example under Unix, Windows NT or Windows 95, Inferno runs as a set of ordinary processes. Instead of mapping its device-control functionality to real hardware, it adapts to the resources provided by the operating system under which it runs. For example, under Unix, the graphics library might be implemented using the X window system and the networking using the socket interface; under Windows, it uses the native Windows graphics and Winsock calls.

Inferno is largely written in the standard C programming language, and most of its components are independent of the many operating systems that can host it.

Inferno provides security of communication, resource control, and system integrity. Each external communication channel may be transmitted in the clear, accompanied by message digests to prevent corruption, or encrypted to prevent corruption and interception. Once communication is set up, the encryption is transparent to the application. Key exchange is provided through standard public-key mechanisms; after key exchange, message digesting and line encryption likewise use standard symmetric mechanisms.

Inferno is secure against erroneous or malicious applications, and encourages safe collaboration between mutually suspicious service providers and clients. The resources available to applications appear exclusively in the name space of the application, and standard protection modes are available. This applies to data, to communication resources, and to the executable modules that constitute the applications. Security-sensitive resources of the system are accessible only by calling the modules that provide them; in particular, adding new files and servers to the name space is controlled and is an authenticated operation. For example, if the network resources are removed from an application's name space, then it is impossible for it to establish new network connections.

Authentication and digital signatures are performed using public key cryptography. Public keys are certified by Inferno-based or other certifying authorities that sign the public keys with their own private key.

Inferno uses encryption for:

- mutual authentication of communicating parties;
- authentication of messages between these parties, and
- encryption of messages between these parties.

The encryption algorithms provided by Inferno include the SHA (Secure Hash Algorithm), MD4 (Message Digest 4), and MD5 (Message Digest 5) secure hashes; Elgamal public key signatures and signature verification; RC4 (symmetric key stream cipher) encryption; DES (Data Encryption Standard) encryption; and public key exchange based on the Diffie–Hellman scheme. The public key signatures use keys with moduli up to 4096 bits, 512 bits by default.

There is no generally accepted national or international authority for storing or generating public or private encryption keys. Thus Inferno includes tools for using or implementing a trusted authority, but it does not itself provide the authority, which is an administrative function. Thus an organization using Inferno (or any other security and key-distribution scheme) must design its system to suit its own needs, and in particular decide whom to trust as a Certifying Authority (CA). However, the Inferno design is sufficiently flexible and modular to accommodate the protocols likely to be attractive in practice.

The certifying authority that signs a user's public key determines the size of the key and the public key algorithm used. Tools provided with Inferno use these signatures for authentication. Library interfaces are provided for Limbo programs to sign and verify signatures.

Generally authentication is performed using public key cryptography. Parties register by having their public keys signed by the CA. The signature covers a secure hash (SHA, Secure Hash Algorithm; MD4, Message Digest 4; or MD5, Message Digest 5) of the name of the party, its public key, and an expiration time. The signature, which contains the name of the signer, along with the signed information, is termed a certificate.

When parties communicate, they use the Station-to-Station (STS) protocol to establish the identities of the two parties and to create a mutually known secret. This STS protocol uses the Diffie–Hellman algorithm to create this shared secret. The protocol is protected against replay attacks by choosing new random parameters for each conversation. It is secured against 'man in the middle' attacks by having the parties exchange certificates and then digitally signing key parts of the protocol. To masquerade as another party an attacker would have to be able to forge that party's signature.

A network conversation can be secured against modification alone or against both modification and snooping. To secure against modification, Inferno can append a secure MD5 or SHA hash (called a digest),

hash(secret, message, messageid)

to each message, where messageid is a 32 bit number that starts at 0 and is incremented by one for each message sent. Thus messages can be neither changed, removed, reordered or inserted into the stream without knowing the secret or breaking the secure hash algorithm.

To secure against snooping, Inferno supports encryption of the complete conversation using either RC4 (symmetric key stream cipher) or DES (Data Encryption Standard) with either DES Cipher Block Chaining (DES-CBC) and Electronic Code Book (DES-ECB).

The strength of cryptographic algorithms depends in part on the strength of the random numbers used for choosing keys, Diffie–Hellman parameters, initialization vectors, etc. Inferno achieves this in two steps: a slow (100 to 200 bit per second) random bit stream comes from sampling the low order bits of a free running counter whenever a clock ticks. The clock must be unsynchronized, or at least poorly synchronized, with the counter. This generator is then used to alter the state of a faster pseudo-random number generator. Both the slow and fast generators were tested on a number of architectures using self correlation, random walk, and repeatability tests.

8.3. THE PEBBLE COMPONENT-BASED OPERATING SYSTEM

Specialized systems or embedded systems run on microcontrollers in cars and microwaves, and on high-performance general purpose processors as found in routers, laser printers, and hand-held computing devices.

Safety is important in mobile code and component-based applications. Although safe programming languages such as Java and Limbo can be used for many applications, hardware memory protection is important when code is written in unsafe programming languages such as C and C++.

High performance cannot be sacrificed to provide safety and flexibility. Systems are chosen primarily for their performance characteristics, and safety and flexibility come in the second place. Any system structure added to support flexibility and safety cannot come at a significant decrease in performance.

To maximize system flexibility, OS Pebble runs as little code as possible in its privileged mode nucleus. If a piece of functionality can be run at the user level, it is removed from the nucleus. This approach makes it easy to replace, layer, and offer alternative versions of operating system services.

Each user-level component runs in its own protection domain, bounded by hardware memory. All communication between protection domains is done by using interrupt handlers, called portals. Only if a portal exists between protection domain A and protection domain B can A invoke a service offered by B. Each protection domain has its own portal table. By restricting the set of portals available to a protection domain, threads in that domain are efficiently isolated from services to which they should not have access.

Portals are the basis for flexibility and safety in Pebble, and the key to its high performance. Specialized, tamper-proof code can be generated for each portal, using a simple interface definition language. The portal code can be optimized for its portal, saving and restoring the minimum necessary state, or encapsulating and compiling out demultiplexing decisions and run-time checks.

Pebble has the same general structure as classical microkernel operating systems such as Mach, Chorus, and Windows NT, consisting of a privileged mode kernel and a collection of user level servers. Pebble's protected mode nucleus is much smaller and has fewer responsibilities than the kernels of these systems, and in that way is much more like the L4 microkernel (second generation microkernel designed and developed by Jochen Liedtke, running on i486 and Pentium CPUs). L4 and Pebble share a common philosophy of running as little code in privileged mode as possible. Where L4 implements IPC (InterProcess Communication) and minimal virtual memory management in privileged mode, Pebble's nucleus includes only code for transferring threads from one protection domain to another and a small number of support functions that require kernel mode.

Liedtke (1995), in his work on L4, espoused the philosophy of a minimal privileged mode kernel that includes only support for IPC (InterProcess communication), and key VM (Virtual Memory) primitives. Pebble goes one step further than L4, removing VM as well (except for TLB fault handling, which is done in software on MIPS).

OS Mach provides a facility for intercepting system calls and servicing them at user level. Pebble's portal mechanism, which was designed for high-performance cross-protection-domain transfer, can be used in a similar way, taking an existing application component and interposing one or more components between the application component and the services it uses.

Pebble's architecture is similar to the nested process architecture of Fluke. Fluke provides an architecture in which virtual operating systems can be layered, with each layer only affecting the performance of the subset of the operating system, which interface it implements. For example, the presence of multiple virtual memory management nesters, which provide demand paging, distributed shared memory, and persistence, would have no effect on the cost of invoking file system operations such as read and write. The Fluke model requires that system functionality be replaced in groups; a memory management nester must implement all of the functions in the virtual memory interface specification. Pebble portals can be replaced piecemeal, which permits finer-grained extensibility.

The Exokernel attempts to exterminate all OS abstractions, leaving the privileged mode kernel in charge of protecting resources, but leaving abstraction of resources to user level application code. As with the Exokernel approach, Pebble moves the implementation of operating system abstractions to user level, but instead of leaving the development of OS abstractions to application writers, Pebble provides a set of OS abstractions, implemented by user-level OS components. Pebble OS components can be added or replaced, allowing alternate OS abstractions to coexist or override the default set.

The Exokernel model attempts to remove all OS abstractions, with the privileged-mode kernel in charge of protecting resources, but leaving resource abstraction to user-level application code. As with the Exokernel approach, Pebble moves the implementation of resource abstractions to user level, but unlike the Exokernel, Pebble provides a set of abstractions, implemented by user-level operating system components. Pebble OS components can be added or replaced, allowing alternate OS abstractions to coexist or override the default set.

Pebble can use the interposition technique to wrap a sandbox around untrusted code. Several extensible operating system projects have studied the use of software techniques, such as safe programming languages and software fault isolation, for this purpose. Where software techniques require faith in the safety of a compiler, interpreter, or software fault isolation tool, a sandbox implemented by portal interposition and hardware memory protection provides isolation at the hardware level, which may be simpler to verify than software techniques.

The Pebble approach to sandboxing is similar to that provided by the Plan 9 operating system. In Plan 9, nearly all resources are modeled as files, and each process has its own file name space. By restricting the name space of a process, it can be effectively isolated from resources to which it should not have access. In contrast with Plan 9, Pebble can restrict access to any service, not just those represented by files.

The Pebble architecture provides low interrupt latency and low-cost intercomponent communication.

(1) *The privileged-mode nucleus is as small as possible. Most executions occur at the user level.*

The privileged-mode nucleus is responsible for switching between protection domains, and it is the only part of the system that must be run with the interrupts disabled. By reducing the length of time the interrupts are disabled, the maximum interrupt latency is reduced.

In a perfect world, Pebble would include only one privileged-mode instruction, which would transfer control from one protection domain to the next. By minimizing the work done in privileged mode, the Pebble's designers reduce both the amount of privileged code and the time needed to perform essential privileged mode services.

(2) *Each component is implemented by a separate protection domain. The cost of transferring control from one protection domain to another should be small enough that there is no performance-related reason to co-locate components.*

Microkernel systems used coarse-grained user level servers, in part because the cost of transferring between protection domains was high. By keeping this cost low, Pebble enables the factoring of the operating system, and application, into smaller components with small performance penalty.

For example, the cost of using hardware memory protection on the Intel x86 can be made extremely small. Pebble can perform a one-way interprotection domain call in 114 machine cycles on a MIPS R5000 processor.

(3) *The operating system is built from fine-grained replaceable components, isolated through the use of hardware memory protection.*

The functionality of the operating system is implemented by trusted user-level components. The components can be replaced, augmented, or layered. For example, Pebble does not handle scheduling decisions, and the user-replaceable scheduler is responsible for all scheduling and synchronization operations.

The architecture of Pebble is based on the availability of hardware memory protection, and it requires a memory management unit.

(4) *Transferring control between protection domains is done by a generalization of hardware interrupt handling, that is, portal traversal. Portal code is generated dynamically and performs portal-specifications.*

Hardware interrupts, interprotection domain calls, and the Pebble equivalent of system calls, are all handled by the portal mechanism. Pebble generates specialized code for each portal to improve run-time efficiency. The portal mechanism provides two important features: abstract communication facilities, which allow components to be isolated from their configuration, and per-connection code specialization, which enables the application of many otherwise unavailable optimizations.

8.3.1. Protection Domains and Portals

Each component runs in its own Protection Domain (PD). A PD consists of a set of pages, represented by a page table, and a set of portals, which are generalized interrupt handlers, stored in the protection domain's portal table. A PD may share both memory pages and portals with other protection domains. Figure 8.1 illustrates the Pebble architecture.

A parent protection domain may share its portal table with its child. In this case, any changes to the portal table will be reflected in both parent and child. Alternatively, a parent protection domain may create a child domain with a copy of the parent's portal table at the time when the child was created. Both copying and sharing portal tables are efficient, since portal tables contain pointers to the actual portal code. No copying of portal code is needed in either case.

A thread belonging to protection domain A (the protection domain in which component A is running) can invoke a service of protection domain B only if A has successfully opened a portal to B. Protection domain B, which exports the service, controls which protection domains may open portals to B, and hence which component scan invoke B's service. Protection domain B

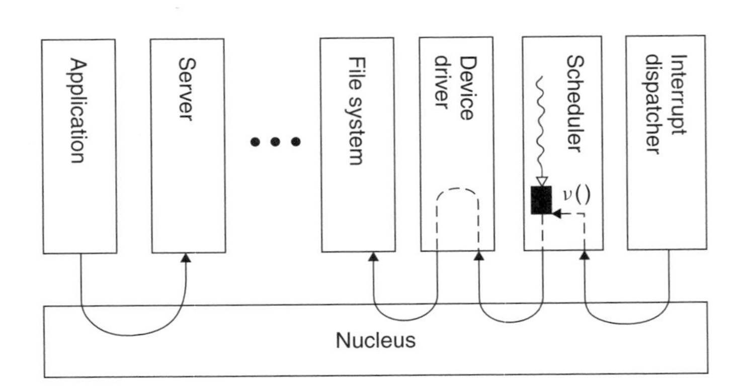

Figure 8.1 Pebble architecture. Arrows denote portal traversals. On the right, an interrupt causes a device driver's semaphore to be incremented, unblocking the device driver's thread.

may delegate the execution of its access-control policy to a third party, such as a directory server or a name-space server.

To transfer control to B, A's thread executes a trap instruction, which transfers control to the nucleus. The nucleus determines which portal A wishes to invoke, looks up the address of the associated portal code, and transfers control to the portal code. The portal code is responsible for saving registers, copying arguments, changing stacks, and mapping pages shared between the domains. The portal code then transfers control to component B. Figure 8.2 shows an example of portal transfer.

When a thread passes through a portal, no scheduling decision is made; the thread continues to run, with the same priority, in the invoked protection domain.

Portals are used to handle both hardware interrupts and software traps, and exceptions. The existence of a portal from PD_A to PD_B means that a thread running in PD_A can invoke a specific entry point of PD_B, and then return. Associated with each portal is code to transfer a thread from the invoking domain to the invoked domain. Portal code copies arguments, changes stacks, and maps pages shared between the domains. Portal code is specific to its portal, which allows several important optimizations to be performed.

Portals are usually generated in pairs. The call portal transfers control from domain PD_A to PD_B, and the return portal allows PD_B to return to PD_A.

Portals are generated when resources are created, for instance, semaphores, and when clients connect to servers, for instance, when files are opened. Interrupt and exception handling portals are created at the system initialization time.

A scheduling priority, a stack, and a machine context are associated with each Pebble thread. When a thread traverses a portal, no scheduling decision is made; the thread continues to run, with the same priority, in the invoked

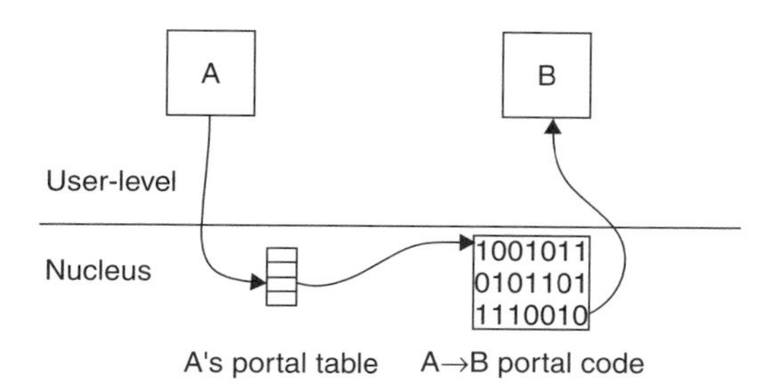

Figure 8.2 Portal transfer. Protection domain A invokes protection domain B via a portal transfer. Protection domain A transfers indirectly through its portal table to the portal code specific to this communication path. The portal code transfers control to protection domain B.

protection domain. Once the thread executes in the invoked domain, it may access all of the resources available in the invoked domain, while it can no longer access the resources of the invoking domain. Several threads may execute in the same protection domain at the same time, which means that they share the same portal table and all other resources.

As part of a portal traversal, the portal code can manipulate the page tables of the invoking and/or invoked protection domains. This most commonly occurs when a thread wishes to map, for the duration of the portal invocation, a region of memory belonging to the invoking protection domain into the virtual address space of the invoked protection domain; this gives the thread a window into the address space of the invoking protection domain while running in the invoked protection domain. When the thread returns, the window is closed.

Such a memory window can be used to save the cost of copying data between protection domains. Variations include:

- windows that remain open to share pages between protection domains;
- windows that transfer pages from the invoking domain to the invoked domain to implement tear-away write;
- windows that transfer pages from the invoked domain to the invoking domain to implement tear-away read.

Although the portal code may modify Virtual Memory (VM) data structures, only the VM manager and the portal manager, which generates portal code, share the knowledge about these data structures. The Pebble nucleus itself is oblivious to those data structures.

Portal code may never block calling threads, and may not contain loops. This is essential to ensure that the portal can be traversed in a small, finite amount of time. If the portal has to block a calling thread (that is, the invoked domain's stacks queue is empty), then the portal code transfers control to the scheduler, inside which the calling thread is waiting for the resource.

Specialized portal code is generated on the fly when a portal is opened. This allows portal code to take advantage of the semantics and trust relationships of the portal. For example, if the caller trusts the callee, the caller may allow the callee to use the caller's stack, rather than to allocate a new one. If this level of trust does not exist, the caller can require that the callee allocate a new stack. Although sharing stacks decreases the level of isolation between the caller and callee, it can improve performance.

Pebble implements a safe execution environment by a combination of hardware memory protection that prevents access to memory outside the protection domain, and by limiting the access to the domain's portal table.

A protection domain may access only the portals it inherited from its parent and new portals that were generated on its behalf by the portal manager. The portal manager may restrict access to new portals in conjunction with the name server. A protection domain cannot transfer a portal it has in its portal table to an unrelated domain. Moreover, the parent domain may intercept all of its child portal calls, including calls that indirectly manipulate the child's portal table.

In Pebble, system services are provided by operating-system server components, which run in the user mode protection domains. Unlike applications, server components are trusted, so they may be granted limited privileges not afforded to application components. For example, the scheduler runs with interrupts disabled, device drivers have device registers mapped into their memory region, and the portal manager may add portals to protection domains (a protection domain cannot modify its portal table directly).

There are advantages in implementing services at user level. First, from a software-engineering standpoint, it is guaranteed that a server component will use only the exported interface of other components. Second, because each server component is only given the privileges that it needs, a programming error in one component will not directly affect other components. If a critical component such as VM fails, the system as a whole will be affected, but a bug in console device driver will not overwrite page tables.

In addition, as user-level servers can be interrupted at any time, this approach has the possibility of offering lower interrupt latency time. Given that server components run at user level (including interrupt-driven threads), they can use blocking synchronization primitives, which simplifies their design. This is in contrast with handlers that run at interrupt level, which must not block, and require careful coding to synchronize with the upper parts of device drivers.

The portal manager is the operating system component responsible for instantiating and managing portals. The portal manager is privileged because it is the only component that is permitted to modify portal tables.

Portal instantiation is a two-step process. First, the server (which can be a Pebble system component or an application component) registers the portal with the portal manager, specifying the entry point, the interface definition, and the name of the portal. Second, a client component requests that a portal with a given name be opened. The portal manager may call the name server to identify the portal and to verify that the client is permitted to open the portal. If the name server approves the access, the portal manger generates the code for the portal, and installs the portal in the client's portal table. The portal number of the newly generated portal is returned to the client. A client may also inherit a portal from its parent.

To invoke the portal, a thread running in the client loads the portal number into a register and traps to the nucleus. The trap handler uses the portal number as an index into the portal table and jumps to the code associated with the portal. The portal code transfers the thread from the invoking protection domain to the invoked protection domain and returns to user level.

Portal interfaces are written using an interface definition language. Each portal argument may be processed or transformed by the portal code. The argument transformation may involve a function of the nucleus state, such as inserting the identity of the calling thread or the current time. The argument transformation may also involve other servers. For example, a portal argument may specify the address of a memory window to be mapped into the receiver's address space. This transformation requires the manipulation of data structures in the virtual memory server.

The design of the portal mechanism presents the following conflict: to be efficient, the argument transformation code in the portal may need to have access to private data structures of a trusted server, that is, the virtual memory system. On the other hand, however, trusted servers should be allowed to keep their internal data representations private.

The solution is to allow trusted servers, such as the virtual memory manager, to register argument transformation code templates with the portal manager. Portals registered by untrusted services are required to use the standard argument types. When the portal manager instantiates a portal that uses such an argument, the appropriate type-specific code is generated as part of the portal. This technique allows portal code to be both efficient by in-lining code that transforms arguments, and encapsulated by allowing servers to keep their internal representations private. Although the portal code that runs in kernel mode has access to server-specific data structures, these data structures cannot be accessed by other servers. The portal manager supports argument transformation code of a single trusted server, the virtual memory server.

8.3.2. Scheduling and Synchronization

Pebble's scheduler implements all actions that may change the calling thread's state, that is, *run* → *blocked* or *blocked* → *ready*. Threads cannot block other threads anywhere except inside the scheduler. In particular, Pebble's synchronization primitives are managed entirely by the user-level scheduler. When a thread running in a protection domain creates a semaphore, two portals that invoke the scheduler (for P and V operations) are added to the protection domain's portal table. The thread invokes P in order to acquire the semaphore. If the P succeeds, the scheduler grants the calling protection domain the semaphore, and returns. If the semaphore is held by another protection domain, the P fails, the scheduler marks the thread as blocked, and then schedules another thread. A V operation works analogously; if the operation unblocks a thread that has higher priority than the invoker, the scheduler can block the invoking thread and run the newly-awakened thread.

The scheduler runs with interrupts disabled, in order to simplify its implementation. Work on the use of lock-free data structures has shown that, with appropriate hardware support, it is possible to implement the data structures used by Pebble's scheduling and synchronization component without locking. Such an implementation would allow the scheduler to run with interrupts enabled which would reduce interrupt latency even further.

Each hardware device in the system has an associated semaphore used to communicate between the interrupt dispatcher component and the device driver component for the specific device.

In the portal table of each protection domain there are entries for the portals that correspond to the machine's hardware interrupts. The Pebble nucleus includes a short trampoline function that handles all exceptions and interrupts. This code first determines the portal table of the current thread and then transfers control to the address that is taken from the corresponding entry in this portal table. The nucleus is oblivious to the specific semantics of the portal that is being invoked. The portal that handles the interrupt starts by saving the processor state on the invocation stack, then it switches to the interrupt stack and jumps to the interrupt dispatcher. This mechanism converts interrupts to portal calls.

The interrupt dispatcher determines which device generated the interrupt, and performs a *V* operation on the device's semaphore. Typically, the device driver would have left a thread blocked on that semaphore. The *V* operation unblocks this thread, and if the now-runnable thread has higher priority than the currently running thread, it gains control of the CPU (Central Processor Unit), and the interrupt is handled immediately. Typically, the priority of the interrupt handling threads corresponds to the hardware interrupt priority in order to support nested interrupts. The priority of the interrupt handling threads is higher than all other threads to ensure short handling latencies. In this way, Pebble unifies interrupt priority with thread priority, and handles both in the scheduler.

Pebble invokes the interrupt dispatcher promptly for all interrupts, including low priority interrupts. However, the interrupt handling thread is scheduled only if its priority is higher than the running thread.

Only a small portion of Pebble runs with interrupts disabled, namely portal code, the interrupt dispatcher, and the scheduler. This is necessary to avoid race conditions due to nested exceptions.

Figure 8.3 shows the interrupt handling in Pebble. Each hardware device in the system is associated with a semaphore that used to communicate between the interrupt dispatcher component and the device driver component for the specific device. Typically, the device driver will have left a thread blocked on the semaphore.

The portal table of each protection domain contains entries for the machine's hardware interrupts. When an interrupt occurs, portal 1 saves the context of the currently running thread, and the contents of the entire register set. Portal 1 then switches the stack pointer to the interrupt stack and calls the interrupt dispatcher, which identifies the device that generated the interrupt. The interrupt dispatcher calls the scheduler to perform a V operation on the device's semaphore via portal 2. This portal saves only a few registers and allows the scheduler to share the same stack as the interrupt dispatcher. The V operation unblocks the handler thread. If the handler thread has a higher priority than the thread that was running at the time when the interrupt was received, the scheduler calls portal 3 with the identity of the handler thread. Portal 3 restores the context of the handler thread, including registers and stack, and the interrupt is handled immediately. Otherwise, the handler thread is added to the ready queue and the scheduler selects resumption of the thread that was running previously by calling portal 3 with the identity of this thread. Portal 3 performs the actual context switch. The scheduler supplies the identity of the next thread to run.

Pebble does not rely on hardware interrupt priorities to schedule interrupt handler threads. The interrupt dispatcher is called promptly for all interrupts, and the Pebble scheduler decides whether or not to run the associated handler thread. Pebble unifies interrupt priority with thread priority, and handles both in the scheduler.

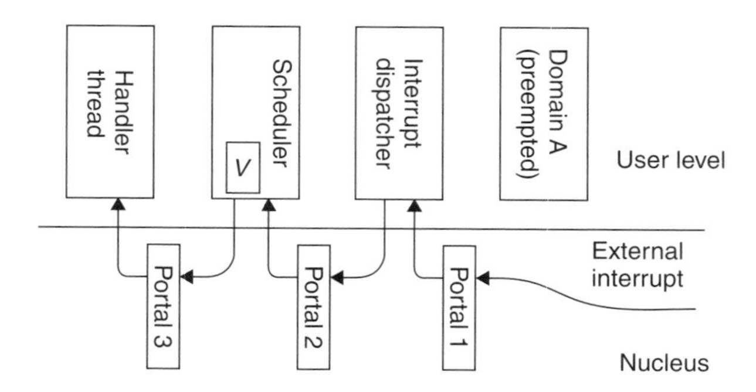

Figure 8.3 Interrupt handling. An interrupt causes a portal call to the interrupt dispatcher, which calls the scheduler to performs a V operation on the device's semaphore. The scheduler wakes up the handler thread that waits on this semaphore.

8.3.3. Implementation

Portal definitions are written using a simple interface definition language. The portal manager dynamically generates specialized code when a portal is created. The interface definition specifies which registers to save, whether to share a stack with the receiving domain, and how to process each argument.

Simple arguments (e.g. integers) are not processed at all; more complex argument types may require more work. For example, an argument may specify the address of a memory window that should be mapped into the receiver's address space, or a capability that must be transformed before being transferred. The transformation code, to be efficient, may need to have access to private data structures of a trusted server (e.g. the virtual memory system or the capability system). On the other hand, however, the trusted servers should be allowed to keep their internal data representations private.

The solution is to allow trusted services to register argument transformation code templates with the portal manager. When the portal manager instantiates a portal that uses such an argument, the code template is used when generating the portal. This technique allows portal code to be both efficient (by in-lining code that transforms arguments) and encapsulated (by allowing servers to keep their internal representations private). Although portal code that runs in kernel mode has access to server-specific data structures, these data structures cannot be accessed by other servers.

In some cases the amount of work done is so small that the portal code itself can implement the service. A short-circuit portal does not actually transfer the invoking thread to a new protection domain, but instead performs the requested action in line, in the portal code. Examples include simple system calls to get the current thread's ID (identifier) and to obtain the time of day. The TLB (Translation Lookaside Buffer) miss handler, which is in software on the MIPS (Millions of Instructions Per Second) architecture, is also implemented as a short-circuit portal.

Pebble's design includes support for capabilities, which are the abstract tokens that represent access rights. The capability manager, a trusted user-level server, keeps track of the capabilities available to each protection domain. Additionally, it registers a capability argument type and associated transformation code with the portal manager. When a capability is passed through a portal, the portal code adds the capability to the receiving protection domain's capability list, and transforms the sending protection domain's external representation of the capability to that of the receiving domain. The standard capability operations include revocation, use-once, nontransferability, reduction in strength, etc.

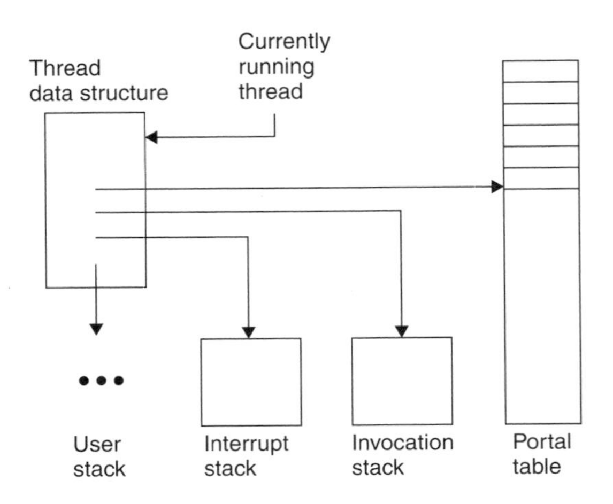

Figure 8.4 Pebble nucleus data structures.

The Pebble nucleus maintains a number of data structures, which are illustrated in Figure 8.4. Each thread is associated with a thread data structure, which contains pointer to the thread's current portal table, user stack, interrupt stack, and invocation stack.

The *user stack* is a regular stack used by the user mode code.

The *interrupt stack* is used whenever an interrupt or exception occurs while the thread is executing. The interrupt portal switches to the interrupt stack, saves state on the invocation stack and calls the interrupt dispatcher server.

The *invocation stack* keeps track of portal traversals and processor state. The portal call code saves the invoking domain's state, and the address of the corresponding return portal, on the invocation stack. The portal return code restores the state from the invocation stack.

The *portal table pointer* in the thread data structure is the portal table of the domain in which the thread is executing. This table is changed by the portal call and restored by the portal return.

The *virtual memory manager* is responsible for maintaining the page tables, which are accessed by the TLB miss handler and by the memory window manipulation code in portals. The virtual memory manager is the only component that has access to the entire physical memory. Pebble does not support demand-paged virtual memory.

Pebble implementation uses the MIPS tagged memory architecture. Each protection domain is allocated a unique ASID (Address Space Identifier), which avoids TLB and cache flushes during context switches. Portal calls and returns also load the mapping of the current stack into TLB entry 0 to avoid a certain TLB miss.

Pebble components run in separate protection domains in user mode, which requires careful memory allocation and cache flushes whenever a component must commit values to the physical memory. For example, the portal manager must generate portal code so that it is placed in contiguous physical memory.

The portal code that opens a memory window updates an access data structure that contains a vector of counters, one counter for each protection domain in the system. The vector is addressed by the ASID of the corresponding domain. The counter keeps track of the number of portal traversals into the corresponding domain that passed this page in a memory window. This counter is incremented by one for each portal call, and is decremented by one for each portal return. The page is accessible if the counter for the domain is greater than zero. Because the same page may be handed to the same domain by multiple concurrent threads, the counters are used to maintain page access rights.

The page table contains a pointer to the corresponding access data structure, and only shared pages have a dedicated access data structure.

The TLB miss handler consults the counter vector to verify the access rights to the memory window page. The portal code does not load the TLB with the mapping of the memory window page. This design saves time if the shared window is passed to another domain without being touched by the current domain. The portal return code must remove the corresponding TLB entry when the counter reaches zero.

The portal call may implement stack sharing, which does not require any stack manipulation. The invoked domain uses the current thread's stack.

If the portal call requires a new stack, it obtains one from the invoked domain's stack queue. The invoked protection domain must preallocate one or more stacks and notify the portal manger to place them in the domain's stack queue. The portal call de-queues a new stack from the invoked domain's stack queue. If the stack's queue is empty, the portal calls the scheduler and waits until a stack becomes available. The portal return enqueues the released stack back into the stack queue. If there are any threads waiting for the stack, the portal return calls the scheduler to pick the first waiting thread and allow it to proceed in its portal code.

The portal than calls the interrupt dispatcher after an interrupt switches the stack to the interrupt stack, which is always available in every thread.

The Pebble nucleus and the essential components (interrupt dispatcher, scheduler, portal manager, real-time clock, console driver, and the idle task) can fit into about 70 pages (8 kb each). Pebble does not support shared libraries, which cause code duplication among components. Each user thread has three stacks (user, interrupt, and invocation) which require three pages, although the interrupt and invocation stacks could be placed on the same page to reduce memory consumption. In addition, fixed size pages inherently waste memory. This could be alleviated on segmented architectures.

An important aspect of component-based system is the ability to interpose code between a client and its servers. The interposed code can modify the operation of the server, enforce safety policies, enable logging and error recovery services, or even implement protocol stacks and other layered system services.

Pebble implements low-overhead interposition by modifying the portal table of the controlled domain. Since all interactions between the domain and its surroundings are implemented by portal traversals, it is possible to place the controlled domain in a comprehensive sandbox by replacing the domain's portal table. All of the original portals are replaced with portal stubs, which transfer to the interposed controlling domain. The controlling domain intercepts each portal traversal that takes place, performs whatever actions it deems necessary, and then calls the original portal. Portal stubs pass their parameters in the same way as the original portals, which is necessary to maintain the semantics of the parameter passing (e.g. Windows). Portal stubs are regular portals that pass the corresponding portal index in their first argument. The controlling domain does not have to be aware of the particular semantics of the intercepted portals; it can implement a transparent sandbox by passing portal parameters verbatim.

The top diagram of Figure 8.5 illustrates the configuration of the original portal table without interposition, where the domain calls its servers directly. The bottom diagram shows the operation of portal interposition. In this case, all the portals in the controlled domain call the controlling domain, which makes the calls to the servers.

However, one-time modification of the controlled domain's portal table is not sufficient. Many servers create new portals dynamically in their client's portal table, and then return an index to the newly created portal back to the client. Since the controlling domain calls the server, the server creates new portals in the controlling domain's table. The controlling domain is notified by the portal manager that a new portal was created in its portal table. The notification portal completes the process by creating a portal stub in the controlled domain's table with the same index as it has in the controlling domain table.

The portal stub calls the controlling domain and passes the parameters in the same way as the original portal. In this way, the controlling domain implements a robust sandbox around the controlled domain, without actually understanding the semantics of the controlled domain portals.

The controlled domain cannot detect that its portals are diverted nor can it thwart the interposition in any way. This mechanism is similar to the Unix

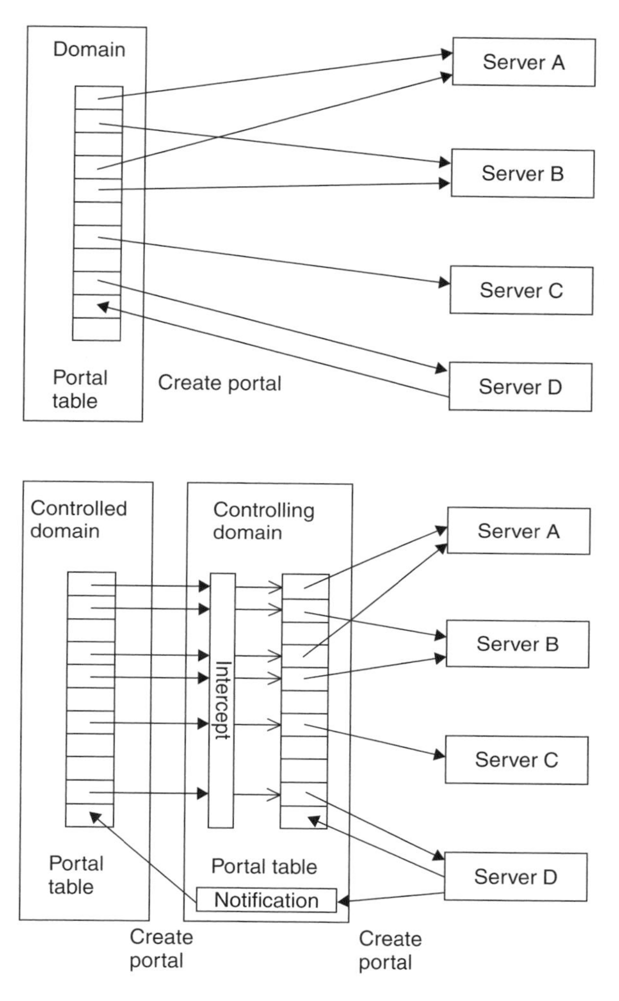

Figure 8.5 Original portal configuration (above) and with portal interposition (below).

I/O redirection, in which a child process accesses standard file descriptor (e.g. 0, 1 and 2), which is redirected by the parent process. Portal interposition is more comprehensive than the Unix I/O redirection, since all interactions between the controlled domain and its environment are controlled. The interposition can be recursive: a controlling domain interposes the portals of a child domain, which does the same to its child, and so on.

Pebble provides a new engineering trade-off for the construction of efficient component-based systems, using hardware memory management to enforce protection domain boundaries, and reducing the cross domain transfer time by synthesizing custom portal code. Pebble enhances flexibility by maintaining a private portal table for each domain. This table can be used to provide different implementations of system services, servers and portal interposition for each domain. Portal interposition allows running untrusted code in a robust sandbox with an acceptable overhead while using unsafe programming languages such as C.

Having a small nucleus with minimal functionality enhances system modularity, while it enables nonstop systems to modify their behavior by integrating new servers on-the-fly.

Pebble is much faster than Open BSD (Berkeley Software Distribution) for a limited set of system-related micro-benchmarks. Pebble efficiency does not stem from clever, low-level, highly optimized code; rather it is a natural consequence of custom portal synthesis, judicious processor state manipulations at portal traversals, encapsulating state in portal code, and direct transfer of control from clients to their servers without scheduler intervention.

Pebble can be used to build flexible, safe, and high performance systems.

8.3.4. Embedded Applications

Pebble operating system architecture is used as a platform for high-end, embedded, communicating devices constructed from reusable software components. A component-based approach to building applications uses software components with clean, abstract interfaces, which allow code to be combined and reused in many different ways. The cost of development can be amortized by using many diverse applications.

The approaches to isolate the components from one another are as follows:

- to provide no protection between components;
- to provide software protection;
- to provide hardware protection.

Each method has its drawbacks. With no protection, a component can compromise the integrity of the system, as it has access to all data and resources of the system. Software protection typically requires that the system be written in a special, safe programming language, which may not be acceptable to all developers. Hardware protection schemes have traditionally exhibited poor performance, due to the cost of switching protection domains when performing an interprotection-domain call.

In Pebble architecture, the operating system services are implemented by a collection of fine-grained, replaceable user-level components. The techniques applied to operating system components are also used by component-based applications running on Pebble, and applications share in the performance benefit provided by these techniques. The performance improvements are significant: for example, on Pebble, a one-way interprotection domain call takes about 120 machine cycles, which is within an order of magnitude of the cost of performing a function call; an equivalent call when running Open BSD (Berkeley Software Distribution), on the same hardware, takes about 1000–2000 machine cycles.

Under the Pebble operating system, an application can dynamically configure the services provided by the system, and safely load new, untrusted components, written in an unsafe programming language, into the system while the system is running. Moreover, old servers may be retired gracefully when new versions of the service are introduced without disrupting the operation of the system. This capability is essential for high-availability systems that must operate continuously.

Communication devices, such as PDA-cell phone hybrids, set-top boxes, and routers require this type of dynamic method of configuration and the ability to run untrusted code safely. This approach is valuable for building embedded systems.

Component-based systems that use software protection schemes are typically written in a type-safe byte-coded programming language, such as Java, and the Limbo programming language of the Inferno operating system. Components run in a single hardware protection domain, but the run-time environment implements, in effect, a software protection domain. These systems are designed to meet the following goals:

(1) to provide an architecture-independent distribution format for code;
(2) to ensure that resources, such as memory, are returned to the system when they are no longer needed;
(3) to ensure that the component does not view or modify data to which it has not been granted access.

In case of Java, these goals are satisfied by (1) the machine-independent Java byte-code, (2) the garbage collector provided by Java run-time environments, and (3) the run-time Java byte-code verifier.

Java byte-code offers a hardware-architecture-neutral distribution format for software components. However, such an architecture-neutral format could also be used for untrusted code. Most compiler front-ends generate a machine-independent intermediate form, which is then compiled by a machine-specific back-end. Such an intermediate form could be used as a distribution format for components written in any programming language, trusted or untrusted.

With the software protection, having all software-protected components in the same address space makes it hard to find a buggy component that is not caught by the type system or the garbage collector.

Hardware protection schemes run each component in a separate hardware protection domain. As an example, a traditional operating system such as Unix, could be thought of as a hardware-protected, component-based system, where the components are programs, and the protection is provided by the operating system working with the hardware memory management unit, such components can be composed using Unix pipes.

Typically, hardware schemes do not provide an architecture-independent distribution format for code since components are distributed in the machine language of the target hardware. Resources, such as memory, are returned to the system when no longer needed through careful bookkeeping: the system keeps track of the resources assigned to each component (process), and when the component (process) terminates, the resources are returned to the system. The component does not view or modify data to which it has not been granted access by using hardware memory protection: each component is run in a separate address space. If a component attempts to view or modify data outside its address space, a hardware trap is invoked and the component (process) is terminated.

By running multiple Java virtual machines on top of hardware protection, the components can be separated in a way that makes it easier to identify buggy components.

Software schemes are the only option when hardware protection is unavailable, such as low-end processors without memory management units. For this reason, the designers of the Inferno operating system and Limbo programming language chose a software protection scheme.

Hardware schemes are used when component code is written in an unsafe programming language, such as C or C++. Although Java provides many facilities unavailable in C and C++, there are often good reasons for running code written in an unsafe programming language. A component may include legacy code that would be difficult or costly to reimplement in a safe programming language. A component may also include a hand-tuned assembler code that uses hardware specific features, for example, a computation-intensive algorithm, such as an MPEG (Motion Pictures Experts Group) decoder, which uses special hardware instructions.

The garbage collection offered by systems such as Java assures programmers and software testers that all allocated resources are eventually freed. Storage leaks are usually hard to find, and automated garbage collection simplifies the task of writing robust code. However, when building applications have fixed latency requirements, the free memory pool may become empty at any time, and any memory allocation could trigger garbage collection, which makes estimating the cost (in terms of time) of memory allocation, a stochastic, rather than deterministic, process. In a real-time embedded system, the uncertainty introduced by the presence of a garbage collector may not be beneficial. This is the reason why some systems, such as the operating system Inferno, ensure that garbage collection does not delay critical functions.

The cost of making transfers between components under a hardware protection scheme is much higher than it is under a software protection scheme. In the Pebble operating system, the cost of cross-component communication under a hardware scheme is within one order of magnitude of the cost of a function call.

Portals can be used to model code and data. A set of portals can be used to represent an open file descriptor. In Pebble, an open call creates three portals in the invoking protection domain, which are for read, write, and seek, on the corresponding file. A read call transfers directly to the appropriate routine, thus, no run-time demultiplexing is needed to determine the type of underlying object, and the appropriate code for a disk file, socket, etc., will be invoked. Additionally, a pointer to its control block can be embedded in the portal code and passed directly to the service routine, thus there is no need to perform a run-time validation of the file pointer. The portal code cannot be modified by the client, and the control block pointer passed to the server can be trusted to be valid. Thus, the server can access the particular file immediately. There is no need for a separate file descriptor table; the data normally associated with the tables is found in the dynamically generated portal code.

Bruno *et al.* (1999) measured the operation of Pebble for several microbenchmarks on three different test hardware platforms, named LOW, MID and HIGH, representing low-, medium- and high-end embedded system configurations. All three platforms included MIPS processors from QED (Quantum Effect Design) RM5230 and RM7000, and IDT (Integrated Device Technology) R5000. All motherboards were manufactured by Algorithmics, a developer of systems for embedded applications.

The LOW platform is representative of low-cost, hand-held devices, which have a single-level cache hierarchy and small memory. The MID platform is representative of more powerful appliances, such as a set-top box, which contain a more powerful processor, two-level cache hierarchy and larger memory. The HIGH platform is representative of high-end systems, which contain top-of-the-line processors with large caches.

The cache in all targets can be accessed in a single machine cycle, and does not cause any pipeline delay. Access to higher levels of the memory hierarchy causes a delay.

The operations are simple communication or synchronization building blocks:

- The measured time of short-circuit portal to return the identity of the calling domain. This is the equivalent of the UNIX null system call. A short-circuit portal performs the action and returns to the caller immediately without a context switch.
- The measured time of an interprotection domain call between two domains, which is implemented by a portal traversal. The portal passes four integer parameters in the machine registers. No parameter manipulation is performed by the portal code. The portal allocates a new stack in the target domain and frees the stack on return. The one leg of interprotection domain call time is constant for a chain of interprotection domain calls through a sequence of domains.
- The measured thread yield operation, in which the current thread calls the scheduler and requests the next thread to be run with a higher or equal priority. There is one active thread in each domain. The time is a single context switch time (total time divided by total number of yields by all threads).
- The measured time to pass a token around a ring of n threads, each running in a separate domain. Each thread shares one semaphore with its left neighbor and one semaphore with its right neighbor. The thread waits on its left neighbor. Once the left semaphore is released, the thread releases its right semaphore and repeats the process. There is one active thread in each domain. The time is that taken to pass the token once, which is the time for a single pair of semaphore acquire/release.

The performance degrades with the number of active threads, which is expected due to more frequent cache misses. Performance degrades the most with LOW platform, and the least with HIGH platform, which has a large cache to hold the entire working set of the test.

The HIGH platform is not significantly faster than the LOW and MID platforms for short-circuit portal, interprotection domain call, and semaphore acquire/release thread, although it has a dual-issue pipeline and much larger caches. This is because these tests do not cause too many cache misses, and the portal code is dominated by sequences of load and store instructions that cannot be executed in parallel.

When an interrupt is generated, there are two factors that control the delay before the interrupt is handled. First, there can be a delay before the system is notified that an interrupt has been received, if the processor is running with interrupts disabled. Second, there may be a delay between the time the interrupt is delivered and the time the device is serviced. The sum of these two delay components is the interrupt latency.

The first delay component, *DF*, is bounded by the length of path through the system where interrupts are disabled. The length, in this context, refers to the amount of time it takes to process the instruction sequence. Intuitively, it can be expected to be proportional to the length, in instructions, of the code path. In particular, the delay *DF* is determined by the portal code and by the scheduler, which are the only frequently used portions of the system that run with interrupts disabled.

The second delay component, *DS*, is bounded by the minimum amount of time required to deliver the interrupt to the interrupt handler. Thus the interrupt latency will range from $[DS, DS + DF]$, provided that interrupt handling does not generate many additional cache misses, such as in the MID and HIGH platforms.

The interrupt latency is measured by computing the difference between the time that an interrupt was generated and the time that a user thread that waited for this interrupt actually woke up. To accomplish this, Bruno *et al.* (1999) had the measurement thread sleep for some randomly chosen duration, and compared the time at which it is woken up with the time it expected to be woken up. The difference between the two is the interrupt latency.

This test is very precise: each of the test platforms includes a high-resolution timer that is incremented every other processor cycle, and the ability to schedule timer interrupts with the same granularity.

To estimate interrupt latencies under various loads, Bruno *et al.* (1999) ran the measurement thread concurrently with a background task that repeatedly performed a specific operation. Different operations exercise different interrupt-disabled paths of the system, and hence have different interrupt latency characteristics. The background threads tested were:

- A background thread that spins in a tight loop on a particular variable in user mode. The idle task can be preempted at any time. The value reported for this case is an estimate of the lower bound of the interrupt latency.
- A background task that calls that implemented by a short-circuit portal routine repetitively. Interrupts are disabled while executing this portal.
- A background thread that repeatedly performs an interprotection domain call to the protection domain in which the measurement thread is running. The portal code associated with this portal is only transferring control, and the call returns immediately. Interrupts are disabled during each leg of the

interprotection domain call return, and are enabled when executing in the caller and called domains.

- A background thread that repeatedly calls the scheduler to yield control. As there is one active thread in the system, the scheduler returns to the calling thread. Interrupts are disabled during the portal call to the scheduler, inside the scheduler, and during the portal return to the thread.
- A pair of background threads, which pass a token both ways. Each thread runs in separate protection domain. Interrupts are disabled during the semaphore operations.

The interrupt latency is bounded by the sum of a platform-specific constant plus a time which is proportional to the longest interrupt-disabled path in the background task. The platform-specific constant is the minimal interrupt response time, which is the median value of the idle test. The measurements indicate that the interrupts are served immediately.

The maximal interrupt latency on the MID and HIGH platforms is very close to the 99th percentile on these platforms, which means that the system performance is highly predictable. However, the maximal interrupt latency on the LOW platform is up to 60 % higher than the 99 % percentile latency. This is the result of the small cache size of LOW, which causes excessive cache misses due to infrequent background events, such as the timer interrupt.

The interrupt latencies for the MID and HIGH systems are quite close, and much lower than those for LOW. Although the cache architectures of the two differ, both MID and HIGH have more effective caches than LOW.

The cost of hardware protection of components and system services is very low and that interrupt latency on Pebble is quite low, with a lower bound of 1200–1300 cycles (6.1 μs to 9.0 μs) depending on the target architecture.

8.4. EMBEDDED OPERATING SYSTEM ENERGY ANALYSIS

Tan *et al.* (2002a) recognized the need to provide embedded software designers with feedback about the effect of different OS services on energy consumption early in the design cycle. They presented a systematic methodology for performing energy analysis and macro-modeling of an embedded OS. Their energy macro-models provide software architects and developers with an intuitive model for OS energy effects, since they directly associate energy consumption with OS services and primitives that are visible to the application software. This methodology consists of:

- an analysis stage to identify a set of energy components, called energy characteristics, which are useful to the designer in making OS-related design trade-offs;
- a subsequent macro-modeling stage, where the data is collected for the identified energy components and macro-models for them are automatically derived.

The methodology is validated by deriving energy macro-models for two embedded OSs, C/OS and Linux OS.

Embedded operating systems form a critical part of a wide range of complex embedded systems, and provide benefits ranging from hardware abstraction and resource management, to real-time behavior. The energy effects of the OS have great bearing on the energy efficiency of the overall embedded system. Most of the embedded OS-related investigations center around performance issues. Also, OS-related issues concerning energy consumption of embedded software are studied. The energy consumption overhead of the OS is called OS energy characterization. Also, the effects of using different OSs or different uses of an OS on energy consumption are important.

An embedded system consists of many components, one of which may be a microprocessor that hosts all the software tasks. The software tasks use an OS as the run-time engine. A conceptual diagram for this system is shown in Figure 8.6. An example software for this embedded system has the task configuration shown in the figure. Task T1 through task T4 perform background processing such as managing the memory, controlling the LCD, etc. Task T5 looks out for new data coming from the A/D converter, performs some preprocessing, and passes the data to task T4. Given this multi-task specification, there are different ways to implement the system. In particular, for task T5, there are two options:

- Task T5 is implemented as an actual software process. After performing preprocessing, it passes the data to task T4 through interprocess communication (IPC).
- Task T5 is implemented as an Interrupt Service Routine (ISR) or signal handler for task T4. The ISR or signal handler is activated periodically using a hardware timer. Since tasks T4 and T5 share the same process space, passing data from task T5 to task T4 does not incur any IPC overhead.

The second implementation has lower overhead in terms of delay and energy, compared with the first implementation. However, the first implementation is a easier to program as compared with the second. Therefore,

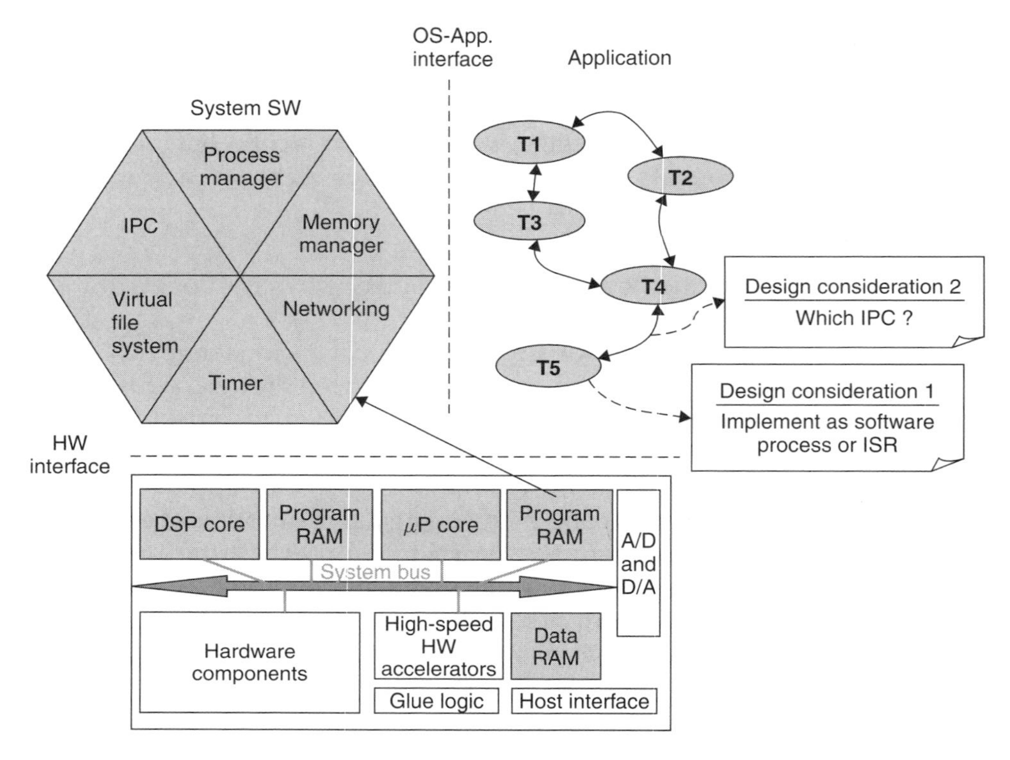

Figure 8.6 Building blocks of an embedded system.

some trade-off has to be made. Apart from that, choosing the actual mechanism for IPC also requires a trade-off between various objectives. A decision can be made when the energy overhead is quantified. To quantify the energy overhead of first over second implementation, both implementations are run and compared. Another way is to use energy characteristic data for the chosen operating system to compare them more efficiently. The energy characteristic data can be provided in the form of energy macro-models.

Figure 8.7 illustrates the overall methodology for OS energy characterization. The following steps are involved:

(1) Energy effect analysis step, in which the essential components of the embedded OS are characterized.
(2) Directed test programs that isolate these components from each other are generated.
(3) The test program is compiled and linked with the OS, and fed into an energy simulation tool. This step requires a low-level energy simulation framework that is capable of executing the application software

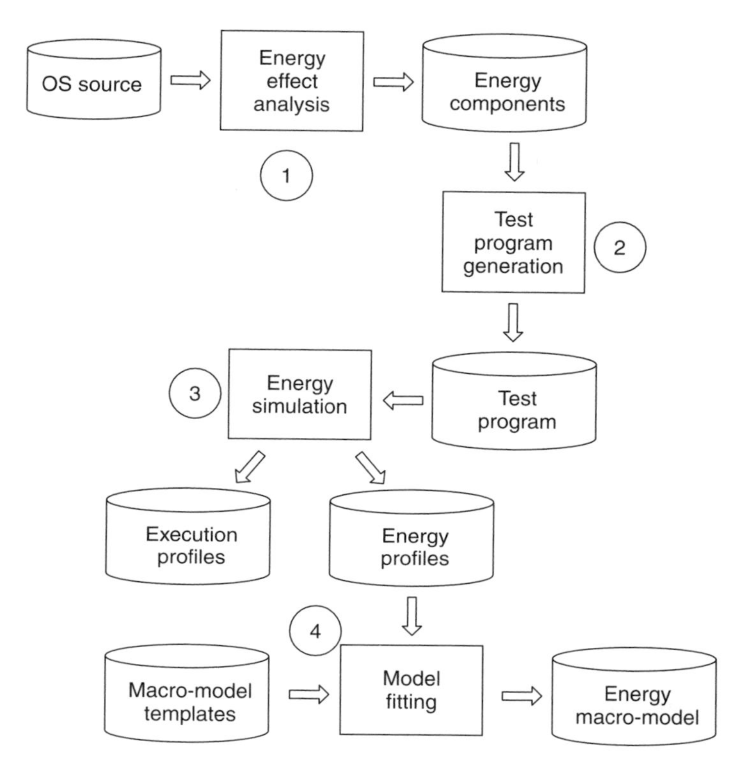

Figure 8.7 A flow diagram for energy characterization of an operating system.

together with the OS and reporting the energy consumption on a function instance basis.

(4) The execution and energy profiles generated from the previous step are subject to further analysis and model fitting to obtain the energy macro-models.

Step 1 is the energy analysis stage, whereas Steps 2, 3, and 4 belong to the energy macro-modeling stage.

For the purpose of energy analysis, an OS can be seen as a multi-entry multi-exit program. Some of the entry–exit pairs belong to system-call interfaces, whereas others belong to implicit paths of execution within the OS, that is, they do not directly correspond to system calls in the application code. Implicit execution paths can be triggered by interrupts. The entry–exit pairs belonging to the system call interfaces are called SCEEP (system call entry–exit pair), and the entry–exit pairs belonging to implicit execution paths are called IEEP (implicit entry–exit pair).

An SCEEP can be overloaded with a few groups of paths because the system function it represents is overloaded with many modes of operations. For example, the read system function in the Linux OS can be used to read from a file, from the network, or from a terminal. These different modes of operation correspond to very different groups of paths for the SCEEP of the read function. Similarly, IEEP could also be overloaded in the sense that different groups of paths may be traversed, depending on the state of the OS.

The objective of OS energy analysis is to identify a useful set of SCEEPs and IEEPs, and to classify the paths between them into groups amenable to macro-modeling. The energies consumed while traversing these paths are the energy characteristics of the OS. The energy characteristic data are classified into two categories, namely the explicit group and the implicit group.

The energy data that is directly related to the OS primitives, functions. The explicit energy basically relates to the energy consumed while traversing the paths between SCEEPs. An OS typically comprises many SCEEPs. A comprehensive set of explicit energy-characteristic data should, in principle, cover all the SCEEPs. However, in most practical applications, a selected set of SCEEPs should be sufficient to provide useful data for the designers. For system functions that are overloaded with multiple modes of operation, each mode of operation should have its own energy macro-model, even though all the modes of operation share the same SCEEP.

Making energy macro-models available for system functions allows the designers to choose among possible alternatives, that is, system functions, efficiently. Therefore, the significance of these energy macro-models lies not in their ability to provide absolute estimates for actual energy consumption, but in their ability to facilitate comparison among different alternatives.

Implicit energy characteristics are not directly related to any OS primitive. The energy is not incurred when exercising an OS primitive, but comes as a result of running the OS engine. This energy basically relates to the energy consumed while traversing the paths between IEEPs. Similar to the SCEEP, there are usually a few groups of paths for a single IEEP, each of them requiring a separate energy macro-model. The energy consumption along the different groups of paths between the IEEP is characteristic of an OS, and they are called implicit energy characteristics. Some of them are as follows:

- Timer interrupt energy, which is the energy overhead incurred by the timer interrupt tied to the scheduler.
- Scheduling energy, which is the energy overhead of performing rescheduling in the preemptive scheduler.
- Context switch energy, which is the energy overhead incurred when a context switch occurs. A call to the scheduler may not always result in

a context switch, hence, the context switch and scheduling energy are characterized separately.

- Signal handling energy: A signal depicts an OS emulation of a low-level interrupt. Since it is commonly used, its energy overhead should be characterized.

Making energy macro-models available for the above energy components allows the designers to efficiently compute the relative energy cost of different software architectures.

Tan *et al.* (2002a) adopted a white-box analysis and black-box measurement philosophy for OS energy macro-modeling. White-box analysis refers to the fact that they identified and analyzed the energy components by studying the internal operation of the OS. Black-box measurement refers to the fact that they measured the energy components by devising experiments that isolate the OS energy by only instrumenting the application code. Black-box measurement and macro-modeling enable the energy models to be used without any knowledge of the OSs internal implementation.

The energy consumption of an OS primitive, that is system function, can be characterized by repeatedly calling it in a test program. Care must be taken to extract the pure energy cost of the system function, isolating it from the implicit energy. For example, in measuring the cost of a message queue read, one must make sure that the energy data collected does not include context switch energy. This is achieved by arranging to have both read and write parts of the experiment in the same software process. Moreover, spurious data resulting from timer interrupt, rescheduling and preemptive context switch must be isolated and eliminated. For example, to avoid a preemptive context switch, the test program must not execute for more than 200 milliseconds in the case of Linux OS.

There are a few key parameters that should be measured to obtain the implicit energy characteristics. Measuring these parameters is not as straightforward as for explicit energy characteristics.

Context switch energy is a good measure of the chosen OSs ability to perform multi-tasking with energy efficiency. Due to its importance, two different approaches are presented for obtaining values for this key parameter. Whether both approaches result in similar energy values is an indicator of their accuracy.

The first approach is to arrange two separate experiments, A and B. Experiment A consists of two tasks, connected by two separate IPC channels going in opposite directions, with a single byte repeatedly being passed back and forth through the two IPC channels. Since two tasks are involved in the activity (reading and writing of the IPC channels), context switches occur repeatedly. Experiment B consists of a similar set-up. However, in this case, both the IPC channels are entirely in a single task. As reading and writing of the IPC channels occur, no context switch is involved. By comparing the differences in the energy consumptions between these two experiments, the context switch energy can be isolated.

The second approach is quite different from the first. In this case, the test program consists of a function that does nothing, expecting to be preempted by the OS. This function is called a large number of times by the main test program, and is preempted by the OS during some of these calls. The energy consumption of this function reveals the energy incurred by the timer interrupt, rescheduling, and context switch. Knowing the underlying scheduling mechanism of the OS, the origin of each energy cluster can be reduced. This approach is a white-box analysis and black-box measurement. The first cluster is the nominal energy consumption of the function. The second cluster is due to the timer interrupt, which arrives with a period 5 milliseconds. Every 10 milliseconds, rescheduling occurs, which results in a third energy cluster. This cluster shows significant dispersion, pointing to the fact that the rescheduling algorithm used in the OS is dynamic. The fourth cluster, which has a low count, is attributed to calls to the function during which an actual context switch occurs. To extract the context switch energy from the histogram, the difference between the fourth and the first energy clusters needs to be calculated. This value is very close to the value obtained using the first approach.

Knowing that the second cluster should be attributed to the timer interrupt, Tan *et al.* (2002a) obtain the timer interrupt energy to be the difference between the second cluster and first. Similarly, the scheduling energy is the energy difference between the third cluster and first.

Signal handling energy needs to be extracted using another experiment. In this experiment, Tan *et al.* (2002a) reuse the same function. However, the main test program also sets up an alarm that generates a signal periodically. An alarm signal handler that does nothing (just return) is also established in the main test program. While the function is called repeatedly by the main test program, some invocations will be interrupted by the alarm. Knowing that this setup is the same as the previous set up except for signal handling, the extra cluster is attributed to signal handling.

8.5. SUMMARY

Pebble provides a new engineering trade-off for the construction of efficient component-based systems, using hardware memory management to enforce protection domain boundaries, and reducing the cross domain transfer time by synthesizing custom portal code. Pebble enhances flexibility by maintaining a private portal table for each domain. This table can be used to provide different implementations of system services, servers and portal interposition for each domain. Portal interposition allows running untrusted code in a robust sandbox with an acceptable overhead while using unsafe programming languages such as C.

Pebble architecture provides an efficient operating system that is easy to modify and debug, and hardware protection for running components written in any programming language. Components communicate via portals and run in user mode with interrupts enabled. Through the use of portals, which are specialized to the specific interrupt or communication they are to handle, Pebble is able to compile out run-time decisions and lead to better performance than other operating system implementations. In addition, a portal can be configured to save and restore only a subset of the machine state, depending on the calling conventions of, and the level of trust between, the client and server. The low interrupt latency provided by Pebble architecture makes it well suited for embedded applications.

OS energy characterization targets embedded operating systems that are monolithic and run on a single processor. This group of OSs includes Linux, C-Linux, C/OS, eCos, etc. Although the overall methodology should be applicable across multiple OSs, the details of test program generation is nevertheless implementation dependent. Tan *et al.* (2002a) used Linux OS as the target OS and leveraged the approach for C/OS. They focused only on the OS energy characterization. Though they have pointed out some possible uses of the energy characteristic data, systematically use of this information in a system-level software energy reduction framework is needed.

PROBLEMS

Learning Objectives

After completing this chapter you should be able to:

- demonstrate understanding of the operating systems for embedded applications;
- explain what the Inferno operating system is;
- demonstrate understanding of the Pebble component-based operating system;

- explain what the Pebble operating system for embedded applications is;
- discuss what is meant by embedded operating system energy analysis and macro-modeling.

Practice Problems

Problem 8.1: How is the communication between protection domains done?
Problem 8.2: What are the characteristics of the portal code?
Problem 8.3: What is the role of the privileged-mode nucleus?
Problem 8.4: What is the cost of transferring control from one protection domain to another?
Problem 8.5: How is the operating system implemented?
Problem 8.6: How is the control transferred between protection domains?
Problem 8.7: What is the portal manager?
Problem 8.8: What are the approaches to isolate components from one another?
Problem 8.9: What are the design goals of component-based systems?

Practice Problem Solutions

Problem 8.1:

All communication between protection domains is done by using interrupt handlers, named portals. Only if a portal exists between protection domain A and protection domain B can A invoke a service offered by B. Each protection domain has its own portal table. By restricting the set of portals available to a protection domain, threads in that domain are efficiently isolated from services to which they should not have access.

Problem 8.2:

Portals are the basis for flexibility and safety in Pebble, and the key to its high performance. Specialized, tamper-proof code can be generated for each portal, using a simple interface definition language. The portal code can be optimized for its portal, saving and restoring the minimum necessary state, or encapsulating and compiling out demultiplexing decisions and run-time checks.

Problem 8.3:

The privileged-mode nucleus is as small as possible. Most executions occur at the user level.

The privileged-mode nucleus is responsible for switching between protection domains, and it is the only part of the system that must be run with the interrupts disabled. By reducing the length of time the interrupts are disabled, the maximum interrupt latency is reduced.

In a perfect world, Pebble would include only one privileged-mode instruction, which would transfer control from one protection domain to the next. By minimizing the work done in privileged mode, the Pebble's designers reduce both the amount of privileged code and the time needed to perform essential privileged mode services.

Problem 8.4:

Each component is implemented by a separate protection domain. The cost of transferring control from one protection domain to another should be small enough that there is no performance-related reason to co-locate components.

Microkernel systems used coarse-grained user level servers, in part because the cost of transferring between protection domains was high. By keeping this cost low, Pebble enables the factoring of the operating system, and application, into smaller components with small performance penalty.

Problem 8.5:

The operating system is built from fine-grained, replaceable components, isolated through the use of hardware memory protection.

The functionality of the operating system is implemented by trusted user-level components. The components can be replaced, augmented, or layered. For example, Pebble does not handle scheduling decisions, and the user-replaceable scheduler is responsible for all scheduling and synchronization operations.

The architecture of Pebble is based on the availability of hardware memory protection, and it requires a memory management unit.

Problem 8.6:

Transferring control between protection domains is done by a generalization of hardware interrupt handling, that is, portal traversal. Portal code is generated dynamically and performs portal specifications.

Hardware interrupts, interprotection domain calls, and the Pebble equivalent of system calls are all handled by the portal mechanism. Pebble generates specialized code for each portal to improve run-time efficiency. The portal mechanism provides two important features: abstract communication facilities, which allow components to be isolated from their configuration, and per-connection code specialization, which enables the application of many otherwise unavailable optimizations.

Problem 8.7:

The portal manager is the operating system component responsible for instantiating and managing portals. The portal manager is privileged because it is the only component that is permitted to modify portal tables.

Problem 8.8:

The approaches to isolate the components from one another are as follows.

- to provide no protection between components;
- to provide software protection;
- to provide hardware protection.

Problem 8.9:

Component-based systems that use software protection schemes are typically written in a type-safe, byte-coded programming language, such as Java, and the Limbo programming language of the Inferno operating system. Components run in a single hardware protection domain, but the run-time environment implements, in effect, a software protection domain. These systems are designed to meet the following goals:

(1) to provide an architecture-independent distribution format for code;
(2) to ensure that resources, such as memory, are returned to the system when they are no longer needed;
(3) to ensure that the component does not view or modify data to which it has not been granted access.

9

Network Support for Embedded Applications

9.1. INTRODUCTION

Bluetooth enables seamless voice and data communication via short-range radio links, and allows users to connect a wide range of devices easily and quickly, without the need for cables, thus expanding communications capabilities for mobile computers, mobile phones, and other mobile devices. Considering a wide range of computing and communication devices such as PDAs, notebook computers, pagers, and cellular phones with different capabilities, Bluetooth provides a solution for access to information and personal communication by enabling collaboration between devices in proximity to each other where every device provides its inherent function based on its user interface, form factor, cost and power constraints. Furthermore, Bluetooth technology enables a vast number of new usage models for portable devices. The development of a short-range Radio Frequency (RF) solution enables the notebook computer to connect to different varieties of cellular phones and other notebook computers. The RF solution also removes many of the wires required for audio and data exchange in cellular handset.

The Bluetooth radio transmission uses a packet-switching protocol with a FHSS (Frequency Hopping Spread Spectrum). The hop frequency is 1600 hops per second, the frequency spectrum is divided into 79 hops of 1 MHz bandwidth each. The frequency hopping scheme is combined with fast

Wireless Sensor Network Designs A. Hać
© 2003 John Wiley & Sons, Ltd ISBN: 0-470-86736-1

ARQ (Automatic Repeat Request), CRC (Cyclic Redundancy Check) and FEC (Forward Error Correction). A binary radio frequency modulation and simple link layer protocols reduce the complexity and the costs of the radio chip.

Bluetooth provides a nominal data rate of 1 Mbit/s for a piconet. One piconet consists of one master and up to seven slaves. The master–slave principle is used to initiate and control the traffic between devices in a piconet. The master is responsible for defining and synchronizing the frequency hop pattern in its piconet. A single Bluetooth unit may send/receive at a maximum data rate of 721 kbit/s or a maximum of three voice channels of 64 kbit/s each, with continuous variable slope delta modulation (CVSD). Both a Synchronous Connection Oriented (SCO) link and an Asynchronous Connectionless (ACL) link for each master–slave pair are supported. Within the same Bluetooth radio range, separate and independent piconets may be formed. These may build up into scatternets to allow for a higher number of Bluetooth devices being active and/or for a higher aggregate bandwidth.

The essence of ubiquitous computing is the creation of environments saturated with computing and communication in an unobtrusive way. WWRF (Wireless World Research Forum) and ISTAG (Information Society Technologies Advisory Group) envisage a vast number of various intelligent devices embedded in the environment, sensing, monitoring and actuating the physical world, communicating with each other and with humans.

The main features of the IEEE 802.15.4 standard are network flexibility, low cost, and low power consumption. This standard is suitable for many applications in the home that require low-data-rate communications in an *ad hoc* self-organizing network.

Wireless sensor networks are used in a wide range of different applications where numerous sensor nodes are linked to monitor and report distributed event occurrences. In contrast to traditional communication networks, the single major resource constraint in sensor networks is power, due to the limited battery life of sensor devices. Data-centric methodologies can be used to solve this problem efficiently. In data-centric storage (DCS), data dissemination framework, all event data is stored by type at designated nodes in the network and can later be retrieved by distributed mobile access points in the network. Resilient Data-Centric Storage (R-DCS) is a method for achieving scalability and resilience by replicating data at strategic locations in the sensor network. This scheme leads to significant energy savings in reasonably large-sized networks and scales well with increasing node density and query rate. R-DCS realizes graceful performance degradation in the presence of clustered as well as isolated node failures, hence making the sensor net data robust.

9.2. BLUETOOTH ARCHITECTURE

Bluetooth enables combined usability models based on functions provided by different devices. In a connection between a computing device like a PDA (Personal Digital Assistant) and a communication device like a cellular phone, by using Bluetooth, and a second connection between the cellular phone and a cellular base station providing connectivity for both data and voice communication, the PDA maintains its function as a computing device and the telephone maintains its role as a communication device. Each device provides a specific function efficiently, yet its function is used separately, and the devices can be used independently of each other. When the devices are close to each other they provide a useful combined function. This function and connectivity model based on a combination of wireless access technologies each matched to different device capabilities and requirements, enables ubiquitous and pervasive wireless communication as in the following examples:

- In the three-in-one phone scenario the user can use the same phone in any place: at the office, the phone functions as an intercom; at home, it functions as a portable phone; and outdoors, it functions as a mobile phone.
- The user can use the e-mail while the notebook is still in the briefcase. When the notebook receives an e-mail, the user gets an alert on the mobile phone. The user can also browse all incoming e-mails and read selected e-mails in the mobile phone's window.
- The automatic background synchronization keeps the user information current. Automatic synchronization of data on the desktop, notebook, personal digital assistant (PDA), and mobile phone allows the user entering the office to have the address list and calendar in the notebook to be automatically updated to agree with the one in the desktop, or vice versa. The user can collect a business card on the phone and add it to the address list on the user's notebook PC.

The following system requirements are needed:

- To handle both voice and data, the protocol must support good quality real-time voice. Voice quality is important to both end-users who are accustomed to it, and for speech recognition engines whose accuracy depends on it.
- To be able to establish *ad hoc* connections. The dynamic nature of mobility makes it impossible to make any assumptions about the operating

environment. Bluetooth units must be able to detect other compatible units and establish connections to them. A single unit must be able to establish multiple connections in addition to accepting new connections while being connected. Ignoring a new connection requests while being connected is confusing to the user and deemed unacceptable, especially in supporting unconscious computing while retaining the ability to perform interactive operations.

- To withstand interference from other sources in an unlicensed band. The Bluetooth radio operates in the unlicensed 2.4 GHz band where many other RF radiators exist. The challenge is to avoid significant degradation in performance when other RF radiators, including other personal area networks nearby, are in operation.
- To be used worldwide. The challenge here is very regulatory in nature with many governments having their own set of restrictions on RF technology. And while the 2.4 GHz band is unlicensed through most parts of the world, it varies in range and offset in a number of different countries.
- To have a similar amount of protection compared as a cable. In addition to the radio's short-range nature and spread spectrum techniques, Bluetooth link protocols also provide authentication and privacy mechanisms.
- To have a small size to accommodate integration into a variety of devices. The Bluetooth radio module must be small enough to permit integration into portable devices. Wearable devices in particular, such as mobile phones, headsets, and smart badges have little space to spare for a radio module.
- To have negligible power consumption compared with the device in which the radio is used. Many Bluetooth devices will be battery powered, which requires that the integration of the Bluetooth radio should not significantly compromise the battery lifetime of the device.
- To allow for ubiquitous deployment of the technology. There is specification defining the radio, physical, link, and higher level protocols and services necessary to support the usage models.

Figure 9.1 outlines the application framework in the context of the radio and protocol stack. The radio is used to send and receive modulated bit streams. The Baseband (BB) protocol defines the timing, framing, packets, and flow control on the link. The Link Manager (LM) assumes responsibility for managing connection states, enforcing fairness among slaves, power management, and other management tasks. The Logical Link Control (LLC) handles multiplexing of higher level protocols, segmentation and reassembly of large packets, and device discovery. Audio data is mapped directly on to the

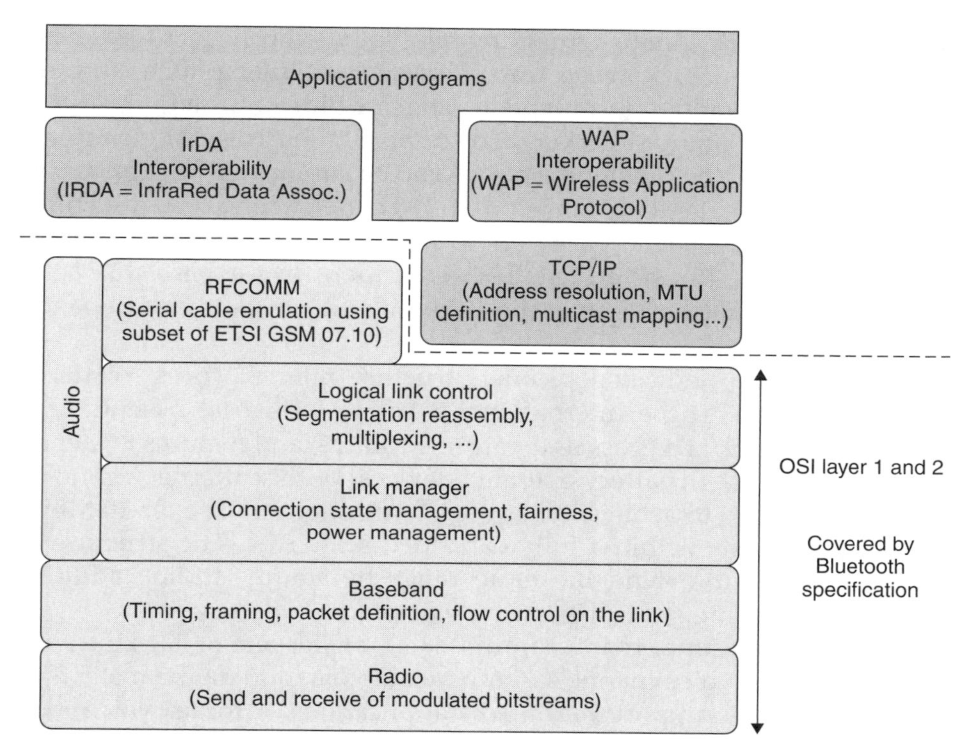

Figure 9.1 Application framework for Bluetooth.

baseband while audio control is layered above the logical link control. Above the data link layer, RFCOMM, which is an interface that allows an application to treat a Bluetooth link in a similar way as if it were communicating over a serial port, and network level protocols, provide different communication abstractions. Other parts of the Bluetooth specification discuss interoperability with other protocols and protocol stacks. Defining TCP/IP over Bluetooth requires that bridging, address resolution, MTU (Maximum Transmission Unit) definition, and multicast/broadcast mappings are solved. To accelerate the number of wireless-specific applications, Bluetooth allows interoperability with higher layer IrDA (Infrared Data Association) and WAP (Wireless Application Protocol) protocol stacks. For example, IrOBEX (Infrared Object Exchange) defines a transport-independent format and session protocol for object exchange and is used as the basis for a variety of applications from exchanging files and business cards to synchronizing address book and calendar schedules.

Bluetooth is specified and designed with emphasis on robustness and low cost. Its implementation is based on a high-performance, low cost, integrated radio transceiver. Bluetooth targets mobile users who need to establish a link, or small network, between their computer, cellular phone, and other peripherals. The required and nominal range of Bluetooth radio is set to 10 meters (with 0 dBm output power). To support other uses, for example the home environment, the Bluetooth chip set can be augmented with an external power amplifier to extend the range (up to 100 meters with a +20 dBm output power). Auxiliary baseband hardware to support, for example, four or more voice channels can also be added. These additions to the base chip set are fully compatible with the nominal specification and may be added depending on the application.

Bluetooth uses a dedicated piconet structure referred to as 'scatternet'. Bluetooth operates in the international 2.4 GHz Industrial, Scientific, and Medical (ISM) band, at a gross data rate of 1 Mbit/s, and features low energy consumption for use in battery operated devices. With scatternet technology, it has been possible to achieve an aggregate throughput of over 10 Mbits/s or 20 voice channels within a fully expanded scatternet. The structure also makes it possible to extend the radio range by simply adding additional Bluetooth units acting as bridges at strategic places.

A single unit can support a maximum data transfer rate of 721 kbits/s or a maximum of three voice channels. A mixture of voice and data transfer is also possible in order to support multimedia applications. A robust voice coding scheme with a rate of 64 kbits/s per voice channel is used. To sustain these transfer rates in busy radio environment, a packet-switching protocol with frequency hopping and advanced coding techniques is employed. Bluetooth features a graceful degradation of both voice and data transfer rates in busy RF environments.

In the Bluetooth network, all units are peer units with identical hardware and software interfaces distinguished by a unique 48-bit address. At the start of a connection, the initializing unit is temporarily assigned as a master. This assignment is valid only during this connection. The master initiates the connection and controls the traffic on the connection. Slaves are assigned a temporary 3-bit member address to reduce the number of addressing bits required for active communication.

The Bluetooth network supports both point-to-point and point-to-multipoint connections. A piconet is a network formed by a master and one or more slaves. Each piconet is defined by a different frequency-hopping channel. All units participating in the same piconet are synchronized to this channel.

To achieve the highest possible robustness for noisy radio environments, Bluetooth uses a packet-switching protocol based on a frequency hopping scheme with 1600 hops per second. The entire available frequency spectrum is used with 79 hops of 1 MHz bandwidth. This frequency hopping gives a reasonable bandwidth and the best interference immunity by utilizing the entire available spectrum of the open 2.4 GHz ISM band. Virtual channels are defined using pseudo-random-hop sequences.

The frequency hopping scheme is combined with fast Automatic Repeat Request (ARQ), Cyclic Redundancy Checks (CRC), and Forward Error Correction (FEC) for data. For voice a continuous variable slope delta modulation (CVSD) scheme is used. All of this results in a very robust link for both data and voice.

To save power and minimize radio interference problems, an RSSI (Received Signals Strength Indicator) with a 72 dB dynamic range is employed. The RSSI measures the signal received from different units and adapts the RF output power to the exact requirement in each instance. That is, with a mouse or headset, the output power can be limited to a 1-meter range, whereas a handset may need a range of 100 meters or more.

When first establishing a network or adding components to a piconet, the units must be identified. Units can be dynamically connected and disconnected from the piconet at any time. Two available options lead to connection times of typically 0.64 and 1.28 seconds, respectively. This applies when the unit address is known and not more than about 5 hours have elapsed since the previous connection. A unit does not need to be connected at all times since only a typical delay of under 1 second is required to start a transaction. Hence, when not in use, the unit can be in a sleep state (STANDBY) most of the time where only a Low Power Oscillator (LPO) is running. This is beneficial for battery operation.

Before any connections are made, all units are in standby mode. In this mode, an unconnected unit will only listen to messages every 1.28 seconds or 2.56 seconds depending on the selected option. Each time a unit wakes up, it will listen on one of 32 hop frequencies defined for this unit.

The connect procedure is initiated by one of the units, the master. A connection is made either by a PAGE message if the address is already known, or by the Inquiry message followed by a subsequent PAGE message if the address is unknown. In the initial PAGE state, the paging unit (which is the master) will send a train of 16 identical page messages on 16 different hop frequencies, defined for the unit to be paged (the slave). The train covers half the sequence of frequencies in which the slave can wake up. It is repeated 128 or 256 times (1.28 or 2.56 seconds) depending on the needs of the paged unit. If no response is received after this time, the master transmits a train of 16 identical page messages on the remaining 16 hop frequencies in the wake-up sequence. The maximum delay before the master reaches the slave is twice 1.28 seconds or 2.56 seconds, if a periodicity of 1.28 seconds was chosen for paging, and the maximum delay is 5.12 seconds with 2.56 seconds periodicity. A trade-off between access delay and power savings exists due to the available choices.

The hop frequencies in the first page train are based on the master's slave clock estimate. The train includes the estimated wake-up hop, and eight hops before and seven hops after this hop. As a result, the estimate can be ±7 hops in error and still the master reaches the slave with the first page train. Because the estimate is updated at each connection establishment, the acquisition delay is shorter when a shorter time has elapsed since the units were last connected. With a Low Power Oscillator (LPO) inaccuracy better than ±250 ppm, the first train is still valid after a lapse of at least 5-hours with no connection.

For a time period of at least 5 hours since the last connection, the average acquisition times are 0.64 s and 1.28 s, respectively. If the first train does not cover the slave's wake-up frequency, then the second train does, and the average acquisition delays are 1.92 s and 3.84 s.

The Inquiry message is typically used for finding public printers, fax machines, and similar equipment with an unknown address. The Inquiry message is very similar to the page message but may require one additional train period to collect all the responses.

If no data needs to be transmitted, the units may be put on Hold where only an internal timer is running. When units leave the Hold mode, data transfer can be restarted instantaneously. Units may thus remain connected, without data transfer, in a low power mode. Hold is typically used when connecting several piconets. It could also be used for units where data needs to be sent very infrequently and low power consumption is important. A typical application would be a room thermostat which may need to transfer data only once every minute.

Two more low-power modes are available, the Sniff mode and the Park mode. If the modes are listed in the increasing order of power efficiency, then the Sniff mode has the higher duty cycle, followed by the Hold mode with a lower duty cycle, and finishing with the Park mode with the lowest duty cycle.

Once a Bluetooth unit has been connected to a piconet, it may communicate by means of two link types between any two members of the piconet forming a master–slave pair. The two link types supported are: Synchronous Connection Oriented (SCO) link, and Asynchronous (or isochronous) Connectionless (ACL) Link.

Different link types may apply between different master–slave pairs of the same piconet and the link type may change arbitrarily during the session. The link type defines what type of packet can be used on a particular link.

On each link type, 16 different packet types can be used. The packets differ in function and data-bearing capabilities. For full duplex transmissions, a Time Division Duplex (TDD) scheme is used. Each packet is transmitted in a different hop channel than the previous packet.

An SCO link is a point-to-point, full-duplex link between the master and a slave. This link is established once by the master and kept alive until being released by the master. The SCO link is typically used for a voice connection. The master reserves the slots used for the SCO link on the channel.

The ACL link makes a momentary connection between the master and any of the slaves for the duration of one frame (master-to-slave slot and slave-to-master slot). No slots are reserved. The master can freely decide which slave to address and in which order. The member sub-address in the packet header determines the slave. A polling scheme is used to control the traffic from the slaves to the master. The link is intended for asynchronous or isochronous data. However, if the master uses this link to address the same slave at regular intervals, it becomes a synchronous link. The ACL link supports both symmetric and asymmetric modes. In addition, modes have been defined with or without FEC, and with or without CRC and ARQ.

9.3. BLUETOOTH INTEROPERABILITY WITH THE INTERNET AND QUALITY OF SERVICE

The main properties of Bluetooth wireless communication technology are:

- low-cost, low-power radio transceiver chip;
- implemented in hardware on small chips (0.5 square inches);
- the low price of Bluetooth module;
- a low nominal range of Bluetooth radio (10 meters) for saving battery power;
- extended range with external power amplifier (100 meters);
- operating in the globally available and unlicensed 2.4 GHz ISM (Industrial, Scientific and Medical) frequency band.

Bluetooth radio transmission uses a packet-switching protocol with a FHSS (Frequency Hopping Spread Spectrum). The hop frequency is 1600 hops per second, the frequency spectrum is divided into 79 hops of 1 MHz bandwidth each. The frequency hopping scheme is combined with fast ARQ (Automatic Repeat Request), CRC (Cyclic Redundancy Check) and FEC (Forward Error Correction). A binary radio frequency modulation and simple link layer protocols reduce the complexity and the costs of the radio chip.

Bluetooth provides a nominal data rate of 1 Mbit/s for a piconet. One piconet consists of one master and up to seven slaves. The master–slave principle is used to initiate and control the traffic between devices in a piconet. The master is responsible for defining and synchronizing the frequency hop pattern in its piconet. A single Bluetooth unit may send/receive at a maximum data rate of 721 kbit/s or a maximum of three voice channels of 64 kbit/s each, with CVSD (Continuous Variable Slope Delta Modulation). Both a Synchronous Connection Oriented (SCO) link and an Asynchronous Connectionless (ACL) link for each master–slave pair are supported. Within the same Bluetooth radio range, separate and independent piconets may be formed. These may build up into scatternets to allow for a higher number of Bluetooth devices being active and/or for a higher aggregate bandwidth.

The state machine for establishing Bluetooth network connections is presented in Figure 9.2. When not in use, the Bluetooth unit stays in a sleep state (Standby) with a Low Power Oscillator (LPO) still running. One unit (the

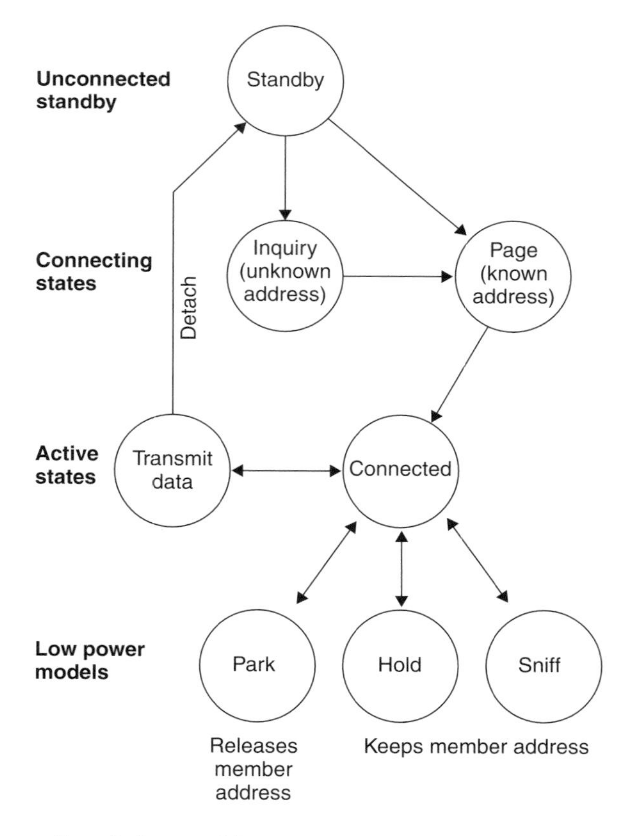

Figure 9.2 Connection state machine.

master) then initiates the connect procedure by sending either a PAGE or an Inquiry message. If the address is already known (e.g. most addresses in an office environment are already known for daily use), the PAGE message is sent. If the address is unknown (e.g. a Bluetooth unit tries to find a public printer with an unknown address), the unit changes to the Inquiry state trying to get response by possibly active units within radio distance.

Once connected, the unit is able to transmit and receive data. For saving battery power, three low power modes are available: Sniff, Hold, and Park (in increasing order of power efficiency):

- In the Sniff mode, both master and slave periodically sleep and sniff for certain time intervals which have been previously negotiated.
- The Hold mode can be used when no data needs to be transmitted for long time intervals (e.g. several minutes). An internal timer determines when the unit will be reactivated.
- The Park mode releases the 3-bit member address. This mode can be chosen when the unit does not participate in data transmission but wants to be synchronized with the frequency hopping.

Possible user situations and networking scenarios of the Bluetooth technology are as follows:

(1) In cordless desktop, which is the wire replacement, the Bluetooth radio provides a simple way of connecting all peripherals to the desktop PC without the need to use a cable.
(2) In the Internet bridge, which is the access to public networks, the Bluetooth radio bridges the gap between portable devices and a public network via an access point or gateway. For example, a cellular phone may be the gateway between Bluetooth devices and the network for accessing the Internet or the telephone network. Connecting a notebook to the Internet, or the Instant postcard, by connecting a camera to the mobile phone, and sending photos or video clips to remote places, are possible applications.
(3) In the three-in-one phone, where a mobile phone with Bluetooth device may work in three modes: (i) at home it works as a portable phone, (ii) in the office it works as an intercom, and (iii) otherwise it works as a cellular phone.
(4) In the interactive conference, which constitutes personal *ad-hoc* networking, a set of mobile hosts forms a wireless network without any additional networking hardware or cable support.

In a broad or abstract sense, the scenarios for the use of Bluetooth are categorized as follows:

- In the office environment, Bluetooth devices and technology are used in professional networking and communication services to bridge the gap between portable devices and a sophisticated (possibly wired) backbone of additional equipment and services.
- In the home environment, Bluetooth technology is used to connect a variety of different devices without cable, for example, cordless desktop, portable phone, remote control, etc.
- In public environments, Bluetooth capable user devices (e.g. a PDA) establish access to local information services in public areas like an airport, railway station, etc.
- In the location independent *ad hoc* networks, the Bluetooth technology is used to connect devices independently and out of range of the fixed networking equipment.

The application framework of Bluetooth achieves interoperability with IrDA (Infrared Data Association) and WAP (Wireless Application Protocol) and other application programs that use Bluetooth technology and protocols.

The interoperability of Bluetooth protocols and protocol interfaces with the Internet protocol family (IP, TCP, UDP) is addressed in the application framework of Bluetooth as shown in Figure 9.1. The WAP approach uses TCP/UDP and IP as one option to get access to Internet based services as WWW (World Wide Web). Furthermore, the TCP/UDP/IP protocol stack allows a variety of applications with a high degree of flexibility to operate on Bluetooth-capable devices over Bluetooth wireless technology (e.g. notebook using Bluetooth Internet bridge with access to IP-based LAN (Local Area Network) and access to all IP-based services).

To achieve Bluetooth interoperability with the Internet protocol family, the TCP/IP over Bluetooth requires that bridging, address resolution, MTU (Maximum Transmission Unit) definition, and multicast/broadcast mappings are solved. For instance, a Bluetooth radio chip in a notebook that provides access to the Internet via a mobile phone. As Internet services are based on the TCP/IP protocol stack, the notebook has to implement this protocol stack on top of Bluetooth. Typically, IP addresses are assigned to hosts connected to the Internet either by globally unique addresses or dynamically chosen local addresses. This leads to the question of does the notebook have its own IP address and how can this IP address be mapped to the Bluetooth's 3-bit address assignment. Aspects of addressing of foreign IP addresses have to be considered. If the notebook does not have its own IP address, local address assignment and address resolution have to be defined.

Another example is the connection of Bluetooth units to selected Internet services via LAN access points. Let us consider a railway station with several LAN access points providing access to Internet services like city information, train route scheduling, etc. Once connected to a LAN access point, the passenger may walk along the railway platform leaving the base station's radio distance of 10 meters. The passenger would appreciate a continuous service, possibly provided by several access points along the platform. Here, the question arises, how the handoff of Bluetooth devices between piconets will be managed from the view of the network protocol IP and higher layer protocols (e.g. the connection set-up between slave and master Bluetooth devices requires a control message exchange, which may require a maximum of 2.56 s or 5.12 s).

The aspects of addressing foreign IP addresses have to be defined for IP over Bluetooth. A possible solution to this is the adaptation of Mobile IP with respect to Bluetooth environments. IP mobility support has been defined in RFC (Request For Comment) 2002 by the mobile IP working group of the Internet Engineering Task Force (IETF).

A network providing mobile IP support for its roaming portable computers has to establish an entity called 'home agent', whereas networks providing access to the Internet for portables with different network addresses than its own, have to establish an entity called foreign agent.

On arrival at a new network, a mobile host contacts the local foreign agent, which supplies it with a care-of address, which may be the address of the foreign agent itself. Then the mobile's home agent is informed that all IP datagrams destined to the mobile host must be forwarded (tunneled) to the new care-of address in order to reach it.

A LAN which offers Bluetooth LAN access points and accepts communication with foreign Bluetooth devices may use concepts of mobile IP for address assignment and providing access to local IP services.

Cellular IP specifies a protocol that makes routing of regular IP datagrams to moving mobile hosts in a local network possible. Cellular IP provides local mobility and handoff support for frequently moving hosts, which means that mobile hosts can migrate inside a cellular IP network with little disturbance to active data flows. Cellular IP is only intended for local area networks and metropolitan area networks. Mobile IP is used between different cellular IP networks.

Mobile hosts connecting to a cellular IP network are able to retain their IP address. Cellular IP makes it possible to route IP packets to that IP address regardless of its location in the cellular IP network and without being influenced by a device's foreign IP address. Thus, hosts inside the cellular IP network are identified by their IP address, but these IP addresses have no location significance. Hosts outside the cellular IP network do not need any changes and they are unaware of the mobile host's location.

Cellular IP was specifically developed for supporting frequently moving hosts, but can also support rarely moving hosts and even static hosts. The concepts of cellular IP may be used in a Bluetooth environment with its low radio range of 10 meters.

The applications may use Quality of Service functions of the IP protocol family for multimedia video/audio streaming over WWW/IP, IP telephony, etc.

The modification of the Resource Reservation Protocol (RSVP) enables resource reservation in wireless networks. The idea is to set up passive reservations to neighboring cells in a wireless architecture. These passive reservations do not consume bandwidth, but they are used to set up reservation states. If the mobile moves to an adjacent cell, the new reservation can be established much faster by using the previously set up reservation states. This concept may work for Bluetooth multimedia applications.

A wireless subnetwork within a heterogeneous end-to-end communication path may drastically degrade the performance of end-to-end protocols like TCP. The error recovery and congestion control mechanisms are not appropriate to cope with bit errors on wireless links, and communication pauses or increased delays during handoff.

9.4. IMPLEMENTATION ISSUES IN BLUETOOTH-BASED WIRELESS SENSOR NETWORKS

Wireless sensor networks use small devices equipped with sensors, microprocessors and wireless communication interfaces. Different applications, ranging from personal health care to environmental monitoring and military applications, are used in such networks. Various wireless technologies, like simple RF, Bluetooth, UWB (Ultra Wide Band) or infrared can be used for communication between sensors.

Various sensors are used as part of different devices (temperature sensors in home or car heating system, smoke alarms, etc.) or as stand-alone devices connected to a network, usually to monitor industrial processes, equipment or installations.

The advancements in MEMS (Micro-Electrical-Mechanical Systems) technology, wireless communications, and electric components, have enabled development of small, low-power and low-cost devices, called smart sensor nodes, capable of performing various sensing tasks, processing data and communicating over wireless connections. Such devices, when organized into a network, present a powerful platform that can be used in many applications, like health monitoring, security systems, detection of chemical agents in air and water, etc.

Wireless sensor networks comprise a number of small devices equipped with a sensing unit, microprocessor, wireless communication interface and power source. In contrast to the traditional sensor networks that are carefully planned and deployed, wireless-sensor networks can be deployed in an *ad-hoc* manner. This deployment requires communication protocols that are able to organize the network automatically, without the need for human intervention.

Beside self-organization capability, another important feature of wireless sensor networks is collaboration of network nodes during the execution of the task. In contrast to the traditional sensor networks where all sensor data is gathered at a server and then analyzed and fused, data processing and fusion is performed by the smart nodes themselves. Each node processes raw measurement data in order to decrease amount of data sent over wireless links and then forwards only relevant parts to nodes responsible for data fusion.

The data-centric nature of the network is yet another specific characteristic of wireless sensor networks. As deployment of smart sensor nodes is not planned in advance and the position of nodes in the field is not determined, it is possible that some sensor nodes are placed such that they either cannot perform the required measurement or the probability of error is high. This is why a redundant number of smart nodes observing the same phenomenon is deployed in the field. These nodes then communicate, collaborate and share data, thus ensuring more accurate results. Each sensor observes its own view of the phenomenon, and when the views from a number of sensors are combined, a better picture of the phenomenon is compiled. Thus, it is more reasonable for a user to send a data request to all sensors monitoring the phenomenon than to send it to one specific sensor node. Using a multicast routing protocol to send messages to all relevant nodes requires a unique addressing scheme in the network. However, due to the sheer number of sensors and user requirements (the user needs information about the phenomenon as a whole, or does not need information about the phenomenon from a particular sensor), the data-centric approach is used where sensors are designated using a description of data they can provide instead of using a unique identifier. Messages are directed to nodes using routing protocols that can find the route based on the data description contained in the message.

Power efficiency is one of the main requirements for all protocols and algorithms used in sensor networks. As power resources of each node are limited, and required lifetime for many scenarios is measured in months or even years, it is of paramount importance to design a system in such a way as to ensure power savings whenever possible.

From the user point of view, querying and tasking are two main services provided by wireless sensor networks. Queries are used when the user requires only the current value of the observed phenomenon. As wireless sensor networks are data-centric networks, the user does not query a specific node for the information it might provide, but defines required data (type, location, accuracy, time, etc.) and requests it from all nodes that can provide the answer. For example, a user can request the temperature in a specific region, or needs to know the location of all sensors where chemical agents are present and their level is above a certain threshold.

To execute a task is a more complex operation and is used when a phenomenon has to be observed over a longer period of time. For example, a user can ask a sensor network to detect a specific type of vehicle in the area and monitor its movement. To execute the task, different types of sensors have to collaborate: seismic to detect motion, video and audio to detect type of vehicle, etc. Information about the vehicle trajectory is forwarded to the user. Both queries and tasks are injected into the network by the gateway which also collects the replies and forwards them to the users.

Smart sensor nodes scattered in the field collect data and send it to users via a gateway using multiple hop routes as shown in Figure 9.3.

The main functions of a gateway are:

- Communication with the sensor network, where short-range wireless communication is used (Bluetooth, UWB, RF, IR, etc.) to provide functions

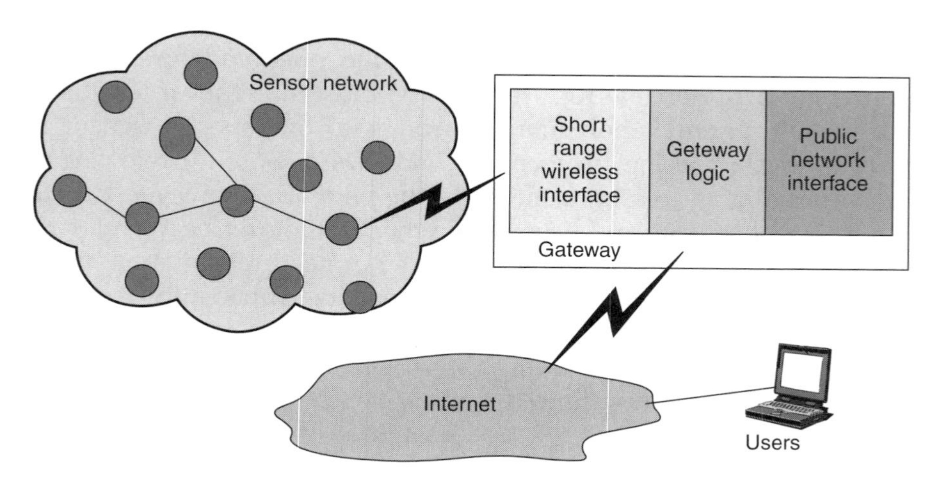

Figure 9.3 A wireless sensor network.

like discovery of smart sensor nodes, generic methods for sending and receiving data to and from sensors, routing, etc.;

- Gateway logic, which controls gateway interfaces and data flow to and from a sensor network. It also provides an abstraction level with the API (Application Programming Interface) that describes the existing sensors and their characteristics. Gateway logic provides functions for uniform access to sensors regardless of their type, location or network topology, injects queries and tasks and collect replies;
- Communication with the users occurs through a gateway. The gateway communicates with the users and the other sensor networks over the Internet, wide area networks, satellite or a short-range communication technology.

A hierarchy of gateways can be built to connect gateways to a backbone and then to provide a higher-level gateway that is used as a bridge to the other networks and users.

The applications of wireless sensor networks include:

- *Health monitoring*. Wireless sensor networks can be used in various ways to improve or enhance health-care services. Monitoring of patients, health diagnostics, drug administration in hospitals, telemonitoring of human physiological data, and tracking and monitoring doctors and patients inside a hospital, are some of the possible scenarios.
  Various sensors (blood pressure, heart monitoring, etc.) can be attached to the patient's body to collect physiological data that can be either stored locally (on a PDA or home PC) or forwarded directly to the hospital server or to the physician. There are several advantages of such monitoring: it is more comfortable for patients, doctors can have 24-hour access to patients and can better understand the patient's condition, and the incurred expenses are lower than when such tests are performed at a hospital. Wearable sensors can also be used to track patients and doctors in the hospital or to monitor and detect behavior and health condition of elderly persons and children.
- *Environmental monitoring*. Fire detection, water pollution monitoring, tracking movements of birds, animals or insects, detection of chemical and biological agents are some of the examples of environmental applications of wireless sensor networks.
  For example, numerous smart sensor nodes with temperature sensors on board can be dropped from an aircraft over a remote forest. After a successful landing, these devices will self-organize the network and will

monitor the temperature profile in the forest. As soon as a fire starts, that information, along with the location of the fire, is transferred to the command center that can act before the fire spreads to cover a large area.

- *Military and security*. Military applications vary from monitoring soldiers in the field, to tracking vehicles or enemy movement.
  - Sensors attached to soldiers, vehicles and equipment can gather information about their condition and location to help planning activities on the battlefield.
  - In the case of nuclear or biological attacks, sensor fields can gather valuable information about the intensity, radiation levels or type of chemical agents without exposing people to danger.
  - Seismic, acoustic and video sensors can be deployed to monitor critical terrain and approach routes, or reconnaissance of enemy terrain and forces can be carried out.
- *Industrial safety*. Similar to personal health care scenarios, wireless sensor networks can be used to monitor buildings, bridges or highways. In such scenarios, thousands of various sensors are deployed in and around monitored objects, and relevant information is gathered and analyzed in order to assess condition of an object after a natural or other disaster. Similarly, sensors can be used to monitor the status of different machines in factories, along with air pollution or fire monitoring.
- *Other applications*. Home automation, smart environments, environmental control in office spaces, detecting car thefts, vehicle monitoring and tracking, and interactive toys are examples of other possible applications.

Research issues are numerous and range from hardware issues to design of efficient communication protocols and distributed data-processing algorithms. All solutions have to be power conscious as well as fault tolerant, scalable, robust, with low production cost, etc.

Wireless sensor networks require low-power, low-cost devices that accommodate a powerful processor, a sensing unit, wireless communication interface, and power source in a robust and tiny package. These devices have to work autonomously, to require no maintenance, and to adapt to environment. For example, the MEMS technology enables production of very small sensing units with low power consumption.

Physical layer issues range from power efficient transceiver design to modulation schemes.

MAC layer protocols have to support self-organization of a distributed network and to ensure fair medium access and collision avoidance. Different power modes have to be supported to enable nodes to save energy resources when possible, but without affecting network performance. Changes in network topology due to node malfunction or mobility have to be taken into account and dealt with automatically.

On the network level, routing protocols are required for dissemination of user queries and tasks. Since a data-centric approach is used, the existing routing protocols for special networks cannot be used and new solutions, capable of routing messages based on data attributes, are required.

Another important requirement for routing protocols in sensor networks is collaboration with data-aggregation algorithms. Data aggregation is required to avoid network implosion, which may occur when many nodes answer the same query and send replies towards the gateway, and overlap problems in data-centric routing.

Based on predefined methods, responsible nodes analyze gathered data and combine it into a set of meaningful information that is forwarded to the user. Data aggregation reduces the amount of network load while preserving validity and amount of information. For certain applications it can be important to know the source of information (position) and in such cases that information has to be forwarded as well. An example of a data aggregation process is shown in Figure 9.4.

At the application level, a framework for attribute-based query definition, task building and their execution at each node, as well as collection of replies, is required.

Sensor network management protocol has to support control of individual nodes, network configuration updates, location information data exchange, network clustering, and data aggregation rules. The sensor network gateway has to provide tools and functions for presentation of network topology, services, and characteristics to the user, and to connect the network to other networks and users.

Low-cost, low-power Bluetooth modules meet the requirements of wireless sensor networks. *Ad-hoc* connection establishment capability, reasonable

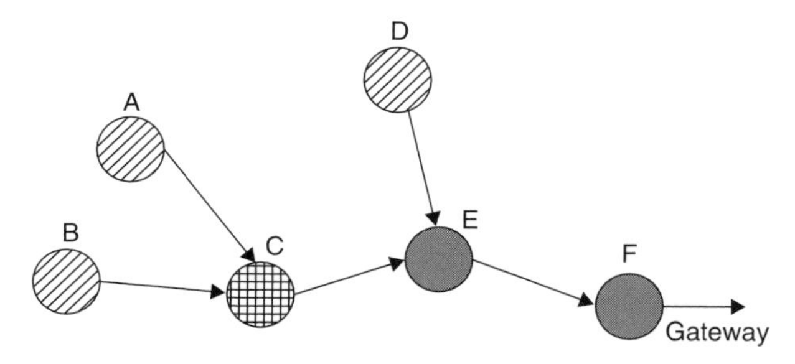

Figure 9.4 An example of data aggregation.

throughput (up to 721 kbit/s in uplink and 56 kbit/s downlink), usage of frequency-hopping schemes with TDD (Time Division Duplex) to minimize the impact of interference in the ISM band of 2.4 GHz, existence of different power-saving modes along with its availability and standardized specification are the main advantages of Bluetooth over other wireless technologies. However, there are several issues, like connection establishment delay and networking functionality, that have to be solved before Bluetooth can be deployed in large sensor networks.

A Bluetooth device has to be a member of a piconet to be able to communicate with other devices. A piconet is a collection of up to eight devices that frequency-hop together. Each piconet has one master, usually the device that initiated establishment of the piconet, and up to seven slave devices. The master's Bluetooth address is used for definition of the frequency-hopping sequence. Slave devices use the master's clock so as to synchronize their clocks so as to be able to hop simultaneously.

When a device wants to establish a piconet, it has to perform Inquiry to discover other Bluetooth devices, which have to perform inquiry scanning at the same time, within the range. Inquiry procedure is defined so as to ensure that two devices will, after some time, visit the same frequency at the same time. When that happens, required information is exchanged (Bluetooth address, and clock of the device that will be master of the piconet) and devices can use paging procedure to establish connection. Time required for communication establishment can be rather lengthy, taking on average around 5 seconds (minimum is 0.00375 seconds and maximum is 12.8 seconds to 33.28 seconds). This delay can be the limiting factor for applications that require instant connection establishment.

When more than seven devices need to communicate, then one or more devices can be put into the Park mode. Bluetooth defines three low power modes: Sniff, Hold, and Park. When a device is in the park mode then it disassociates from the piconet, but still maintains timing synchronization with the piconet. The master of the piconet periodically broadcasts beacons inviting the slave to rejoin the piconet or to allow the slave to request to rejoin. The slave can rejoin the piconet only if there are fewer than seven slaves already in the piconet. Otherwise, the master has to park one of the active slaves first. All these actions cause delays and for some applications this can be unacceptable, for example, a process control that requires immediate response from the command center.

Another option is to build a scatternet. A scatternet consists of several piconets connected by devices participating in multiple piconets. These devices can be slaves in all piconets, or master in one piconet and slave in other piconets. Using scatternets, a higher throughput is available and multi-hop connections between devices in different piconets are possible. However, hardware does not support this functionality for several reasons:

- Bluetooth specification gives no way for a slave to demand park, hold or sniff mode, but can only request it from the master so there is no guarantee that the slave will be allowed to leave one piconet and join the other
- Each time a device switches between piconets it might lose up to two slots for communication due to difference in the piconets' clocks
- Scheduling switches between piconets to maintain communication links with devices uninterrupted, is very difficult.

A possible solution, before scatternet is supported by Bluetooth hardware, is to perform switching between piconets on the application level.

This scatternet building mechanism assumes that all nodes in the network are peer nodes. A mitigating circumstance for sensor networks is that a gateway can be used to direct establishment of the scatternet. In this centralized approach it is possible to generate close-to-optimal network topology and to solve scheduling, bandwidth allocation and routing problems.

The additional requests for scatternets in sensor networks complicate scatternet building. Various sensor types produce different amount of data (video sensor and temperature sensor, for example). If too many high-output sensors are connected to the same branch in the scatternet it can cause link congestion or buffer overflow in intermediate nodes. Hence, parameters like number of sensors, amount of data generated by sensor per measurement, and buffer size, have to be taken into account during building scatternet topology.

The main goal of an implementation is to build a hardware platform and generic software solutions that can serve as the basis and act as a testbed for wireless sensor network protocols. The implementation supports dedicated deployment of sensors, where sensor characteristics are automatically collected and presented in a structured way, and there are no limits in terms of sensor type and number of sensors, generic functions for querying sensors and collecting replies, and the basis for attribute-based routing is provided. Software architecture is designed in such a way that new protocols can be added easily without affecting current functionality.

The sensor network consists of several smart sensor nodes and a gateway. Each smart node can have several sensors and is equipped with a microcontroller and a Bluetooth radio module. Gateway has two wireless interfaces: Bluetooth for communication with sensors, and the network protocol for communication with users. Gateway and smart nodes are members of one piconet, thus a maximum of seven smart nodes can exist simultaneously in the network.

The smart sensor node comprises of three functional blocks: sensing, data processing, and communication. One or more sensors can be attached to the microcontroller. Temperature, heart monitor, and smart fabric sensors can be used.

A microcontroller is responsible for the smart sensor node logic. An application is developed that gathers data from sensors and controls the Bluetooth module and communication with the gateway. It also stores sensors' profiles and data.

The gateway plays the role of the piconet's master in the sensor network. It controls establishment of the network, gathers information about the existing smart sensor nodes and sensors attached to them, and provides access to them.

A set of core services is developed to take care of common procedures and services required by all layers. Logging, scheduling, event subscription, and services required for automatic application starting and restarting, are supported.

The sensor network communication interface handles communication with sensors and can also control connections to a mobile phone, or to local users that are using Bluetooth to access the network. Depending on the available hardware resources, more than one sensor network can be attached to the gateway by using Bluetooth or any other communication interface.

The sensor network abstraction layer and its API are independent of the underlying communication technology and provide information about and access to all available sensors in the network.

Smart sensor node discovery is the first procedure that is executed upon the gateway initialization. Its goal is to discover all sensor nodes in the area and to build a list of sensor's characteristics and network topology. Afterwards, it is executed periodically to facilitate addition of new or removal of the existing sensors.

When the gateway is initialized, it performs the Bluetooth Inquiry procedure. When a new Bluetooth device is discovered, its major and minor device classes are checked (these parameters are obtained along with the Bluetooth address and other parameters). These parameters are set by each smart node to define type of the device and types of attached sensors. If discovered device is not smart node, it is discarded. Otherwise, the SDP (Service Discovery Protocol) is invoked and the service database of the discovered smart node is searched for sensor services and for the serial port profile connection parameters. Once a connection string is obtained from the device, the Bluetooth link is established and data exchange with smart node can start.

Using two types of message, a gateway can request either list of sensors attached to the particular smart sensor node, or sensor data. Reply messages have a very flexible structure and can relay information about any number or type of sensor. Sensor profiles are defined by sensor vendors. As all relevant information about a sensor is contained in its profile (sensor type, measuring unit, accuracy, manufacturer, calibration date, etc.), gateway can automatically build knowledge of a sensor network and its characteristics, i.e. sensors can be deployed in an *ad-hoc* fashion.

Bluetooth links are maintained as long as the gateway and smart sensor node are in the range. When an event happens on the sensor side, the sensor can send information about it to the gateway. However, the power resources are used on maintenance of communication link, and it is not possible to have more than seven smart nodes in one piconet.

Gateway's abstraction layer uses sensor profiles to create a list of objects that represents each sensor in the network. Each object provides methods that enable sending and receiving of data to and from a sensor. Specifics of actual data transmission are hidden from users.

Applications can access sensor objects by using queries that describe the data. By comparing data description in the query and profiles associated with each sensor, the gateway determines which sensors can have an answer, and sends data requests to them using methods provided by each object.

If larger network and scatternet topology are used, then similar functionality needs to be provided by each master in the scatternet, and appropriate attribute-based routing solutions are required, to disseminate queries. Data aggregation rules also need to be defined.

Bluetooth is a possible choice for data communication in sensor networks. Good throughput, low-power, low-cost, standardized specification and hardware availability are Bluetooth advantages, while slow connection establishment and lack of scatternet support are deficiencies.

9.5. LOW-RATE WIRELESS PERSONAL AREA NETWORKS

Various in-home applications can be classified as Internet connectivity, multi-PC connectivity, audio/video networking, home automation, energy conservation, and security. They are characterized by different requirements for bandwidth, cost, and installation procedure. With the growth of the Internet, the focus is on meeting the requirement for shared high-speed connectivity.

Other applications, such as home automation, security, and gaming, have relaxed throughput requirements. These applications cannot handle the complexity of large protocol stacks that impact power consumption and utilize many computational resources.

As example, a temperature sensor at a window may need to report its temperature only a few times per hour, be inconspicuous, and have a very low selling price. This application can use a low-throughput, low-cost wireless communications link. The use of wires (for communication or power) is impractical because of the use of a window. Also, the wired installation cost would exceed by several times the cost of the sensor. In this case, extremely low power consumption is needed, since frequent battery replacement is impractical. Both IEEE 802.11 and Bluetooth devices require battery replacement several times per year, which is impractical if many windows are involved in the application.

The IEEE 802.15.4 standard defines a Low-Rate Wireless Personal Area Network (LR-WPAN) which has ultra-low complexity, cost, and power for low data-rate wireless connectivity among inexpensive fixed, portable, and moving devices. The IEEE 802.15.4 standard defines the physical (PHY) layer and Media Access Control (MAC) layer specifications.

IEEE 802.15.4 is useful in a wide variety of applications, including industrial control and monitoring; public safety, including sensing and location determination at disaster sites; automotive sensing, such as tire pressure monitoring; smart badges and tags; and precision agriculture, such as the sensing of soil moisture, pesticide, herbicide, and pH levels. In home automation and networking applications, there are several possible market sectors: PC peripherals, including wireless mice, keyboards, joysticks, low-end PDAs, and games; consumer electronics, including radios, televisions, VCRs (Video Cassette Recorders), CDs, DVDs, remote controls, and a truly universal remote control to control them; home automation, including Heating, Ventilation, and Air Conditioning (HVAC), security, lighting, and the control of objects such as curtains, windows, doors, and locks; health monitoring, including sensors, monitors, and diagnostics; and toys and games, including PC-enhanced toys and interactive gaming between individuals and groups. The maximum required data rate for these applications ranges from 115.2 kb/s for PC peripherals to less than 10 kb/s for home automation. The consumer electronics message latency ranges from approximately 15 milliseconds for PC peripherals to 100 milliseconds or more for home automation applications.

The IEEE 802.15.4 standard encompasses the layers up to and including portions of the Data Link Layer (DLL). Higher-layer protocols are at the discretion of the individual applications utilized in an in-home network environment.

The network layer is responsible for topology construction and maintenance, naming and binding services, which incorporate the necessary tasks of addressing, routing, and security. These services are challenging to implement in a wireless in-home network because of the importance of energy conservation. A network layer implementation built on the energy conscious IEEE 802.15.4 standard should also conserve energy. Network layers built on the standard are self-organizing and self-maintaining, to minimize total cost to the consumer.

The IEEE 802.15.4 standard supports multiple network topologies, including both star and peer-to-peer networks. The topology depends on the design where applications, such as PC peripherals, may require the low-latency connection of the star network, while others, such as perimeter security, may require the large area coverage of peer-to-peer networking. Multiple address types, including both physical (i.e. 64-bit IEEE) and short (i.e. 8-bit network-assigned) are provided.

The IEEE 802 splits the DLL into two sublayers, the MAC and the Logical Link Control (LLC). The LLC is standardized in 802.2 and is common among the 802 standards such as 802.3, 802.11, and 802.15.1, while the MAC sublayer is closer to the hardware and may vary with the physical layer implementation. Figure 9.5 shows how IEEE 802.15.4 fits into the Open Systems Interconnection (OSI) reference model. The IEEE 802.15.4 MAC provides services to an IEEE 802.2 type I LLC through the Service-Specific Convergence Sublayer (SSCS), or a proprietary LLC can access the MAC services directly without going through the SSCS. The SSCS ensures compatibility between different LLC sublayers and allows the MAC to be accessed through a single set of access points. By using this model, the 802.15.4 MAC provides features not utilized by 802.2, and therefore allows more complex network topologies.

The features of the IEEE 802.15.4 MAC are association and disassociation, acknowledged frame delivery, channel access mechanism, frame validation, guaranteed time-slot management, and beacon management. The MAC sublayer provides two services to higher layers that can be accessed through two Service Access Points (SAPs). The MAC data service is accessed through

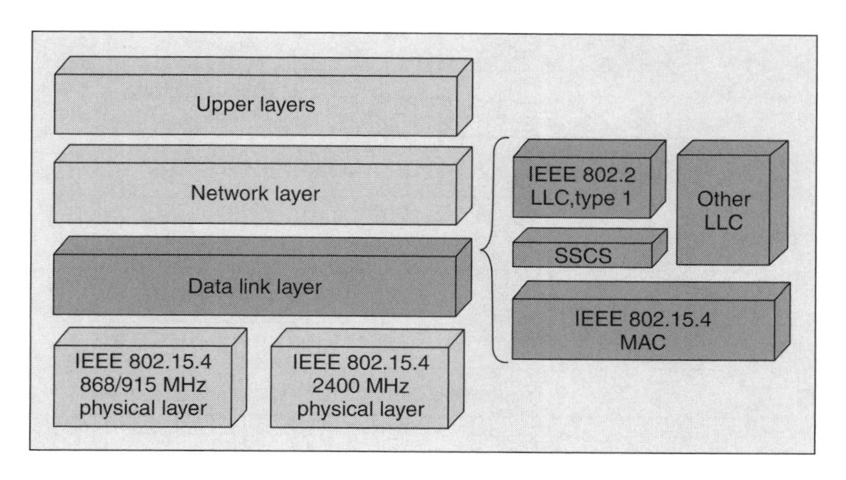

Figure 9.5 The IEEE 802.15.4 and OSI layered network model.

the MAC common part sublayer (MCPS-SAP), and the MAC management services are accessed through the MAC Layer Management Entity (MLME-SAP). These two services provide an interface between the SSCS or another LLC and the physical layer.

The MAC management service has 26 primitives. Compared to the IEEE 802.15.1 (i.e. Bluetooth), which has about 131 primitives and 32 events, the IEEE 802.15.4 MAC has very low complexity, making it suitable for its intended low-end applications, but at the cost of a smaller feature set than IEEE 802.15.1 (for instance, IEEE 802.15.4 does not support synchronous voice links).

The MAC frame structure is very flexible to accommodate the needs of different applications and network topologies while maintaining a simple protocol. The general format of a MAC frame is shown in Figure 9.6. The MAC frame is called the MAC Protocol Data Unit (MPDU) and is composed of the MAC Header (MHR), MAC Service Data Unit (MSDU), and MAC Footer (MFR). The first field of the MAC header is the frame control field. It indicates the type of MAC frame being transmitted, specifies the format of the address field, and controls the acknowledgment.

The frame control field specifies how the rest of the frame is built and what it contains. The size of the address field may vary between 0 and 20 bytes. For instance, a data frame may contain both source and destination

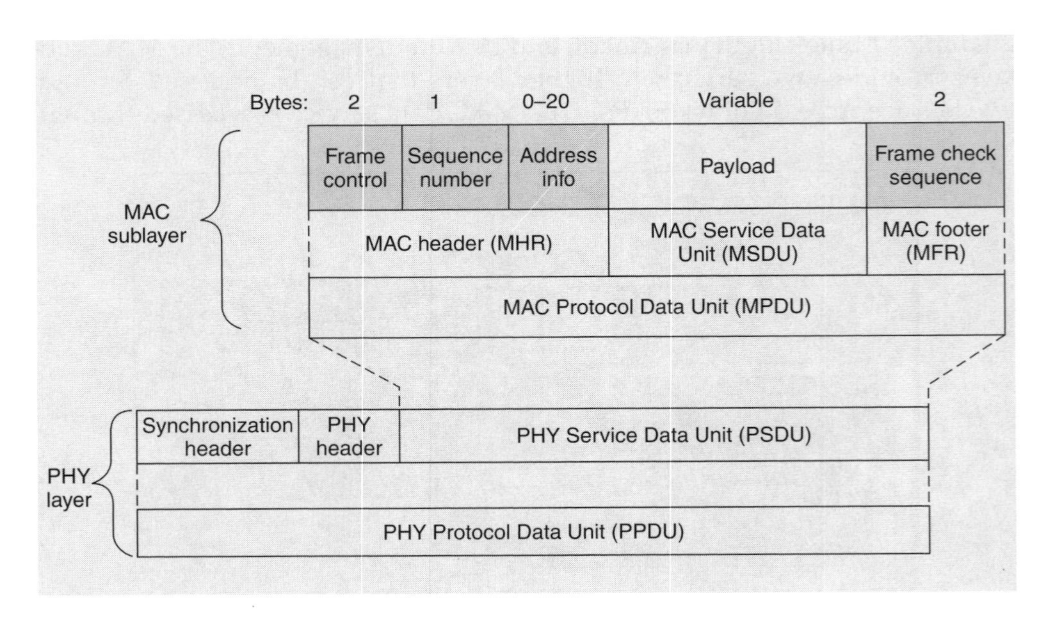

Figure 9.6 The MAC frame format.

information, while the return acknowledgment frame does not contain an address information. On the other hand, a beacon frame may only contain source address information. In addition, short 8-bit device addresses, or 64-bit IEEE device addresses, may be used. This flexible structure helps to increase the efficiency of the protocol by keeping the packets short.

The payload field has variable length; however, the complete MAC frame may not exceed 127 bytes. The data contained in the payload is dependent on the frame type. The IEEE 802.15.4 MAC has four different frame types: the beacon frame, data frame, acknowledgment frame, and MAC command frame. Only the data and beacon frames contain information sent by higher layers. The acknowledgment and MAC command frames originate in the MAC and are used for MAC peer-to-peer communication.

Other fields in a MAC frame are the sequence number and Frame Check Sequence (FCS). The sequence number in the MAC header matches the acknowledgment frame with the previous transmission. The transaction is considered successful only when the acknowledgment frame contains the same sequence number as the previously transmitted frame. The FCS helps verify the integrity of the MAC frame. The FCS in an IEEE 802.15.4 MAC frame is a 16-bit International Telecommunication Union – Telecommunication Standardization Sector (ITU-T) Cyclic Redundancy Check (CRC).

Some applications may require dedicated bandwidth to achieve low latencies. To accomplish these low latencies, the IEEE 802.15.4 LR-WPAN can operate in an optional superframe mode. In a superframe, a dedicated network coordinator, called the PAN coordinator, transmits superframe beacons at predetermined intervals. These intervals can be as short as 15 milliseconds or as long as 245 seconds. The time between two beacons is divided into 16 equal time-slots independent of the duration of the superframe. A device can transmit at any time during a slot, but must complete its transaction before the next superframe beacon. The channel access in the time-slots is contention based; however, the PAN coordinator may assign time-slots to a single device requiring dedicated bandwidth or low-latency transmissions. These assigned time-slots are called Guaranteed Time-Slots (GTS) and together form a contention-free period located immediately before the next beacon. The size of the contention-free period may vary depending on demand by the associated network devices; when GTS are employed, all devices must complete their contention-based transactions before the contention-free period begins. The beginning of the contention-free period and duration of the superframe are communicated to the attached network devices by the PAN coordinator in its beacon.

Depending on the network configuration, an LR-WPAN may use one of two channel-access mechanisms. In a beacon-enabled network with superframes, a slotted Carrier Sense Multiple Access with Collision Avoidance (CSMA-CA) mechanism is used. In networks without beacons, unslotted or standard CSMA-CA is used. When a device wants to transmit in a nonbeacon-enabled network, it first checks if another device is currently transmitting on the same channel. If this is the case, the device may back off for a random period, or indicate a transmission failure if it is unsuccessful after some retries. Acknowledgment frames confirming a previous transmission do not use the CSMA mechanism since they are sent immediately following the previous packet.

In a beacon-enabled network, any device wishing to transmit during the contention access period waits for the beginning of the next time slot and then determines if another device is currently transmitting in the same slot. If this is the case, the device backs off for a random number of slots, or it indicates a transmission failure after some retries. In a beacon-enabled network, acknowledgment frames do not use CSMA.

Successful reception and validation of data, or MAC command frame, is confirmed with an acknowledgment. If the receiving device is unable to handle the incoming message, the receipt is not acknowledged. The frame control field indicates whether or not an acknowledgment is expected. The acknowledgment frame is sent immediately after successful validation of the received frame. Beacon frames sent by a PAN coordinator and acknowledgment frames, are never acknowledged.

The IEEE 802.15.4 standard provides for three levels of security: no security of any type (e.g. for advertising kiosk applications); access control lists (noncryptographic security); and symmetric key security, employing AES-128 (advanced encryption standard 128-bit cryptographic keys). To minimize the cost for devices that do not require security, the key distribution method (e.g. public key cryptography) is not specified in the standard but may be included in the upper layers of the application.

IEEE 802.15.4 offers two physical layer options that can be combined with the MAC to enable a broad range of networking applications. Both physical layers are based on Direct Sequence Spread Spectrum (DSSS) methods that result in low-cost digital IC (Integrated Circuit) implementation, and both share the same basic packet structure for low-duty-cycle, low-power operation. The fundamental difference between the two physical layers is the frequency band. The 2.4 GHz physical layer specifies operation in the 2.4 GHz ISM band, which has nearly worldwide availability, while the 868/915 MHz physical layer specifies operation in the 868 MHz band in Europe and 915 MHz ISM band in the United States. The international availability of the 2.4 GHz band offers advantages in terms of larger markets and lower manufacturing costs. On the other hand, the 868 MHz and 915 MHz bands offer an alternative to the growing congestion and other interference (for instance, microwave ovens) associated with the 2.4 GHz band, and longer range for a given link availability due to lower propagation losses.

A second distinguishing physical layer characteristic is the transmission rate. The 2.4 GHz physical layer provides a transmission rate of 250 kb/s, while the 868/915 MHz physical layer offers rates of 20 kb/s and 40 kb/s, for its 869 MHz and 915 MHz bands, respectively. The higher rate in the 2.4 GHz physical layer is attributed largely to a higher-order modulation scheme, in which each data symbol represents multiple bits. The different transmission rates can be exploited to achieve a variety of different goals, for example, the low rate of the 868/915 MHz physical layer can be translated into better sensitivity and larger coverage area, thus reducing the number of nodes required to cover a given physical area, while the higher rate of the 2.4 GHz physical layer can be used to attain higher throughput, lower latency, or lower duty cycle.

Twenty-seven frequency channels are available across the three bands as shown in Figure 9.7. The 868/915 MHz physical layer supports a single channel between 868.0 and 868.6 MHz, and 10 channels between 902.0 and 928.0 MHz. Due to the regional support for these two bands, it is unlikely that a single network will use all 11 channels. However, the two bands are considered close enough in frequency that similar hardware can be used for both, thus lowering manufacturing costs. The 2.4 GHz physical layer supports 16 channels between 2.4 and 2.4835 GHz with ample channel spacing (5 MHz) aimed at easing the filter requirements for transmitting and receiving.

Since the home is likely to contain multiple types of wireless network vying for the same frequency bands, as well as unintentional interference

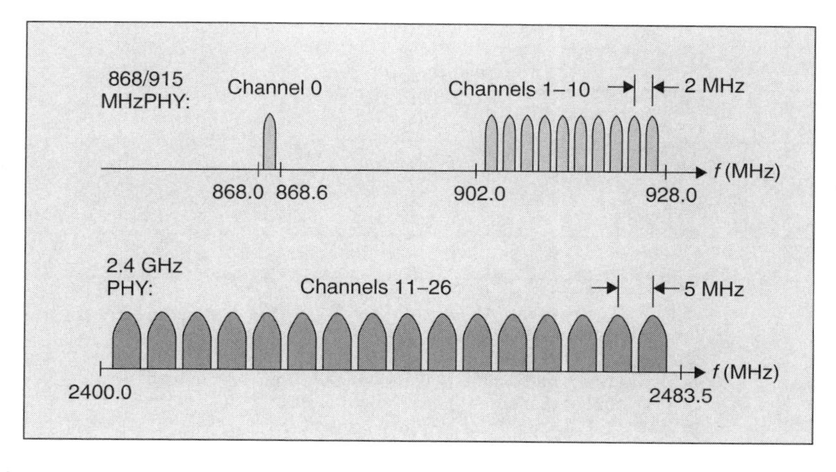

Figure 9.7 The IEEE 802.15.4 channel structure.

from appliances, the ability to relocate within the spectrum is an important factor. The standard allows dynamic channel selection, although the specific selection algorithm is left to the network layer. The MAC layer includes a scan function that steps through the list of supported channels in search of a beacon, while the physical layers contain several lower-level functions, such as receiver energy detection, link quality indication, and channel switching, which enable channel assessment and frequency agility. These functions are used by the network to establish its initial operating channel and to change channels in response to a prolonged outage.

To maintain a common simple interface with the MAC, both physical layers share a single packet structure as shown in Figure 9.8. Each packet, or Physical layer Protocol Data Unit (PPDU), contains a synchronization header (preamble plus start of packet delimiter), a physical layer header to indicate the packet length, and the payload, or Physical layer Service Data Unit (PSDU). The 32-bit preamble is designed for acquisition of symbol and chip timing, and in some cases may be used for coarse frequency adjustment. Channel equalization is not required for either physical layer due to the combination of small coverage area and relatively low chip rates.

Within the physical layer header, 7 bits are used to specify the length of the payload (in bytes). This supports packets of length 0–127 bytes, although due to the MAC layer overhead, zero-length packets do not occur in practice. Typical packet sizes for home applications such as monitoring and control of security, lighting, air conditioning, and other appliances are about 30–60 bytes, while more demanding applications such as interactive games and computer peripherals, or multitop applications with more address overhead, may require larger packet sizes. Adjusting for the transmission rates

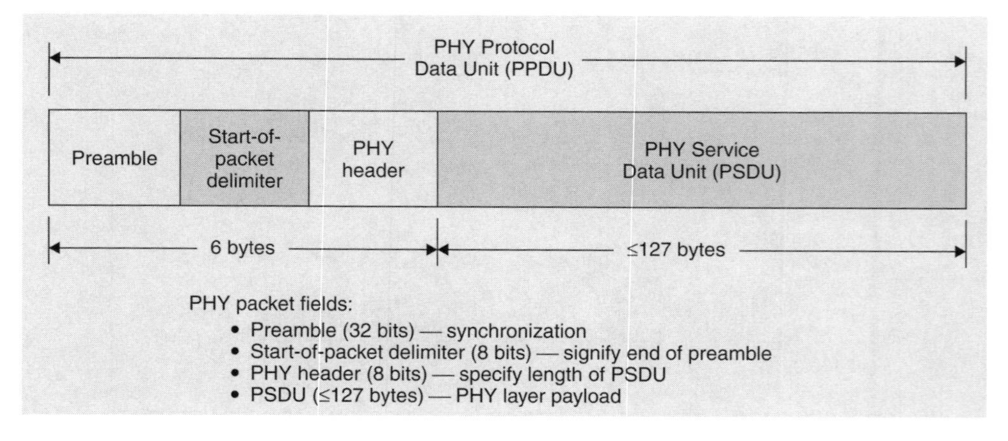

Figure 9.8 The IEEE 802.15.4 physical layer packet structure.

in each band, the maximum packet durations are 4.25 milliseconds for the 2.4 GHz band, 26.6 milliseconds for the 915 MHz band, and 53.2 milliseconds for the 868 MHz band.

The 868/915 MHz physical layer uses a simple DSSS approach in which each transmitted bit is represented by a 15-chip maximal length sequence (m-sequence). Binary data is encoded by multiplying each m-sequence by +1 or −1, and the resulting chip sequence is modulated onto the carrier using Binary Phase Shift Keying (BPSK). Differential data encoding is used prior to modulation to allow low-complexity differentially coherent reception.

The 2.4 GHz physical layer employs a 16-ary quasi-orthogonal modulation technique based on DSSS methods (with similar properties, e.g. processing gain). Binary data are grouped into 4-bit symbols, and each symbol specifies one of 16 nearly orthogonal 32-chip Pseudo-Noise (PN) sequences for transmission. PN sequences for successive data symbols are concatenated, and the aggregate chip sequence is modulated onto the carrier using Minimum Shift Keying (MSK), which is equivalent to Offset Quadrature Phase Shift Keying (O-QPSK) with half-sine pulse shaping. The use of a nearly orthogonal symbol set simplifies the implementation in exchange for a relatively small performance penalty (less than 0.5 dB).

IEEE 802.15.4 specifies receiver sensitivities of −85 dBm for the 2.4 GHz physical layer and −92 dBm for the 868/915 MHz physical layer. These values include sufficient margin to cover manufacturing tolerances as well as to permit very low-cost implementation approaches. In each case, the best devices may be of the order of 10 dB better than the specification.

The achievable range is a function of the receiver sensitivity and the transmit power. The standard specifies that each device shall be capable of transmitting at least 1 mW, but depending on the application needs, the actual transmit power may be lower or higher, within regulatory limits. Typical devices (1 mW) are expected to cover a range of 10–20 meters; however, with good sensitivity and a moderate increase in transmit power, a star network topology can provide complete home coverage. For applications allowing more latency, mesh network topologies provide an alternative for home coverage since each device needs only enough power (and sensitivity) to communicate with its nearest neighbor.

Devices operating in the 2.4-GHz band must accept interference caused by other services operating in this band. This is compatible with IEEE 802.15.4 applications, which have relatively low quality of service (QoS) requirements, do not require isochronous communication, and may be expected to perform multiple retries on occasion, to complete packet transmissions. A primary requirement of IEEE 802.15.4 applications is excellent battery life achieved in the standard by the use of low transmit power and very low duty-cycle operation. IEEE 802.15.4 devices may be sleeping as much as 99.9 % of the time they are operational, and employ low-power spread spectrum transmissions, thus they are good neighbors in the 2.4 GHz band.

The IEEE 802.15.4 standard targets residential and industrial customers. LR-WPAN is designed as an enabler technology. IEEE 802.15.4 is complementary to other wireless networking technologies by occupying the lower end of the power consumption and data throughput space.

9.6. DATA-CENTRIC STORAGE IN WIRELESS SENSOR NETWORKS

In most communication networks, naming of nodes for low-level communication leverages topological information. An example of this is the Internet (point-to-point communication model) where IP addresses are assigned to each node, and these serve as unique node identifiers in IP routing. Such a naming scheme is not very efficient in a sensor network scenario, since the identity of individual sensor nodes is not as important as the data associated with them. Data-centric models for sensor networks allow the sensor data itself (as contrasted to sensor nodes) be named, based on attributes such as event type or geographic location. In particular, data-centric routing and data-centric storage is energy efficient in sensor networks.

Data-Centric Storage (DCS) is a data-dissemination paradigm for sensor networks. In DCS, data is stored, according to event type, at corresponding sensor-net nodes. All data of a certain event type (e.g. temperature measurements) are stored at the same node. A significant benefit of DCS is that queries for data of a certain type can be sent directly to the node storing data of that type, rather than flooding the queries throughout the network (unlike data-centric routing proposals). DCS is based on the low-level routing functionality provided by the GPSR (Greedy Perimeter Stateless Routing) geographic routing algorithm, and on distributed hash-table functionality provided by peer-to-peer lookup algorithms. DCS offers reduced total network load and very good network usage.

Replication of control and data information in a DCS framework is the primary mechanism for reducing data retrieval traffic and increasing resilience to node failures. The storage of data of a particular type occurs at one of several replica nodes in the network assigned to this type, and storage of control and summary information pertaining to this type at geographically distributed monitor nodes in the network. By increasing the number of nodes where data can be stored for each event-type, as well as maintaining summary and control information at several nodes, the average cost of storing data and querying data is decreased.

Resilient Data-Centric Storage (R-DCS) is a method of achieving scalability and resilience by replicating data at strategic locations in the sensor network. R-DCS outperforms other schemes in terms of scaling to a large number of nodes and a large number of queries. In the case where nodes in the sensor network are unreliable and experience random failures, R-DCS does not experience a dramatic increase in the number of messages sent as the node failure rate is increased. It also maintains a high query success rate in this scenario. Hence this scheme realizes graceful performance degradation in the presence of node failures.

Wireless sensor networks have certain unique features which must be accounted for in any data dissemination methodology designed for such networks. Sensor devices have significantly higher processing capabilities and storage capabilities than available bandwidth (this is in contrast to wired networks, where an explosion of available bandwidth has led to a drastic reduction in its relative cost). The reason for this difference is that sensors typically have limited battery life. Hence they must use low-power (and consequently, low-bandwidth) wireless communication techniques to conserve battery power. In a typical scenario, it has been estimated that 3000 instructions could be executed for the same energy cost of sending a bit 100 meters by radio. This framework encourages the use of computational techniques to reduce the total communication overhead in the network.

The gateway through which sensor networks communicate with the external world (e.g. a monitoring terminal or the Internet) is an access point. 'Access path' refers to the set of data paths from the sensor nodes to the access point. The access points usually have a higher communication load than other sensor-net nodes. In a high-traffic scenario, such access points can become a bottleneck in the sensor-net (hotspot). The peak amount of traffic flowing through these access points ought to be minimized.

The energy constraints of wireless sensor networks are more efficiently achieved by using an attribute-based naming system rather than topological naming schemes (e.g. IP). Such attributes could be predefined to reduce the overhead during communication. For example, all sensor data in an environmental sensing network can be classified as being of types temperature, pressure and humidity. All such data can be named by including these predefined event attributes in the data itself.

When an event occurs, the sensors record and store the event data locally, and name this data based on its attributes. The low-level output from sensors (observations) is named, based on the attributes of the associated data. This data can be handled in a number of different ways. The three methods involve substantially different assumptions and cost–benefit trade-offs.

- *External Storage* (ES) in which all event data is stored at an external storage point for processing;
- *Local Storage* (LS) in which all event information is stored locally (at the detecting node), and
- *Data-Centric Storage* (DCS) in which all event data is stored by event-type within the sensornet at designated nodes.

Queries are used to retrieve event information from the sensornet. It is important to consider the ratio of query traffic to event-detection traffic while designing a sensor network. Each of the approaches has different relative costs for query and event traffic. Hence, depending on the function of a sensor network, one of the approaches may be more useful than the others.

DCS uses a Distributed Hash-Table (DHT) and offers the following interface:

- The Put(dataName, dataValue) primitive to store the value of the data corresponding to a certain event at the sensor-net node corresponding to the dataName (which serves as the key in the DHT and is typically based on the event type). The name of the data is typically based on the relevant event type.
- The Get(dataName) primitive to retrieve the value of the data stored at the node corresponding to the given dataName.

DCS uses the GPSR (Greedy Perimeter Stateless Routing) geographic routing algorithm for low-level routing. It then builds a DHT on top of GPSR. Unlike the commonly used shortest path technique, geographic routing uses the relationship between geographic position and connectivity in a wireless network. Since GPSR needs knowledge about the geographic coordinates of sensornet nodes to route messages, it is assumed that these nodes know their location through the use of localization methodologies. GPSR uses greedy forwarding to forward packets to nodes that are always progressively closer to the destination. The sender includes the approximate coordinates of the final destination while sending the packet. In regions where such a greedy path does not exist, GPSR recovers by forwarding in perimeter mode, wherein a packet traverses successively closer faces of a planar subgraph of the network connectivity graph, until reaching a node closer to the destination, where greedy forwarding resumes. If N is the number of nodes in the network, then to go from one random location to another requires $O(\sqrt{N})$ packet transmissions. In contrast, a flooding algorithm sends a packet to the entire sensor-net and requires $O(N)$ packet transmissions.

DHT is used to hash the name of a certain event to a key (dataName) which is a location somewhere within the boundaries of the sensor-net. The put(dataName, dataValue) primitive sends a packet with the given payload into the sensor-net which is routed towards the location dataName. The get(dataName) primitive is routed to the node closest to the dataName location, which then transmits a packet to the node originating the query with the corresponding data. In a sensor-net with completely stationary and reliable nodes, this approach is sufficient.

In order to make DCS resilient to node failures and mobility, there are certain extensions to the basic scheme. The storage node for an event type periodically routes a refresh message to all nodes which had transmitted event data to this node. Regular GPSR routing returns these refresh messages to the storage node along the network perimeter. In the intermittent time interval, the nodes in the sensor network could be displaced from their original locations. If a new node is closer to the location of the original storage node than the original storage node itself, then this new node will become the storage node. Timer based algorithms ensure that in case a storage node dies in this fashion, the new storage node automatically starts generating refresh messages. This process is called the Perimeter Refresh Protocol and is used to accomplish replication of (key, value) pairs and their consistent placement at the appropriate home nodes when the network topology changes. To protect against node failure, all nodes that receive a refresh message, will cache the data contained in it (local replication).

The Structured Replication in DCS (SR-DCS) scheme achieves load balancing in the network. SR-DCS uses a hierarchical decomposition of the key space and associates each event-type e with a hierarchy depth d. It hashes each event type to a root location. For a hierarchy depth d, it then computes $(4^d - 1)$ images of the root. When an event occurs, it is stored at the closest image node. Queries are routed to all image nodes, starting at the root and continuing through the hierarchy. SR-DCS significantly improves the scalability of DCS, and is useful for frequently detected events. However, that SR-DCS does not involve actual replication of data. It only stores one copy of any event-data at the closest image node. If all nodes in a certain location fail simultaneously (clustered failures), SR-DCS might not be able to recover the data stored at these nodes. The root node is a single point-of-failure in the sense that if the root for event-type e fails, one might not be able to issue any queries for event data corresponding to this type.

The extensions to DCS are used to achieve the following:

- minimize query-retrieval traffic hence saving energy consumption;
- increase data availability, ensuring that event information is not lost even with multiple node failures.

The original version of DCS has all events of the same type stored in one sensor-net node. It is evident that if there are too many events of a particular type, then this storage node will become a bottleneck point (hotspot) in the network. The data-dissemination scheme, Resilient Data-Centric Storage (R-DCS), overcomes these issues by a two-level replication strategy (control and data).

In R-DCS, the coordinate space of the sensornet field is partitioned into Z zones. The set of available zones is denoted as $z_j : j = 1, \ldots Z$. This zoning can be done on the basis of geographical boundaries, as shown in Figure 9.9. The zones can contain sensor nodes operating in three possible modes:

(1) *Monitor mode* Each zone has one monitor node for each event-type. The monitor node stores and exchanges information in the form of a monitoring map for each event-type. The monitoring map includes control and summary information in the following fields:
   - list of zones containing replica nodes (for forwarding event data and queries);
   - list of zones containing monitor nodes (for facilitating map exchange);

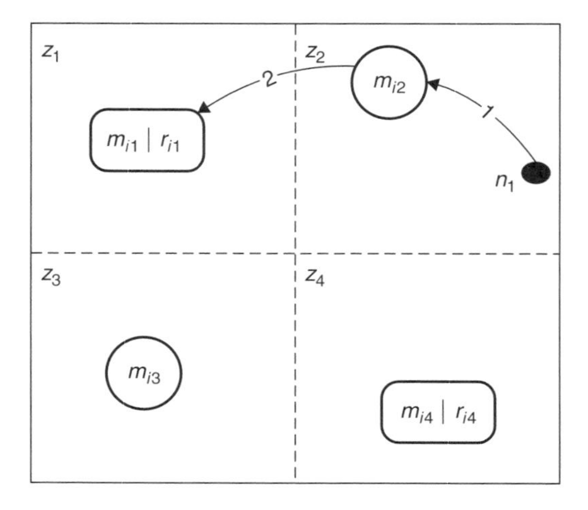

Figure 9.9 Event storage in R-DCS.

- event summaries (for facilitating summary-mode queries). The exact nature of event summaries depends on the event-type. For example, in a sensornet designed for temperature monitoring, the summary information could contain the number of events detected and the average temperature reading for each zone.
- *Bloom filters* (for enabling attribute-based queries). Event-data is organized in the form of a set of attributes and their values. In the temperature monitoring case, for example, these attributes could be event-time and temperature. A user might want to access all temperature measurements conducted between 10 a.m. and 11 a.m., or all temperature readings between 50 and 70. Bloom filters offer an efficient way to support attribute-based queries. Note that this field is optional – it is required only to support attribute-based queries.

(2) *Replica mode* Each zone has at most one replica node for each event type. The replica node, if present, is always the same as the monitor node. In addition to performing the functions of a monitor node, the replica node actually stores event data for the given event type.

(3) *Normal mode* All nodes that are not monitor or replica nodes operate in this mode. A normal node may originate or forward (i.e. route) event-data, but is not involved in storing any event data or control information.

Let E denote the number of event types in the sensornet. Let M_i be the total number of monitor nodes for each event type e_i. Let R_i be the number of replica nodes for each event-type e_i. The following system constraints must be satisfied for each $i = 1, \ldots, E$:

(1) $R_i \leqslant M_i \leqslant Z$ – This holds because each zone may have at most one replica node and one monitor node, and all replica nodes are monitor nodes as well. Under normal operations without clustered node failures (a majority of nodes failing in one zone), there will be one monitor node per zone: $M_i = Z$.

(2) $R_i \geqslant 1$ – There must be at least one replica node in the network, since all event-data for each event type is stored at the respective replica nodes.

For the distributed hash-table, a hash function H is used, which is a function of event type e_i and zone z_j. If event type e_i in zone z_j hashes to a location $(x_{ij}, y_{ij}) \equiv H(e_i, z_j)$, then a sensor node m_{ij} geographically closest to (x_{ij}, y_{ij}) is the monitor node for event type i within zone j. Depending on local decision rules, this monitor node may also serve as a replica node r_{ij}. For the purpose of load balancing, it is desirable that the function $H(e_i, z_j)$ be chosen such that for each zone z_j, different event types e_i hash to distinct nodes.

A sensing node (situated in zone j) sends an event of type e_i to the monitor node in the same zone m_{ij}. If this monitor node is also the replica node r_{ij}, then the event-data is stored at r_{ij}. If not, the data is forwarded to the closest replica node for this event type. The closest replica node can be determined from the information in the list of replica nodes field of the local monitoring map. The target replica node stores the event-data and updates its local copy of the monitoring map. This operation is illustrated in Figure 9.9.

There are three types of queries in an R-DCS system: summary, list or attribute-based as follows.

- *List*: A list query for an event type is a request for all stored data for events of this type. A querying node in zone z_j sends the query for event type e_i to the local monitor (and possibly replica) node m_{ij}. The monitor node then duplicates the query and forwards it to all other active replica nodes $r_{ix} : x = 1, \ldots, Z$ in the sensor net. All active replica nodes reply directly to the querying node with event data. This operation is illustrated in Figure 9.10.
- *Summary*: The querying node requests a summary of event information for an event type. A querying node in zone z_j sends the query for event type e_i to the local monitor node m_{ij}. The monitor node responds with the event summary information from the local monitoring map. This is illustrated in Figure 9.11.
- *Attribute based*: An attribute-based query requests data for all events which match certain constraints on their attribute values. A querying node in zone z_j sends the query for event-type e_i to the local monitor (and possibly replica)

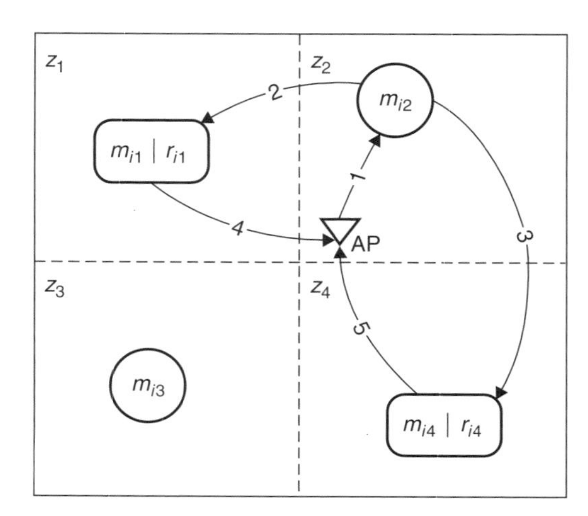

Figure 9.10 Querying in R-DCS (list mode).

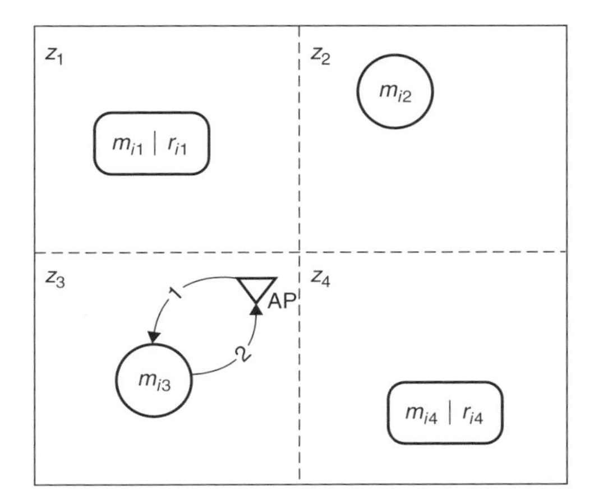

Figure 9.11 Querying in R-DCS (summary mode).

node m_{ij}. The monitor node then duplicates the query and forwards it to all other active replica nodes in the sensor net with Bloom filter matches. All active replica nodes reply directly to the querying node with event-data:

When an event of type e_i occurs in zone j, it updates the local monitoring map in m_{ij}. However, for global consistency of information such as the number of replica nodes for type e_i and event-summary information, these monitoring maps must be exchanged between the respective monitor nodes at periodic intervals. For this purpose, all active monitor nodes for a type e_i form a logical ring as shown in Figure 9.12. Each zone has two adjacent zones.

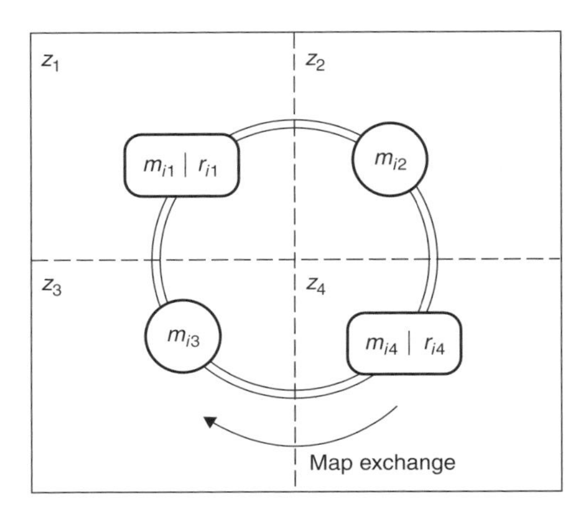

Figure 9.12 Logical ring in R-DCS.

When a monitor node receives a new map, it adds its own local updates (based on events received since the last map update) and forwards it to the next monitor node.

9.7. SUMMARY

Wireless sensor networks use small devices equipped with sensors, microprocessors and wireless communication interfaces. Different applications, ranging from personal health care to environmental monitoring and military applications are used in such networks. Various wireless technologies, like simple RF, Bluetooth, UWB or infrared can be used for communication between sensors.

Wireless sensor networks require low-power, low-cost devices that accommodate powerful processor, a sensing unit, wireless communication interface and power source, in a robust and tiny package. These devices have to work autonomously, to require no maintenance and to adapt to environment. For example, the MEMS technology enables production of very small sensing units with low power consumption.

Sensor network management protocol has to support control of individual nodes, network configuration updates, location information data exchange, network clustering, and data aggregation rules.

The sensor network gateway has to provide tools and functions for presentation of network topology, services, and characteristics to the user and to connect the network to other networks and users.

The IEEE 802.15.4 standard defines a Low-Rate Wireless Personal Area Network (LR-WPAN) which has ultra-low complexity, cost, and power for low-data-rate wireless connectivity among inexpensive fixed, portable, and moving devices. The IEEE 802.15.4 standard defines the physical (PHY) layer and media access control (MAC) layer specifications.

The IEEE 802.15.4 standard targets the residential and industrial markets. and LR-WPAN is designed as an enabler technology. IEEE 802.15.4 is complementary to other wireless networking technologies by occupying the lower end of the power consumption and data-throughput space.

Data-centric storage is a data-dissemination paradigm for sensor networks. In DCS, data is stored, according to event type, at corresponding sensor net nodes. All data of a certain event type (e.g. temperature measurements) are stored at the same node. A significant benefit of DCS is that queries for data of a certain type can be sent directly to the node storing that data, rather than flooding the queries throughout the network (unlike data-centric routing proposals). DCS is based on low-level routing functionality provided by the GPSR (Greedy Perimeter Stateless Routing) geographic routing algorithm, and on distributed hash-table functionality provided by peer-to-peer lookup algorithms. DCS offers reduced total network load and very good network usage.

PROBLEMS

Learning Objectives

After completing this chapter you should be able to:

- demonstrate understanding of the Bluetooth architecture;
- discuss Bluetooth services and QoS issues;
- demonstrate understanding of the Bluetooth-based wireless sensor network implementation;
- explain what home networking with IEEE 802.15.4 is, a standard for low-rate wireless personal area network;
- demonstrate an understanding of codesign and reconfiguration;
- explain what resilient data-centric storage in wireless *ad-hoc* sensor networks is.

Practice Problems

Problem 9.1: How does Bluetooth operate?

Problem 9.2: What connections are supported by the Bluetooth network?

Problem 9.3: What are the links by which a Bluetooth unit communicates?

Problem 9.4: What is an SCO link?

Problem 9.5: What is an ACL link?

Problem 9.6: What is required to achieve the Bluetooth interoperability with the Internet protocol family?

Problem 9.7: How are the sensor networks deployed?

Problem 9.8: What are the main services provided to the user by the wireless sensor network?

Problem 9.9: When does executing of a task occur in a wireless sensor network?

Problem 9.10: What are the main functions of a gateway?

Problem 9.11: How to the routing protocols and data aggregation algorithms collaborate?

Problem 9.12: What are the functions of the sensor network management protocol?

Problem 9.13: What are the functions of the sensor network gateway?

Problem 9.14: How can more than seven devices in a Bluetooth network be accommodated?

Problem 9.15: What are the functional blocks of a smart sensor node?

Problem 9.16: What is the first procedure executed upon gateway initialization?

Problem 9.17: In which applications is the IEEE 802.15.4 standard used?

Problem 9.18: What topologies are supported by the IEEE 802.15.4 standard?

Problem 9.19: What are the sublayers of DLL in the IEEE 802.15.4 standard?

Problem 9.20: What are the features of the IEEE 802.15.4 MAC sub-layer?

Problem 9.21: What are the levels of security provided by the IEEE 802.15.4 standard?

Problem 9.22: What are the physical layer options in the IEEE 802.15.4 standard?

Problem 9.23: What is the transmission rate in the IEEE 802.15.4 standard?

Problem 9.24: What frequency channels are available in the IEEE 802.15.4 standard?

Problem 9.25: How are the energy constraints achieved in wireless sensor networks?

Problem 9.26: What is data-centric storage?

Practice Problem Solutions

Problem 9.1:

Bluetooth uses an *ad-hoc*, piconet structure referred to as scatternet. Bluetooth operates in the international 2.4 GHz ISM band, at a gross data rate of 1 Mbit/s, and features low energy consumption for use in battery operated devices. With scatternet technology, it has been possible to achieve an aggregate throughput of over 10 Mbits/s or 20 voice channels within a fully expanded scatternet. The structure also makes it possible to extend the radio range by simply adding additional Bluetooth units to act as bridges at strategic places.

Problem 9.2:

The Bluetooth network supports both point-to-point and point-to-multipoint connections. A piconet is the network formed by a master and one or more slaves. Each piconet is defined by a different frequency hopping channel. All units participating in the same piconet are synchronized to this channel.

Problem 9.3:

Once a Bluetooth unit has been connected to a piconet it may communicate by means of two link types. That is, between any two members of the piconet forming a master–slave pair. The two link types supported are: Synchronous Connection Oriented (SCO) link, and Asynchronous (or isochronous) Connectionless (ACL) link.

Problem 9.4:

An SCO link is a point-to-point full-duplex link between the master and a slave. This link is established once by the master and kept alive until being released by the master. The SCO link is typically used for a voice connection. The master reserves the slots used for the SCO link on the channel.

Problem 9.5:

The ACL link makes a momentary connection between the master and any of the slaves for the duration of one frame (master-to-slave slot and slave-to-master slot). No slots are reserved. The master can freely decide which slave to address and in which order. The member sub-address in the packet header determines the slave. A polling scheme is used to control the traffic from the slaves to the master. The link is intended for asynchronous or isochronous data. However, if the master uses this link to address the same slave at regular intervals, it becomes a synchronous link. The ACL link supports both symmetric and asymmetric modes. In addition, modes have been defined with or without FEC, and with or without CRC and ARQ.

Problem 9.6:

To achieve the Bluetooth interoperability with the Internet protocol family, the TCP/IP over Bluetooth requires that bridging, address resolution, MTU (maximum transmission unit) definition, and multicast/broadcast mappings are solved.

Problem 9.7:

Wireless sensor networks comprise a number of small devices equipped with a sensing unit, microprocessor, wireless communication interface, and power source. In contrast to the traditional sensor networks that are carefully planned and deployed, wireless sensor networks can be deployed in an *ad-hoc* manner. This deployment requires communication protocols that are able to organize the network automatically, without the need for human intervention.

Problem 9.8:

From the user point of view, querying and tasking are two main services provided by wireless sensor networks. Queries are used when user requires only the current value of the observed phenomenon.

Problem 9.9:

Executing a task is used when a phenomenon has to be observed over a longer period of time. For example, a user can ask a sensor network to detect a specific type of vehicle in the area and to monitor its movements. To execute the task, different types of sensor have to collaborate: seismic to detect motion, video and audio to detect type of vehicle, etc. Information about the vehicle trajectory is forwarded to the user. Both queries and tasks are injected into the network by the gateway which also collects the replies and forwards them to the users.

Problem 9.10:

The main functions of a gateway are:

- communication with the sensor network, where short-range wireless communication is used (Bluetooth, UWB, RF, IR, etc.) to provide functions like discovery of smart sensor nodes, generic methods for sending and receiving data to and from sensors, routing, etc.
- Gateway logic, which controls gateway interfaces and data flow to and from sensor network. It also provides an abstraction level with the API (Application Programming Interface) that describes the existing sensors and their characteristics. Gateway logic provides functions for uniform access to sensors regardless of their type, location or network topology, injects queries and tasks and collects replies;
- Communication with the users occurs through a gateway. The gateway communicates with the users and the other sensor networks over the

Internet, wide area networks, satellite or a short-range communication technology.

Problem 9.11:

An important requirement for routing protocols in sensor networks is collaboration with data aggregation algorithms. Data aggregation is required to avoid network implosion, which may occur when many nodes answer the same query and send replies towards the gateway, and overlap problems in data-centric routing. Based on predefined methods, responsible nodes analyze gathered data and combine it into a set of meaningful information that is forwarded to the user. Data aggregation reduces the amount of network load while preserving validity and amount of information. For certain applications it can be important to know the source of information (position) and in such cases that information has to be forwarded as well.

Problem 9.12:

Sensor network management protocol has to support control of individual nodes, network configuration updates, location information data exchange, network clustering, and data aggregation rules.

Problem 9.13:

A sensor network gateway has to provide tools and functions for presentation of network topology, services, and characteristics to the user and to connect the network to other networks and users.

Problem 9.14:

When more than seven devices need to communicate there are two options. The first one is to put one or more devices into the park state. Bluetooth defines three low power modes: sniff, hold and park. When a device is in the park mode it disassociates from the piconet, but still maintains timing synchronization with it. The master of the piconet periodically broadcasts beacons to invite the slave to rejoin the piconet or to allow the slave to request to rejoin. The slave can rejoin the piconet only if there are fewer than seven slaves already in the piconet. If this is not the case, then the master has to park one of the active slaves first. All these actions cause delays and for some applications it can be unacceptable, for example, process control that requires immediate response from the command center.

The other option is to build a scatternet. Scatternet consists of several piconets connected by devices participating in multiple piconets. These devices can be slaves in all piconets or master in one piconet and slave in other piconets. Using scatternets, higher throughput is available and multi-hop connections between devices in different piconets are possible. However, hardware still does not support this functionality.

Problem 9.15:

The smart sensor node comprises three functional blocks: sensing, data processing, and communication.

Problem 9.16:

Smart sensor node discovery is the first procedure that is executed upon the gateway initialization. Its goal is to discover all sensor nodes in the area and to build a list of sensor characteristics and network topology. Afterwards, it is executed periodically to facilitate addition of new, or removal of existing, sensors.

Problem 9.17:

IEEE 802.15.4 is useful in a wide variety of applications, including industrial control and monitoring; public safety, including sensing and location determination at disaster sites; automotive sensing, such as tire-pressure monitoring; smart badges and tags; and precision agriculture, such as the sensing of soil moisture, pesticide, herbicide, and pH levels.

Problem 9.18:

The IEEE 802.15.4 standard supports multiple network topologies, including both star and peer-to-peer networks. The topology depends on the design where applications, such as PC peripherals, may require low-latency connection of a star network, while others, such as perimeter security, may require large-area coverage of peer-to-peer networking. Multiple address types, including both physical (i.e. 64-bit IEEE) and short (i.e. 8-bit network assigned) are provided.

Problem 9.19:

IEEE 802 splits the DLL into two sublayers, the MAC and the logical link control (LLC).

Problem 9.20:

The features of the IEEE 802.15.4 MAC are association and disassociation, acknowledged frame delivery, channel access mechanism, frame validation, guaranteed time-slot management, and beacon management. The MAC sub-layer provides two services to higher layers that can be accessed through two service access points (SAPs). The MAC data service is accessed through the MAC common part sub-layer (MCPS-SAP), and the MAC management services are accessed through the MAC layer management entity (MLME-SAP). These two services provide an interface between the SSCS or another LLC and the physical layer.

Problem 9.21:

The IEEE 802.15.4 standard provides for three levels of security: no security of any type (e.g. for advertising kiosk applications); access control lists (noncryptographic security); and symmetric key security, employing AES-128 (advanced encryption standard 128-bit cryptographic keys). To minimize the cost for devices that do not require security, the key distribution method (e.g. public key cryptography) is not specified in the standard but may be included in the upper layers of the applications.

Problem 9.22:

The IEEE 802.15.4 offers two physical layer options that can be combined with the MAC to enable a broad range of networking applications. Both physical layers are based on direct sequence spread spectrum (DSSS) methods that result in low-cost digital IC implementation, and both share the same basic packet structure for low-duty-cycle, low-power operation. The fundamental difference between the two physical layers is the frequency band. The 2.4 GHz physical layer specifies operation in the 2.4 GHz ISM band, which has nearly worldwide availability, while the 868/915 MHz physical layer specifies operation in the 868 MHz band in Europe and 915 MHz ISM band in the United States. The international availability of the 2.4 GHz band offers advantages in terms of larger markets and lower manufacturing costs. On the other hand, the 868 MHz and 915 MHz bands offer an alternative to the growing congestion and other interference (for instance, from microwave ovens) associated with the 2.4 GHz band, and longer range for a given link availability due to lower propagation losses.

Problem 9.23:

The 2.4 GHz physical layer provides a transmission rate of 250 kb/s, while the 868/915 MHz physical layer offers rates of 20 kb/s and 40 kb/s, for its 869 MHz and 915 MHz bands, respectively.

Problem 9.24:

Twenty-seven frequency channels are available across the three bands. The 868/915 MHz physical layer supports a single channel between 868.0 and

868.6 MHz, and 10 channels between 902.0 and 928.0 MHz. Due to the regional support for these two bands, it is unlikely that a single network will use all 11 channels. However, the two bands are considered close enough in frequency for similar, if not identical, hardware to be used for both, thus lowering manufacturing costs. The 2.4-GHz physical layer supports 16 channels between 2.4 and 2.4835 GHz with ample channel spacing (5 MHz) aimed at easing the filter requirements to transmit and receive.

Problem 9.25:

The energy constraints of wireless sensor networks are more efficiently achieved by using an attribute-based naming system rather than topological naming schemes (e.g. IP). Such attributes could be predefined to reduce the overhead during communication. For example, all sensor data in an environmental sensing network can be classified as being of types temperature, pressure and humidity. All such data can be named by including these predefined event attributes in the data itself.

Problem 9.26:

Data-centric storage is a data-dissemination paradigm for sensor networks. In DCS, data is stored, according to event type, at corresponding sensor-net nodes. All data of a certain event type (e.g. temperature measurements) are stored at the same node. A significant benefit of DCS is that queries for data of a certain type can be sent directly to the node storing data of that type, rather than flooding the queries throughout the network (unlike data-centric routing proposals). DCS is based on the low-level routing functionality provided by the GPSR (Greedy Perimeter Stateless Routing) geographic routing algorithm, and on distributed hash-table functionality provided by peer-to-peer lookup algorithms. DCS offers reduced total network load and very good network usage.

10

Applications of Wireless Sensor Networks

10.1. INTRODUCTION

Distributed, dynamic, and adaptive, embedded software is used in highly constrained devices. An active message communication model is used to build nonblocking applications and higher level networking capabilities. The TinyOS event-driven approach is used to implement the communication model with very limited storage and the radio channel modulated directly in software in an energy efficient manner. The open, component-based design allows novel relationships between system and application.

To explore the system design techniques underlying this kind of application and the emerging technology of microscopic computing, there is a series of small RF wireless sensor devices, a tiny operating system (TinyOS), and a networking infrastructure for low-power, highly constrained devices in dynamic, self-organized, interactive environments. The severe resource constraints put the hardware platform far beyond reach of conventional operating systems. TinyOS is a simple, component-based operating system, which is primarily a framework for managing concurrency in a storage- and energy-limited context. A collection of modular components is built up by modulating the radio channel and accessing sensors via ADCs (analog-to-digital converters) to an event-driven environmental monitoring application with dynamic network discovery and multi-hop *ad hoc* routing. A nonblocking discipline is carried throughout the design and most components are re-entrant cooperating state machines.

Wireless Sensor Network Designs A. Hać
 ISBN: 0-470-86736-1

In wireless embedded systems, the communication path to the devices is a shared channel, which must be shared effectively in the context of resource constrained processing and *ad-hoc* multi-hop routing. Many applications require that nodes have roughly equal ability to move data through the network, regardless of position within the network topology. The low-level TinyOS communication components are extended with an energy-aware media access control (MAC) protocol and use a simple technique for application specific adaptive rate control.

Rockwell Science Center has created a development environment for Wireless Integrated Networked Sensors (WINS) that includes customizable, sensor-laden networked nodes and both mobile and Internet-hosted user interfaces. The WINS development system allows evaluation of the design, deployment and usage of microsensor networks. It uses multiple sensors, processes sensor data both autonomously and in cooperation with neighboring nodes into information, and communicates this information to users via a variety of network topologies. WINS are self-organizing and establish and maintain the network without user intervention. Minimizing power consumption is a primary concern in WINS development. Each node processes sensor data into information, thereby reducing power-demanding communications requirements. Power minimization allows the design of integrated WINS hardware and the creation of networking protocols specific to the needs of microsensor networks. WINS are applied to area monitoring, surveillance, and security, to networking of personnel and physical assets over large areas, and to machinery and platform health and status monitoring.

In microsensor networks, large numbers of devices (e.g. more than 10) are needed to address issues such as scalability, spatial distribution, frequency reuse, and a number of potential application scenarios. The devices in a testbed system need to have software and networking capabilities that support data collection and algorithm development using results from WINS nodes under field conditions. Microsensor nodes have limitations on computation, memory, and communication resources caused by battery limitations.

Aggregates of sensors are used in collaborative processing tasks for sensor networks such as tracking and localization. Sensor aggregate is defined by the nodes in a network that satisfy a grouping predicate. The parameters of the predicate depend on task and resource requirements. A distributed protocol is needed for constructing sensor aggregates in the context of counting distinct targets in a sensor field. The node processing and communication capabilities allow implementations on resource constrained hardware.

Wireless sensor networks can be applied to real-world habitat monitoring. A set of system design requirements covers the hardware design of the nodes, the design of the sensor network, and the capabilities for remote data access and management. A system architecture uses these requirements for habitat monitoring. The architecture presents monitoring seabird nesting environment and behavior.

10.2. APPLICATION AND COMMUNICATION SUPPORT FOR WIRELESS SENSOR NETWORKS

The embedded software is agile, self-organizing, resource constrained, and communication centric on numerous small devices operating collectively. The applications include:

- ubiquitous computing environments where numerous devices placed on humans and things interact in a context-aware manner;
- dense *in situ* monitoring of life-science experiments;
- condition-based maintenance, and
- disaster management in a smart civil infrastructure.

The mode of operation is concurrency intensive for bursts of activity but otherwise very passive, watching for a significant change or event. In the bursts, data and events are streaming in from sensors and the network, out to the network, and to various actuators. A mix of real-time actions and longer-scale processing must be performed. In the remaining time, the device must shutdown to a very low power state, yet monitor sensors and network for important changes while perhaps restoring energy reserves. Net accumulation of energy in the passive mode and efficiency in the active mode determine the overall performance capability of the nodes.

TinyOS provides convenient abstractions of physical devices and highly tuned implementations of common functions. This goal is especially challenging because of the highly constrained resource context, and the application-specific devices.

A TinyOS application consists of a scheduler and the components. Each component is described by its interface and its internal implementation, in a manner similar to many hardware description languages, such as VHDL [VHSIC (Very High Scale Integrated Circuit) Hardware Description Language] and Verilog (a Cadence Design Systems digital simulation tool). An interface comprises synchronous commands and asynchronous events. The component has an upper interface, which names the commands it implements and the events it signals, and a lower interface, which names the commands it uses and the events it handles. The implementation is written using the interface name space. A component also has internal storage, structured into a frame, and internal concurrency, in the form of very light-weight threads, called tasks. The command, event, and task handlers are declared explicitly in the source. The points where an external command is called, event signaled, or task posted, are also explicit in the static code, as are references to frame storage. A separate application description shows how the interfaces are wired together to form the overall application composition. An event may be delivered to multiple components or multiple components may use the same command. Thus, although the application is modular, the compiler has static information to use in optimizing across the entire application, including the operating system. In addition, the underlying run-time execution model and storage model can be optimized for specific platforms. A typical application graph is shown in Figure 10.1, containing a low-power radio stack, a UART (Universal Asynchronous Receiver Transmitter) serial port stack, sensor stacks, and higher level network discovery, and dedicated routing to support distributed sensor data acquisition. This entire application occupies about three kilobytes.

The TinyOS concurrency model is a two-level scheduling hierarchy, where events preempt tasks, and the tasks do not preempt other tasks. The vast majority of operation is in the form of nonblocking state transitions. Within a task, commands may be called, a command may call subordinate commands, or it may post tasks to continue working logically in parallel with its invocation. By convention, all commands return a status indicating whether the command was accepted, providing a full handshake. Since all components have bounded storage, a component must be able to refuse commands. A command may initiate an operation, for instance, by accessing a sensor or

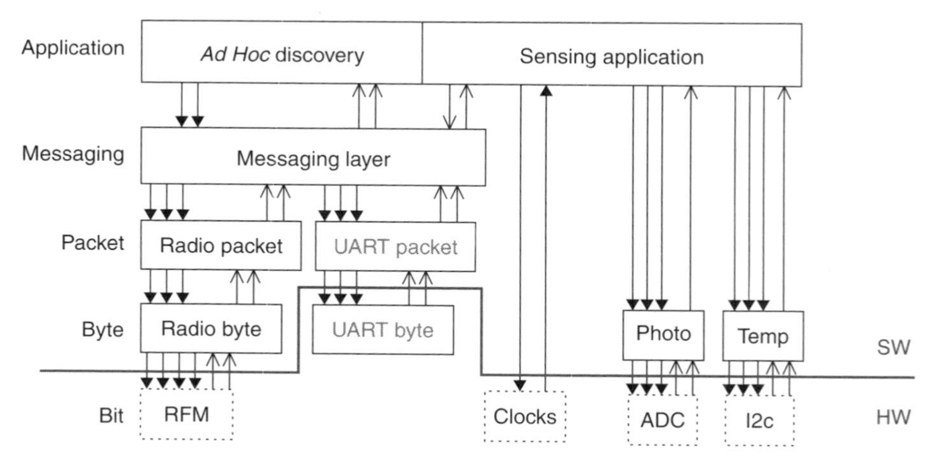

Figure 10.1 A networking application component graph.

sending a message, leaving the operation to be carried out concurrently with other activities, by using either hardware parallelism or tasks.

Events are initiated at the lowest level by hardware interrupts. Events may signal higher level events, call commands, or post tasks. Commands cannot signal events. Thus, an individual event may propagate through multiple levels of components, triggering collateral activity. Whenever the work cannot be accomplished in a small, bounded amount of time, the component should record continuation information in its frame and post a task to complete the work. By convention, the lowest level hardware abstraction components perform enough interrupt processing to re-enable interrupts before signaling the event. Events (or tasks posted within events) typically complete the split-phase operations initiated by commands, signaling to the higher-level component that the operation has completed and perhaps passing it the data.

A nonblocking approach is taken throughout TinyOS. There are no locks, and components never spin on a synchronization variable. A lock-free queue data structure is used by the scheduler. Components perform a phase of an operation and terminate, allowing the completion event to resume their execution. Most components are written essentially as re-entrant state machines. TinyOS is written in the C programming language with conventional preprocessor macros to highlight the key concepts. The TinyOS execution model is implemented on a single shared stack with a static frame per component.

The communication-centric design approach in TinyOS is used to build a networking infrastructure for self-organized, deeply embedded collections of devices.

Active Messages (AM) is a simple, extensible paradigm for message-based communication by using procedure calls. Each message contains the name of a handler to be invoked on a target node upon arrival, and a data payload. The handler function serves the dual purpose of extracting the message from the network and either integrating the data into the computation or sending a response. The AM communication model is event driven and specifically designed to allow a very lean communication stack to process packets directly off the network, while supporting a wide range of applications.

Initiating an active message involves specifying the data arguments, naming the handler, requesting the transmission, and detecting transmission completion. Receiving AM involves invoking the specified handler on a copy of the transmitted data.

The send message command identifies intended recipients, the handler that will process the message on arrival, and the source output message buffer in the local frame. A handler registry is maintained, and the identifier for the named handler is extracted. The status handshake for this command illustrates the general notion of components managing their bounded resources.

The messaging component may refuse the send request, for example, if it is busy transmitting or receiving a message and does not have resources with which to queue the request. The reaction to this occurrence is application specific.

The message arrival event is similar to other events. One key difference is that the active-message component dispatches the event to the component with the associated message handler. Many components may register one or more message handlers. Additionally, the input to the handler is a reference to a message buffer provided by the active-message component.

Managing buffer storage is a difficult problem in a communication stack because the following issues must be addressed:

- encapsulating useful data with transport header and trailer information;
- determining when output message data storage can be reused, and
- providing an input buffer for an incoming message before the message has been inspected, to determine where it goes.

The Tiny active-message layer provides simple primitives for resolving these issues with no copying and very simple storage management.

The message buffer has a defined type in the frame that provides fields for system specific encapsulation, such as routing information and error detection. These fields are used as the packet moves down the stack, rather than following pointers or copying. The application components refer only to the data field or the entire buffer. References to message buffers are the only pointers carried across component boundaries in TinyOS.

Once the send command is called, the transmit buffer is considered to be owned by the network until the messaging component signals that transmission is complete. The mechanism for tracking ownership is application specific.

The message handler receives a reference to a system owned buffer, which is distinct from its frame. The typical behavior is to process information in the message and return the buffer. In general, the handler must return a reference to a free buffer. It could retain the buffer it was given by the system and return a different buffer, which it owns. A common special case of this scenario is a handler that makes a small change to an incoming message and retransmits it. We would like to avoid copying the remainder of the message, however, we cannot retain ownership of the buffer for transmission and return the same buffer to the system. Such a component should declare a message buffer and a message-buffer pointer in its frame. The handler modifies the incoming buffer and exchanges buffer ownership with the system. If its previous transmit buffer is still busy, one of the two operations must be discarded. A component performing reassembly from multiple packets may own multiple such buffers. In any case, run-time buffer-storage management is reduced to a simple pointer swap.

The Tiny active message is used to support dynamic network discovery and multi-hop, *ad-hoc* routing. Discovery can be initiated from any node, but often it is rooted at gateway nodes that provide connectivity to conventional networks. Each root periodically transmits a message carrying its ID and its distance, which is equal to zero, to its neighborhood. The message handler checks whether or not the source is the closest node from which it has heard recently (i.e. in the current discovery phase) and, if so, records the source ID as its multi-hop parent, increments the distance, and retransmits the message with its own ID as the source. The discovery component utilizes the buffer swap.

The packets are routed up the tree as follows. A node transmitting data to be routed specifies a multi-hop forwarding handler and identifies its parent as the recipient. The handler will fire in each of its neighbors. The parent retransmits the packet to its parent, using the buffer swap. Other neighbors simply discard the packet. The data is thus routed hop-by-hop to the root. Reduction operators can be formed by accumulating data from multiple children before transmitting a packet up the tree.

The discovery algorithm is nonoptimal because of redundancy in the outgoing discovery wave front and may be improved by electing cluster leaders or retransmitting the beacon with some probability inversely related to the number of siblings. Alternatively, the discovery phase can be eliminated entirely by piggybacking the distance information on the sensor data messages. When a node hears a packet from a node, which is fewer hops from the base station, it adopts the source as its parent. The root node simply transmits a packet to itself to grow the routing tree. Nodes must also age their current distance to adapt to changes in network topology due to movement or signal propagation changes. These examples illustrate the fundamental communication step upon which distributed algorithms for embedded wireless networks are based: receiving a packet, transforming it, and selectively retransmitting it or not. Squelching retransmission forms an outgoing wave front in discovery and forms a beam on multi-hop routing. In these algorithms the data structure for determining whether to retransmit is a cache of recent packets.

One challenge is to move the message data from the application storage buffer to the physical modulation of the channel without making entire copies, and similarly in the reverse direction. A common pattern that has emerged is a cross-layer data pump. We find this at each layer of the stack in Figure 10.1. The upper component has a unit of data partitioned into subunits. It issues a command to request transmission of the first subunit. The lower component acknowledges that it has accepted the subunit and when it is ready for the next one, it signals a subunit event. The upper handler provides the next unit, or indicates that no more units are forthcoming. This is done by calling the next subunit command within the ready handler. The message layer is effectively a packet pump. The packet layer encodes and frames the packet, pumping it byte-by-byte into the byte layer. On the UART, the byte-by-byte abstraction is implemented directly in hardware, whereas on the radio the byte layer pumps the data bit-by-bit into the radio. Each of these components utilizes the frame, command, and event framework to construct a re-entrant software state machine.

In a multi-hop data collection network, each node transmits its own data from time to time, and listens during the remaining time for data that it needs to forward toward a sink.

Although active transmission is the most power intensive mode, most radios consume a substantial fraction of the transmit energy when the radio is on and does not receive anything. In special networks, a device transmits for short periods of time, but must be continually listening in order to forward data for the neighboring nodes. The total energy consumption of a device is dominated by the RF reception cost.

Power consumption can be reduced by using periodic listening. By creating time periods when transmitting is not permitted, the nodes must listen only part time. This approach works well when the time scale of the invalid periods is quite large relative to the message transmission time. The downside of this approach is that it limits the used bandwidth.

In sensor networks, a node may act as a router or data processing point, and may need to use the radio bandwidth fully. Low-power listening keeps the same listener duty cycle concept, but greatly reduces the time scale.

To further reduce the average power consumption of the network, low power listening can be combined with periodic listening. Running both schemes simultaneously results in listening at reduced power for only a fraction of the time, and the power reductions are multiplicative. These techniques provide a mechanism for trading bandwidth and transmission cost for a reduction in receive power consumption.

The hardware directly connects the central microcontroller to the radio. This places all of the real time requirements of the radio onto the microcontroller, which must handle every bit that is transmitted or received in real time. Additionally, it controls the timing of each bit so that any jitter in the control signals that it generates is propagated to the transmitted signal. The TinyOS communication stack handles these constraints while allowing higher level functions to continue in parallel.

At the base of the component stack is a state machine that performs the bit timing. The RFM (RF Monolithics) component transfers a single bit at a time to or from the RF Monolithics radio. For a correct transmission to occur, the transmitted bit must be placed and held on the TX (data output) line of the radio for exactly one bit time, for instance, 100 microseconds. For reception, the RX (data input) line of the radio must be sampled at the midpoint of the transmission period. The radio provides no support for determining when bit times have completed.

The interface to the RFM component forms a data pump performing a bit-by-bit transfer from a byte-level component to the physical hardware. To start the transmission of data, a command is issued to the RFM component to switch into transmit mode. Then a second command is used to transfer a single bit down to the RFM component. This bit is immediately placed onto the transmit line. After 100 microseconds has passed, the RFM component will signal an event to indicate that it is ready for another bit. The byte-level component's response is to issue another command to the RFM component that contains the next bit. This interaction of signaling an event and receiving the next bit continues until the entire packet is completed. The RFM layer component abstracts the real-time deadlines of the transmission process from the higher layer components.

During transmission, complex encoding must be done on each byte while simultaneously meeting the strict real-time requirements of the bit layer. The encoding operation for a single byte takes longer than the transmission time of a single bit. To ensure that the encoded data is ready in time to meet the bit level transmission deadline, the encoding of the next byte starts prior to the completion of the transmission of the current byte. The TinyOS task mechanism executes the encoding operation while simultaneously performing the transmission of previous data. By encoding data one byte in advance of transmission, the buffering is used to decouple the bit level timing from the byte encoding process.

Data reception takes the same form as transmission, except that the receiver must first detect that a transmission is about to begin and then determine the timing of the transmission. To accomplish this, when there is activity on the radio channel, the RFM layer component is set to sample bits every 50 microseconds, double sampling each byte. These bits are handed up one at a time to the byte-level component. The byte-level component creates a sliding buffer of these bit values that contains the last 18 bits. When the value of the last 18 bits received matches the designated start symbol, the start of a packet is detected. Additionally, the timing of the packet is determined to within half a bit time. Next, the RFM layer samples a single bit after 75 microseconds. This causes the next sample to fall in the middle of the next bit window, half way between where the double sampling would have occurred if the sample period had remained at

50 microseconds. Then the RFM samples every 100 microseconds for the remainder of the packet.

In wireless embedded systems, the communication path to the devices is a shared channel, which must be shared effectively in the context of resource constrained processing and *ad hoc* multi-hop routing. Many applications require that nodes have roughly equal ability to move data through the network, regardless of their position within the network topology. The low-level TinyOS communication components are extended with an energy-aware Media Access Control (MAC) protocol and use a simple technique for application specific adaptive rate control.

The MAC protocols must be performed on microcontroller concurrently with other operations. The RF transceiver lacks support for collision detection, thus the Carrier Sense Multiple Access (CSMA) scheme is used, where a node listens for the channel and only transmits a packet if the channel is idle. The mechanism for clocking in bits at the physical layer is also used for carrier sensing. Thus, the MAC layer is implemented at both the bit and byte level in the network stack. If consecutive sampling of the channel discovers no signal, the channel is deemed idle and a packet transmission is attempted. However, if the channel is busy, a random back off occurs. The entire process repeats until the channel is idle. A simple 16-bit linear feedback shift register is used as a pseudo-random-number generator for the back off period. The radio is turned off during back off. Many applications collect and transmit data periodically, perhaps after detecting a triggering event, thus traffic can be highly correlated. Detection of a busy channel suggests that a neighboring node may indicate that the communication patterns of the nodes are synchronized. The application uses the failure to send as feedback and shifts it sampling phase to potentially desynchronize.

Another common application requirement is roughly equal coverage of data sampling over the entire network. Each node in the network should be able to deliver fair allocation of bandwidth to the base station. With special routing layers, nodes self organize into a spanning forest, where each node originates and routes traffic to a base station. The competition between originated and route-through traffic for upstream bandwidth must be balanced to meet the fairness goal. The capacity of a multi-hop network is limited, and the nodes must adapt their offered load to the available bandwidth, rather than over commit the channel and waste energy in transmitting packets that can never reach the base station. The adaptive transmission control scheme is a local algorithm implemented above the active-message layer and below the application level. The application has a baseline sampling rate that determines maximum transmission rate and transmits a sample with a dynamically determined probability. Upon successful transmission, the probability is increased linearly, whereas upon failure it is decreased multiplicatively. A successful transmission can be indicated by an explicit acknowledgment from the receiver or an implicit acknowledgment when the sender hears its packet being forwarded by its parent. Since implicit acknowledgment is often application specific, the application decides if the transmission was successful and propagates the information down to the transmission control layer. Rejection of application's transmission command at the transmission control level triggers the adaptation.

The TinyOS approach has proven quite effective in supporting general purpose communication among, potentially, many devices that are highly constrained in terms of processing, storage, bandwidth, and energy with primitive hardware support for I/O. The event driven model facilitates interleaving of the processor between multiple flows of data and between multiple layers in the stack for each flow while still meeting the severe real-time requirements of servicing the radio. Since storage is very limited, it is common to process messages incrementally at several levels, rather than buffering entire messages and processing them level by level. However, events alone are not sufficient, and it is essential that an event be able to hand any substantial processing off to a task that will run outside the real-time window. This provides logical concurrency within the stack and is used at every level except the lowest hardware abstraction layer. By adopting a nonblocking, event-driven approach, the traditional threads are not supported, with their associated multiple stacks and complex synchronization.

The component approach yields robust operation despite limited debugging capabilities, and facilitates experimentation. The packet components can be swapped with a simple change to the description graph and temporary components can be interposed between existing components, without changing any of the internal implementations. Moreover, the use of components allows, essentially, an entire subtree of components to be replaced by hardware and vice versa.

The Tiny active message programming model permits experiments with numerous higher level networking layers and fine-grained distributed algorithms. The nodes can be reprogrammed over the network. A node can obtain code capsules from its neighbors or over multihop routes and assemble a complete execution image in its EEPROM secondary store. The node can then use this to reprogram itself. Other examples include a general purpose data logging and acquisition capability, a facility to query nodes by schema, and to aggregate data from a large number of nodes within the network.

Without the traditional layers of abstraction dictating what kinds of capabilities are available, it is possible to foresee many novel relationships between the application and the underlying system. The adaptive transmission control scheme is a simple example; rejection of the send request causes the application to adjust its rate of originating data. The application level forwarding of multi-hop traffic allows the node to keep track of its changing set of neighbors. Moreover, the radio is itself another sensor, since receive signal strength is provided to the ADC. Thus, each packet can be accompanied by signal strength data for use in estimating physical distance or presence of obstructions. The radio is also an actuator, as its signal strength, and therefore cell size, can be controlled.

The lowest layer components are synchronizing all receivers to the transmitter to within a fraction of a bit. Thus, very fine grain time synchronization information can be provided with every packet for control applications.

10.3. AREA MONITORING AND INTEGRATED VEHICLE HEALTH MANAGEMENT APPLICATIONS

Wireless distributed microsensor networks consist of a collection of communicating nodes, where each node incorporates

- one or more sensors for measuring the environment;
- processing capability in order to process sensor data into high value information and to accomplish local control, and
- a radio to communicate information to/from neighboring nodes and to external users.

Ultra-low-power CMOS (Complementary Metal-Oxide Semiconductor) chips can integrate radios for communication, digital computing, and MEMS sensing components, on a single die produced in high volumes for low-cost. This permits large numbers of wireless integrated networked sensors to be easily and rapidly deployed (e.g. air dropped into battle fields or deployed throughout an aircraft or space vehicle) to form highly redundant, self-configuring, dedicated sensor networks. For ease of deployment, the nodes use wireless communications, and are capable of establishing and operating their own network. To prolong battery life, all node and network functions are designed to consume minimal power. Highly capable and ultrareliable systems are built out of large numbers of such nodes that are individually inexpensive and use cooperation between nodes to produce highly reliable, high quality information. WINS node is based on an open modular design using widely available commercial off-the-shelf technology. The wireless microsensor nodes combine sensing capabilities (such as seismic, acoustic, and magnetic) with a commercial digital cordless telephone radio and an embedded commercial RISC microprocessor in a small package. As these networks are designed for low power, embedded signal processing is performed in order to reduce communication requirements. For example, many thousands of bytes of raw time series data from a vibration sensor are reduced to a few bytes of amplitude and frequency information using the on-board processor. Communicating only processed information reduces the power required to convey information by orders of magnitude. WINS nodes are supporting experiments in multi-hop data communication protocols, dynamic cooperative signal processing (e.g. beamforming with randomly spaced nodes), and distributed resource management.

The unique aspects of microsensor networks can be examined with significant numbers of prototype devices explicitly designed for this purpose, as opposed to generic computing platforms. Some of the unique requirements for WINS include:

- small, lightweight form factor;
- robustness to wide temperature ranges and other demanding environmental conditions;
- battery or other stand-alone power sources;
- low power operation and access to internal power control mechanisms;
- a small, low-power radio having sufficient range;
- a real-time execution environment;
- the ability to code in a high level language for rapid algorithm hosting and testing, and
- a reasonable cost.

WINS nodes support battlefield applications, and a variety of vehicle health management and condition-based maintenance applications on industrial, military, and space platforms. For example, a motor and pump test bed for developing component (e.g. bearing), process (e.g. fluid pumping), and system-level (e.g. an overall collection of motors and pumps in a large-scale process) monitoring and diagnostics was constructed at the Rockwell Science Center. This test bed was instrumented with WINS nodes which incorporate acceleration, pressure, and temperature sensors and algorithms for machinery and process diagnostics. The signal processing algorithms running on the individual nodes provide for incipient detection of a wide variety of faults. The wireless networked communications provide for simple installation and cooperative diagnostics among groups of motors, pumps, and valves in the system. A web-based browser allows the entire system, and any component within the system, to be remotely monitored.

Distributed microsensor networks use cooperative processing and low power communication protocols. Scenarios, such as monitoring large areas, buildings, or avenues of approach, are accomplished by positioning the sensors close to the areas of interest in high densities. Close spacing permits low-power sensing and short-range radio links. The nodes can be precisely located or dispersed in random configurations with spatial knowledge (or lack thereof) incorporated in the signal processing and communication algorithms. This versatility makes the nodes suitable for a wide range of applications, for example, perimeter security and reconnaissance, monitoring of machinery and other assets within large industrial plants, monitoring of many subsystems on-board large vehicles.

WINS nodes communicate with the external world via an enterprise level network, such as a factory control network and/or the Internet as shown in Figure 10.2. Two-way communication is provided throughout the system, as each WINS node supports bidirectional, peer-to-peer communications with its neighbors. The WINS nodes can be static or slowly mobile. Multiple portals for transporting information into or out of the sensor network can be established. A portal can be extended by allowing the connection of long-range radios to any one of the nodes or through a gateway to a wired network, such as the Internet, allowing users to monitor and control the network remotely. A WINS user can issue commands through a user interface hosted on a personal

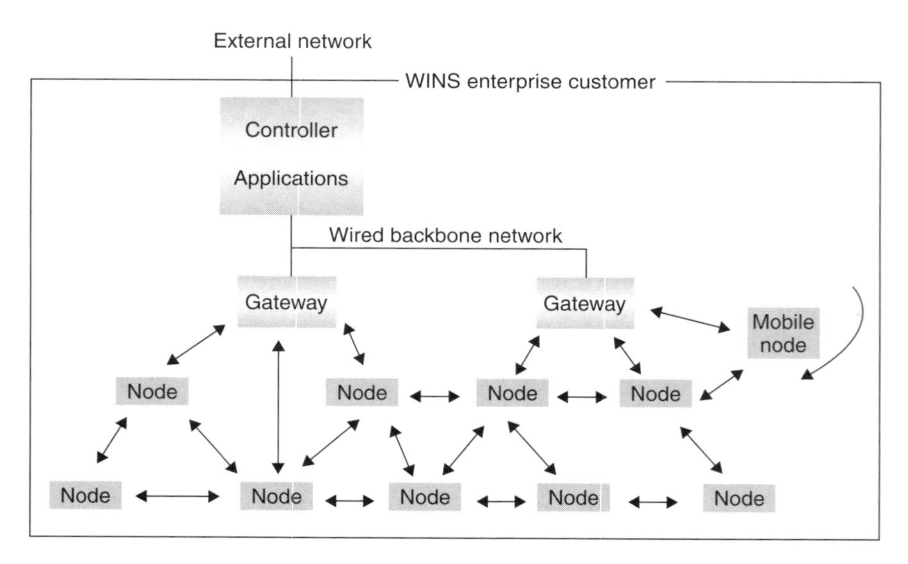

Figure 10.2 A schematic description of a WINS system architecture.

or handheld computer, allowing users to control the network of nodes, for example, setting sensor sensitivity thresholds or to reprogram the nodes over the wireless link. The user interface can display activity at each of the sensors along with their health status (e.g. battery level).

A digital spread spectrum radio in each WINS node provides a robust wireless communication link, and enables data rates of 100 kb/s over ranges in excess of 100 meters. Two-way, peer-to-peer communication among nodes in a small neighborhood supports multi-hop data transfers, avoiding the requirement for all nodes to be in the range of a base station. This feature gives users a very high degree of flexibility in the deployment of the nodes, thus enabling strategic sensor placement in the area of interest without the constraint of line-of-sight communications to a central data collection or gateway site. The WINS concept takes advantage of the fact that short-range radio hops are exponentially more power efficient than larger hops covering the same distance. Power control on each radio is further used to minimize the transmit power needed to communicate with its neighbors.

Networking in a WINS system is distinguished from that in a conventional wireless data network for the following reasons:

- Nodes have limited battery energy, making Time-Division Multiple Access (TDMA) schemes attractive, but requiring special routing schemes optimized for minimal power consumption.
- The sensor nodes may require synchronization for time tagging of data and coherent signal processing that is implemented with power conserving, network time distribution algorithms.
- Nodes may have multiple sensor types (e.g. seismic, acoustic, IR, etc.), each with different coverage, accuracy, and power consumption, and allowing local sensor fusion.
- The generated traffic patterns of WINS are generally predictable, allowing efficient tuning of protocols. While traffic is created by random events (e.g. target detections, user commands), the destinations and, hence, routes are constrained, as are the message volumes and allowed latencies. Detection information is forwarded to portals. There is also data summarization along the network routing path.
- Cooperative processing, such as beamforming, requires dynamic multicast groups of nodes that are closest to the events. Since targets or other phenomena that cause events can be mobile, the set of nodes that are actively sensing them will change, moving the locus of message generators.

The requirement for simple node deployment necessitates that the network of nodes be capable of self-discovery and self-configuration. Self-organizing procedures for boot-up and automatic node incorporation into the network allow nodes to be added to an operational network for improved coverage or replenishment. Mechanisms for recovering from node failures allow the network to be self-healing. The WINS uses a power-efficient, time-division multiple access scheme supporting multi-hop communication. Routing algorithms avoid creating power consumption hotspots that result in sensors in a neighborhood dissipating battery energy much more rapidly than the rest of the network causing partitions when their energy is depleted.

Research into low-power signal processing algorithms is an integral part of the system development effort and, for battlefield applications, is focusing on the following:

- Target detection/classification – WINS nodes run vibration detection algorithms based on energy thresholding. This technique is subject to false alarms leading to consideration of more sophisticated spectral signature algorithms. Low-power algorithms to classify a detected event as an impulsive event (e.g. either a foot-step or gun-shot) or vehicle (e.g. wheeled or tracked, light or heavy) are used.
- On-board sensor fusion – The inclusion of multiple sensors on each of the nodes enables fusion of different sensed phenomenologies, leading to higher quality information and decreased false alarm rates. Algorithms for fusing the seismic, acoustic and magnetic sensors on a single node are used.
- Multi-node sensor fusion – Algorithms utilizing the advantages of a network of spatially separate nodes span a range of cooperative behaviors, each of which trades off detection quality versus energy consumption. Examples of cooperative fusion range from high-level decision corroboration (e.g. voting), to feature fusion, and full coherent beam formation.

10.3.1. Development Platform

The hardware in each microsensor node uses an open, modular design that allows incorporation of a range of sensors. Board interconnection is provided by two 40-pin miniconnectors. The connectors form a system bus that provides power and control lines to the sensor boards, and supports multiple open interfaces. The WINS node consists of a stack of base circuits comprising the processor, radio and power supply, which are coupled with the desired sensors. The hardware components are as follows:

- acoustic sensor;
- DCT (Digital Cordless Telephone) digital spread spectrum radio module;

- StrongARM processor module;
- multiple voltage power supply module;
- seismic sensor module;
- Mark 4 Products geophone (seismic sensor), and
- two standard 9-V batteries

The basic hardware block diagram given in Figure 10.3 shows the connectivity and power distribution between the major modules within the system.

Processor Module: The processor module is built around the Intel StrongArm SA1100 embedded controller. The SA1100 is a general-purpose, 32-bit RISC microprocessor based on the ARM architecture. The processor offers a 16-kb instruction cache, an 8-kb data cache, serial I/O and JTAG (Joint Test Action Group) interface all combined in a single chip. Program and data storage are provided by 128-kb SRAM (Static RAM) and 1-Mb of bootable flash memory. Connection with the sensor modules is easily achieved using the four-wire SPI (Serial Peripheral Interface). An RS232 port is added to the module for connection to external devices. The processor has three states: normal, idle and sleep, that can be controlled to reduce power consumption.

Radio Module: The radio module uses the Conexant Systems, Inc. RDSSS9M Digital Cordless Telephone (DCT) chip-set which implements a 900-MHz

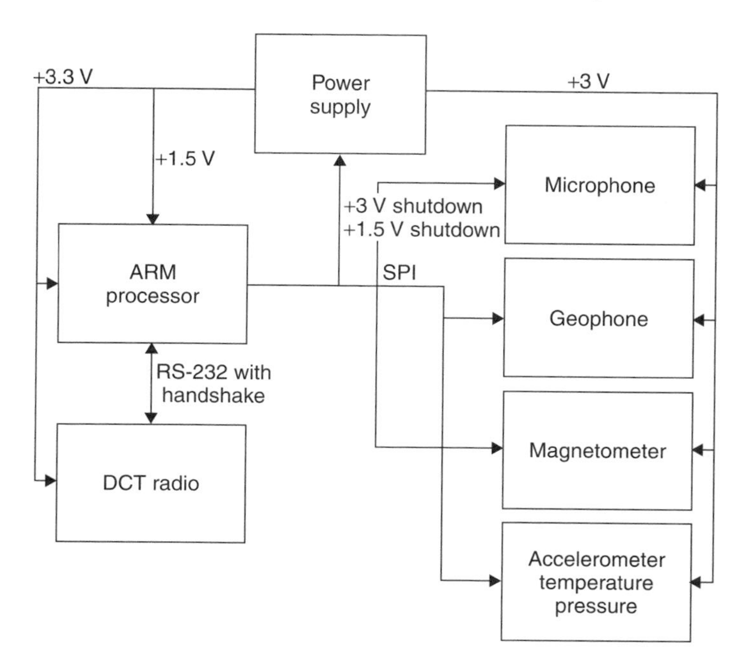

Figure 10.3 A hardware block diagram.

spread spectrum RF communications link. The chipset has an embedded 65C02 microcontroller that performs all control and monitoring functions required for direct sequence spread-spectrum communication (12 chips/bit), as well as data exchange, with the processor module. The radio operates on one of 40 channels in the ISM frequency band, selectable by the controller. Program and data storage are provided with 32-kb SRAM and 1-Mb of bootable flash memory. Embedded firmware is developed to support multiple access networking with minimal ARM processor support. The board also provides a 4-bit ADC for battery voltage monitoring. The RF portion of the radio is packaged as a small multi-chip module, interfaces to a 50 ohm helical antenna, and is capable of operating at multiple transmit power levels between 1 and 100 mW, enabling the use of power-optimized communication algorithms.

Seismic Sensor Module: The seismic sensor board uses a Mark IV geophone designed for low frequency detection of seismic events. The sensitivity of this geophone is about 15 g. The circuit employs an Analog Devices AD 7714 sigma-delta converter that results in a clean, 20-dbit signal from 1 Hz to 400 Hz. The circuit is sufficiently repeatable to allow phase matching between sensor nodes to support cooperative coherent processing, such as beamforming.

Other sensor modules include:

- *Acoustic Sensor*. The acoustic sensor board employs a miniature microphone such as a Knowles BL1785 microphone element having a low frequency cutoff of 4 Hz. The maximum frequency of interest for acoustic sensor applications is selected as 2 kHz. It is desirable to preserve phase information for beamforming applications.
- *Magnetometer*. A magnetometer module employs the Honeywell HMC1001 and has a 10 Hz bandwidth. The rated sensitivity is 27 microgauss making it able to detect 1 lb of iron from 6 feet.
- *Accelerometer*. An accelerometer board is fabricated for use in machinery vibration monitoring. This board includes a high-speed accelerometer sampled at 48 kHz. The WINS accelerometer board also provides inputs for temperature and pressure sensors.

To support experimentation and algorithm development, a flexible software environment in which applications can be written in a high-level language, such as C, while maintaining access to low-level hardware functions, such as power control, is essential. The key WINS software functions are organized in the following layers:

- *Monitor/Hardware Abstraction Layer (HAL)*. The HAL provides routines for initialization, external communication, program loading and debugging, and interrupt processing. A packet protocol interpreter routes packets arriving from either the radio or the external RS-232 to internal tasks. Program loading can occur either through an attached device, or through the radio.

- *Run-time environment.* This real-time kernel on each node provides the low-level distributed WINS network infrastructure. The low-level controls for communication protocols, as well as the sensor drivers, are hosted at this level.
- *System Applications.* Signal processing computations and higher layer network functions (e.g. scheduling, routing) are performed and written in a high-level programming language, such as C. New applications may be downloaded onto sensor nodes that are deployed in the field via the RF network.
- User interface applications hosted on PCs that allow users to perform various tasks and to interact with the sensor network. An interface for communicating with the network through a gateway is supported as well as display and logging of network information.

A real-time, preemptive, multi-tasking kernel is ported to the processor module that is based on the Micro C/OS, and designed to run on top of the HAL. The HAL provides the three critical assembly language functions for the Micro C/OS (real-time timer, context switching, and interrupt handler). The relationship between the applications, the operating system and the HAL is shown in Figure 10.4, with some details of the implementation. The OS is used to schedule the applications, enforce interprocessor communication with the radio and other attached intelligent modules, control higher-level power management, control attached sensor device drivers, and handle network messaging and related protocol functions. The HAL provides a standard view of interrupts to the OS and low-level access to power management, code downloads, Flash memory programming function, and JTAG debugging interface. A standard I/O function is implemented to support C-language debugging and a C-level interface for network messaging.

Figure 10.5 shows the overall software architecture with software entities hosted on the ARM processor module, the DCT module, the sensor modules and a host computer (PC). Support for a long-range radio is also provided. Several development tools reside on the PC and the ARM for use in software coding and debugging. The users obtain the ARM System Developers Toolkit (SDT) or similar compiler in order to develop C-based applications.

A power-efficient TDMA scheme is implemented as the basic WINS link-layer protocol. The TDMA scheme allows nodes to turn off their receiver and/or transmitter when they are not scheduled to communicate. A multi-hop

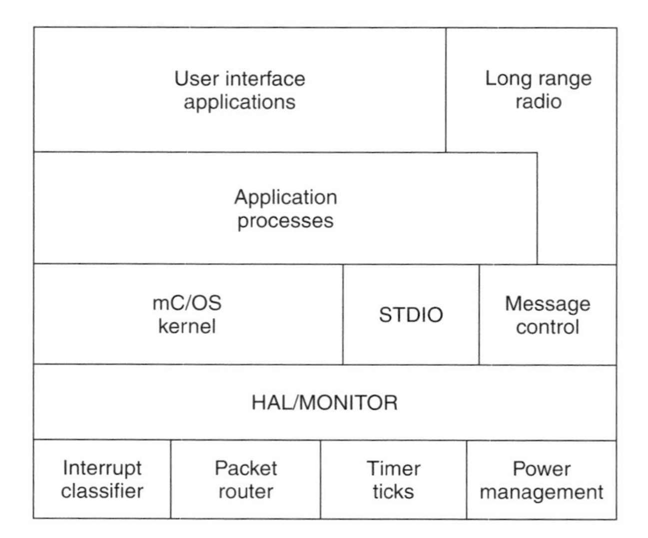

Figure 10.4 Runtime environment components.

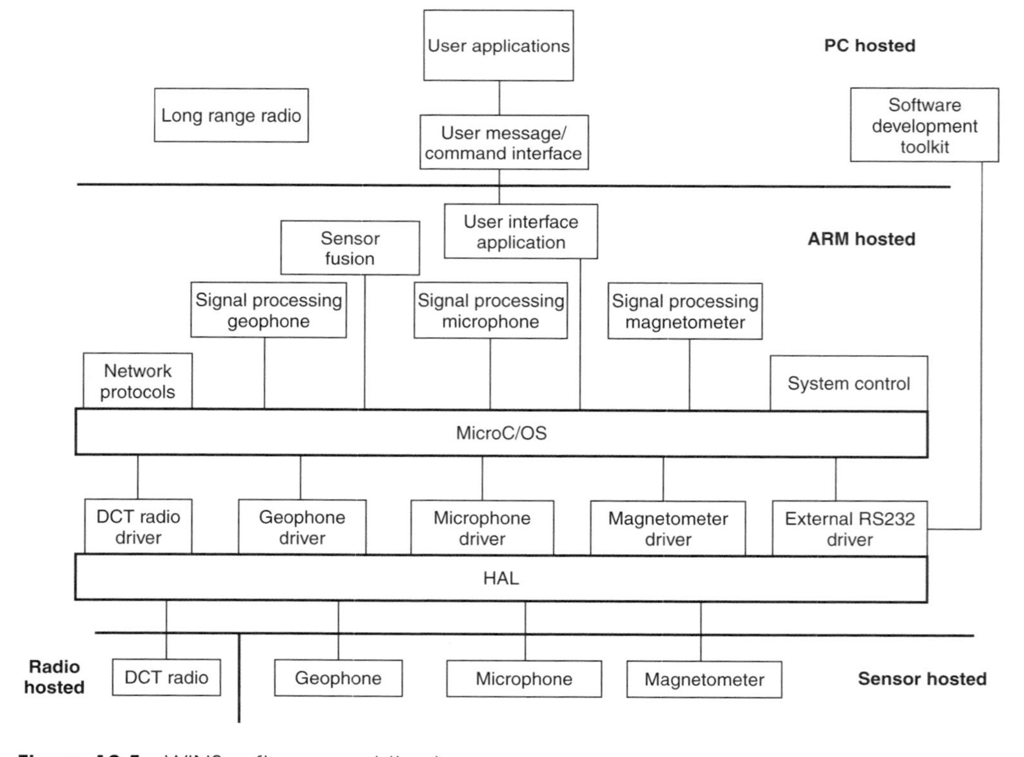

Figure 10.5 WINS software architecture.

routing scheme is also implemented so that information from distant nodes can be forwarded to destination locations. The link layer protocols are built on top of the digital spread spectrum radio broadcast channel and provide a raw data rate of 100 kb/s. Various low-overhead forward error correction schemes are also implemented.

The entire WINS sensor node consumes a peak of 1 W of power, with the processor consuming 300 mW, the radio consuming 600 mW in transmit mode and 300 mW in receive mode, and less than 100 mW consumed by the sensor transducers. Proper control of the system ensures that peak power is rarely required. An essential capability of the devices is that they can be put into idle or sleep modes under low-level software control to increase the system operational lifetime.

10.3.2. Applications

For military users, a primary focus has been area monitoring. WINS employs large arrays of distributed small and capable sensors for both security and surveillance applications. The added feature of robust, self-organizing networking makes WINS deployable by untrained troops in essentially any situation. Distributed sensing has the further advantages of being able to provide redundant and hence highly reliable information on threats as well as the ability to localize threats by both coherent and incoherent processing among the distributed sensor nodes. WINS is used in traditional sensor network applications for large-area and perimeter monitoring and enables every platoon, squad, and soldier to deploy sensor networks to accomplish a myriad of mission and self-protection goals. For the urban terrain, WINS dramatically improves troop safety as they clear and monitor intersections, buildings, and rooftops by providing continuous vigilance for unknown troop and vehicle activity.

The primary challenge facing WINS, or indeed any military area monitoring or security system, is accurate identification of the signal being sensed. Rockwell is developing state-of-the-art vibration, acoustic, and magnetic signal classification algorithms to accomplish this goal. For added assurance, Rockwell has also integrated the latest in CMOS (Complementary Metal-Oxide Semiconductor) visible and uncooled infrared imaging into the WINS architecture. Sensor-cued images of detected threats are rapidly relayed to cognizant personnel and battlefield, commanders for real-time threat ID and prosecution. These capabilities provide unparalleled safety and awareness on the battlefield, and help to realize the goal of zero friendly force casualties, even in the most stressful situations.

WINS systems are specifically tailored to the requirements for monitoring complex machinery and processes. WINS are deployed inside factories and on board ships for continuous health monitoring of equipment. WINS are also being explored for use on aircraft, rotorcraft, and spacecraft as part of an overall integrated vehicle health management system. The primary driver in these applications is to reduce the requirement for human monitoring of equipment and provide detailed and continuous knowledge on the operating state of equipment such that costly, unanticipated downtime is avoided and logistics activities are optimized. The cost savings that can be achieved with on-line equipment monitoring are dramatic. WINS is enabling these applications because it eliminates the high costs (e.g. economic, size, and weight) associated with wire-line networks.

The primary challenge for WINS in machinery and process monitoring is related to the quality of the information produced by both the individual sensors and the distributed sensor network. Nodes located on individual components must not only be able to provide information on the present state of the component (e.g. a bearing or a gear box), but also predict the remaining useful life of the component. Rockwell is developing diagnostic algorithms for machinery vibration monitoring that differ from previous efforts in this area in that they are designed to be generally applicable to broad classes of components, as opposed to being tailored for a specific component. For example, an algorithm that autonomously determines all of the critical frequencies of a bearing (e.g. inner race, outer race, number of balls, operating speed, etc.), and then uses the information to predict remaining useful life, was developed and tested under a large variety of conditions. State-of-the-art bearing diagnostic algorithms either do not provide a comprehensive picture of bearing health, or must have a large amount of bearing-specific data supplied as initial input. This same algorithm is finding general diagnostic application for gear boxes and other rotating machinery.

Distributed collections of WINS nodes located on machine components and/or throughout a process provide information on the overall machine and/or process on which they are deployed. Distributed sensing system enables inferences from individual component data to be used to provide diagnostics for aspects of the system that are not directly being sensed. For example, monitoring bearing vibrations or motor currents can provide information on not only the bearing health, but also the inception and severity of pump cavitation. Pump cavitation, in turn, can provide information on the state of valves located throughout a pumping process. The heart of this capability is model-based diagnostics and Rockwell is a leader in the development of tools and applications for model-based diagnostics.

The dynamically reconfigurable nature of WINS is used in an application of WINS to space vehicle health monitoring in collaboration with the Boeing company. WINS is deployed throughout space vehicles and performs different missions during the different phases of the space flight. For example, during the launch phase, WINS located on various critical components of the spacecraft monitors vibration levels for out-of-compliance signals. During flight and re-entry, the WINS network monitors structural disturbances caused by the significant temperature gradients encountered as different portions of the vehicle are alternately exposed and shadowed from the sun and atmosphere. This is accomplished via coherent collection and processing of vibration and strain data. Upon landing, critical components are once again monitored for out of compliance signals. These data are used to determine those components needing post-flight maintenance or replacement, thus enabling faster turn-around for the space vehicle and thereby dramatically lowering costs.

WINS was applied in an experiment demonstrating the ease with which ultra-low-cost picosatellites can be deployed and communicated with. In collaboration with the Aerospace Corporation, WINS-based picosatellites were launched on two space vehicles. WINS communications and processor modules form the heart of an experiment in which multiple satellites are deployed from a mother satellite. These satellites communicate both with each other and with a ground station. The cost of the WINS-based picosatellites is less than 5000 US dollars and they are slightly larger than a deck of playing cards. A network of Micro-Electro-Mechanical (MEMS) switches serves as the satellite payload, thus demonstrating the efficacy of MEMS in space.

10.4. BUILDING AND MANAGING AGGREGATES IN WIRELESS SENSOR NETWORKS

Designing and operating a sensor network requires forming and managing aggregates of sensors for collaborative processing tasks. Consider the problem of tracking a moving herd of zebra in wildlife habitat management. An example of a collaboration region is defined as the set of seismic sensor nodes that can potentially sense the movement of the animals, i.e. within the propagation range of vibrations from the animal footsteps. Such a group of sensors is an aggregate which collaboratively performs a specific task. For example, these sensors can collectively estimate the size of the herd from the intensity of the vibrations and the speed at which the herd travels from the frequency of the signals. As the herd moves to the next region, a new aggregate of sensors will have to wake up and start to track the animals, and so on. The definition of such collaboration regions depends on the task objectives and resource constraints. For example, some sensors may be on critical paths of routing and their energy reserve is more likely to be depleted than others. These sensors should participate in forming an aggregate only when the expected gain exceeds a threshold. Moreover, the collaboration regions are dynamically defined and updated, as the physical events of interest, environmental conditions, or network topology change. To define and maintain the collaboration regions adaptively is one of the key tasks in sensor network operation.

A decentralized protocol Distributed Aggregate Management (DAM) forms sensor aggregates for a target counting task. The protocol comprises a decision predicate P for each node i to decide if it should participate in an aggregate, and a message exchange scheme M about how the grouping predicate is applied to nodes. A node determines if it belongs to an aggregate based on the result of applying the predicate to the data of the node, as well as an information from other nodes. Aggregates are formed when the process converges. The protocol supports a representative collaborative signal-processing task in sensor networks counting distinct targets in a sensor field. Sensor aggregates defined by multiple interfering targets are considered.

Directed diffusion is an effective mechanism for coordinating information transport in sensor networks. It uses a fine-grain data-level publish-and-subscribe for data sources to advertise data attributes of signals they detect, and for data sinks to express data attributes in which they are interested. The data-source attributes and data-sink interest are propagated and met throughout the network. Routing pathways between the sources and sinks are established as shortest paths in the network connectivity graph.

The DAM protocol can be considered as an example of the next-level-up coordination mechanism that defines regions of sensors in a network. Unlike directed diffusion where data attributes are first class objects, DAM makes grouping predicates first class entities. Directed diffusion forms data routing and aggregation paths, while DAM forms sensor aggregates that are defined by constraints arising from tasks, resources, or geometries of a space.

Geographic routing is a mechanism for routing data to a geographic region instead of to a destination node specified by an address. The destination region must be specified as a rectangle or some other regular geometric object for computational reasons. DAM can form arbitrarily complex regions as long as the network topology permits and the resulting aggregates can be abstracted as geometric objects for use by geographic routing. On the other hand, geographic routing could implement the information exchange within the groups of sensors in DAM.

SQL-style queries such as AVERAGE, MIN, MAX can be used for distributed sensor networks. A SQL-style database system supports an aggregation function and a grouping predicate. For example, the query

SELECT TRUNC(temp/10)

AVERAGE(light) FROM sensors

GROUP BY TRUNC(temp/10)

HAVING AVERAGE(light) $\geqslant$ 50

forms groups of sensors according to the temperature bins, computes average light for each group, and then excludes those groups with light values less than 50. Collaborative signal processing applications must form aggregates specified not just by individual node data, but also by the relations on the data across nodes. In the target counting problem, for example, nodes exchange and compare amplitude detection values in order to form groups belonging to each target.

Approaches to collaborative signal processing address the formation and management of collaboration regions in several application contexts. For example, to track moving vehicles on the street, sensor groups are dynamically formed, with each group responsible for collecting and processing information about one vehicle. These collaboration patterns can be abstracted into a set of generic schemas to support a wide class of applications for sensor networks.

We consider a task of counting multiple targets in a two-dimensional sensor field. Targets can be stationary or moving at any time independent of the states of the other targets.

In addition, the following assumptions are made about this network:

- Targets are point sources of signals. Target signal amplitude attenuates, as a monotonically decreasing function of the distance from the source, according to an inverse distance squared law (e.g. acoustic signal propagation in free space) or exponentially.
- Each sensor has a finite sensing range. Sensors can only sense the amplitude. Signals of two targets sum at a sensor.
- Each sensor can communicate wirelessly with other sensors within a fixed radius larger than the mean internode distance.
- Sensors are time synchronized to a global clock.
- The main limiting factors are the on-board battery power, and the network bandwidth and latency.

The task here is to determine the number of targets in the field, forming an initial count and recomputing the count when targets move, enter, or leave the field. For each distinct target, a sensor leader corresponding to the target is elected. As targets move, new leaders are elected to reflect network changes. Hence, we can obtain a target count by determining the number of leaders elected.

Formally, a sensor network is represented as a graph $G(V, E)$, where V are the vertices representing sensor nodes and E edges representing one-hop connectivity in the network. The counting protocol has the structure (G, T, P, M), with T the targets, P the grouping predicate, and M the messaging schema. The schema M applies P to nodes in the network to compute sensor aggregates $A\{V_1, \ldots, V_n\}$, where $V_i \in V$.

When targets are well separated, sensor aggregates for the targets become islands in the network. In other cases when the influence regions of the targets overlap and each sensor in a region is able to separate signal components for each target, then the network can maintain overlapping sensor aggregates, one for each target.

Sensor networks with diverse target characteristics (velocity, moving patterns, etc.) and limited network resources require a protocol with the following characteristics:

- The protocol is distributed and autonomous for scalability.
- The leader election process converges quickly to allow fast leader re-election for fast-moving objects.
- The protocol is designed so that minimal amount of inter-sensor communication is needed while keeping application semantics intact.
- A reasonable level of fault tolerance is supported.

As the sensors in this network can only sense amplitude, the spatial characteristics of target signals have to be considered when multiple targets are in close proximity of each other.

(1) When the target influence areas are well separated, leader election can be considered as a clustering problem. Otherwise, it becomes a peak counting problem.
(2) Target signal propagation has a large impact on target resolution. The faster the signal attenuates with distance from the source, the easier targets are to discern from their neighbors, based on signal amplitude they emit.
(3) Spacing of sensors is also critical in obtaining correct target count. Sensor density has to be high enough for sampling of target signal amplitude provided by sensors to yield enough information for obtaining correct

target counts. On the other hand, too close a proximity of a sensor to its neighbors makes its measurement redundant and wastes resources.

In the protocol design, sensors are somewhat evenly spaced with a mean inter-sensor distance determined by the target signal attenuation characteristics, sensor sensitivity to target signals and target signal strength.

Leader elections are conducted by sensors exchanging information with their neighbors via one-hop broadcast.

Neighbors of sensor S refer to sensors that are within the transmitting radius of sensor S, i.e. all the sensors that can hear sensor S directly.

Broadcast, in this context, refers to a multicast to all neighbors of a sensor.

There is the minimum signal amplitude that a sensor has to receive from target(s) for it to participate in the leader election process. This value is determined by the protocol designer, and cannot be smaller than the sensor receiving threshold determined by the noise floor.

Protocol period is the time duration of a leader election process. The leader election process runs every protocol period.

Sensor state is a set of parameters that a sensor keeps during each protocol period in order to process packets from other sensors to elect leaders. There is a field indicating if the sensor node participates in the leader election process.

The design criteria are fast convergence and minimal amount of inter-sensor communications. They are achieved by dropping those packets that will not become a leader at the earliest possible stage.

10.5. HABITAT AND ENVIRONMENTAL MONITORING

Habitat and environmental monitoring represent a class of sensor network applications with enormous potential benefits for scientific communities and society. Instrumenting natural spaces with numerous networked microsensors can enable long-term data collection at scales and resolutions that are difficult, if not impossible, to obtain otherwise. The intimate connection with its immediate physical environment allows each sensor to provide localized measurements and detailed information that is hard to obtain through traditional instrumentation. The integration of local processing and storage allows sensor nodes to perform complex filtering and triggering functions, as well as to apply application-specific or sensor-specific data compression algorithms. The ability to communicate not only allows information and control to be communicated across the network of nodes, but allows nodes to cooperate in performing more complex tasks, like statistical sampling, data aggregation, and system health and status monitoring. Increased power efficiency gives applications flexibility in resolving fundamental design tradeoffs, e.g. between sampling rates and battery lifetimes. Low-power radios with well-designed protocol stacks allow generalized communications among network nodes, rather than point-to-point telemetry. The computing and networking capabilities allow sensor networks to be reprogrammed or retasked after deployment in the field. Nodes have the ability to adapt their operation over time in response to changes in the environment, the condition of the sensor network itself, or the scientific endeavor.

The potential impact of human presence when monitoring plants and animals in field conditions includes changing behavioral patterns or distributions. The anthropogenic disturbance can seriously reduce or even destroy sensitive populations by increasing stress, reducing breeding success, increasing predation, or causing a shift to unsuitable habitats. While the effects of disturbance are usually immediately obvious in animals, plant populations are sensitive to trampling by even well-intended researchers, introduction of exotic elements through frequent visitation, and changes in local drainage patterns through path formation.

Disturbance effects are of particular concern in small-island situations, where it may be physically impossible for researchers to avoid some impact on an entire population. In addition, islands often serve as refugia for species that cannot adapt to the presence of terrestrial mammals, or may hold fragments of once widespread populations that have been extirpated from much of their former range.

Seabird colonies are notorious for their sensitivity to human disturbance. Even a visit of several minutes to a cormorant colony can result in up to 20 % mortality among eggs and chicks in a given breeding year. Repeated disturbance will lead to complete abandonment of the colony. In another example, Leach's storm petrels are likely to desert their nesting burrows if they are disturbed during the first 2 weeks of incubation.

Sensor networks represent a significant advance over traditional invasive methods of monitoring. Sensors can be deployed prior to the onset of the breeding season or other sensitive period (in the case of animals), or while plants are dormant or the ground is frozen (in the case of botanical studies). Sensors can be deployed on small islets where it would be unsafe or unwise repeatedly to attempt field studies.

10.5.1. Island Habitat Monitoring

The College of the Atlantic (COA) is field testing *in-situ* sensor networks for habitat monitoring. COA has ongoing field research programs on several remote islands with well established on-site infrastructure and logistical support. Great Duck Island (GDI) (44.09°N, 68.15°W) is a 237 acre island located 15 km south of Mount Desert Island, Maine. The Nature Conservancy, the State of Maine, and the College of the Atlantic hold much of the island in joint tenancy.

Mainwaring *et al.* (2002) are primarily interested in three major questions in monitoring the Leach's storm petrel at GDI:

(1) What is the usage pattern of nesting burrows over the 24–72-hour cycle when one or both members of a breeding pair may alternate incubation duties with feeding at sea?
(2) What changes can be observed in the burrow and surface environmental parameters during the course of the approximately 7-month breeding season (April–October)?
(3) What are the differences in the microenvironments with and without large numbers of nesting petrels?

Each of these questions has unique data needs and suitable data acquisition rates. Presence/absence data is most likely to be acquired through occupancy detection and temperature differentials between burrows with adult birds and burrows that contain eggs, chicks, or are empty. Petrels are unlikely to enter or leave during the light phase of a 24-hour cycle, but measurements every 5–10 minutes during the late evening and early morning are needed to capture time of entry or exit. More general environmental differentials between burrow and surface conditions during the extended breeding season can be captured by records every 2–4 hours, while differences between popular and unpopular sites benefit from hourly sampling, especially at the beginning of the breeding season.

It is unlikely that any one parameter recorded by wireless sensors could determine why petrels choose a specific nest site. However, by making multiple measurements of many variables predictive models can be developed. These models will correlate which conditions seabirds prefer.

Great Duck Island requirements are as follows:

- *Internet access*: The sensor networks at GDI must be accessible via the Internet. An essential aspect of habitat monitoring applications is the ability to support remote interactions with *in-situ* networks.
- *Hierarchical network*: The field station at GDI needs sufficient resources to host Internet connectivity and database systems. However, the habitats of scientific interest are located up to several kilometers farther away.

A second tier of wireless networking provides connectivity to multiple patches of sensor networks deployed at each of the areas of interest. Three to four patches of 100 static (not mobile) nodes is sufficient to start.

- *Sensor network longevity*: Sensor networks that run for 9 months from non-rechargeable power sources are used. Although ecological studies at GDI span multiple field seasons, individual field seasons typically vary from 9 to 12 months. Seasonal changes as well as the plants and animals of interest determine their durations.
- *Management at-a-distance*: The remoteness of the field sites requires the ability to monitor and manage sensor networks over the Internet. The goal is zero on-site presence for maintenance and administration during the field season, except for installation and removal of nodes.
- *Inconspicuous operation*: Habitat monitoring infrastructure must be inconspicuous. It should not disrupt the natural processes or behaviors under study. Removing human presence from the study areas both eliminates a source of error and variation in data collection, as well as a significant source of disturbance.
- *System behavior*: From both a systems and end-user perspective, it is critical that sensor networks exhibit stable, predictable, and repeatable behavior whenever possible. An unpredictable system is difficult to debug and maintain. More importantly, predictability is essential in developing trust in these new technologies for life scientists.
- *In-situ interactions*: Although the majority of interactions with the sensor networks are expected to be via the Internet, local interactions are required during initial deployment and during maintenance tasks, as well as during on-site visits. PDAs serve an important role in assisting with these tasks. They may directly query a sensor, adjust operational parameters, or simply assist in locating devices.
- *Sensors and sampling*: For those particular applications, the ability to sense light, temperature, infrared, relative humidity, and barometric pressure provide an essential set of useful measurements. The ability to sense additional phenomena, such as acceleration/vibration, weight, chemical vapors, gas concentrations, pH, and noise levels is also useful.
- *Data archiving*: Archiving sensor readings for off-line data mining and analysis is essential. The reliable off-loading of sensor logs to databases in the wired, powered infrastructure is an essential capability. The desire to drill-down and explore individual sensors interactively, or a subset of sensors, in near real-time complement log-based studies. In this mode of operation, the timely delivery of fresh sensor data is key. Lastly, nodal data summaries and periodic health-and-status monitoring requires timely delivery.

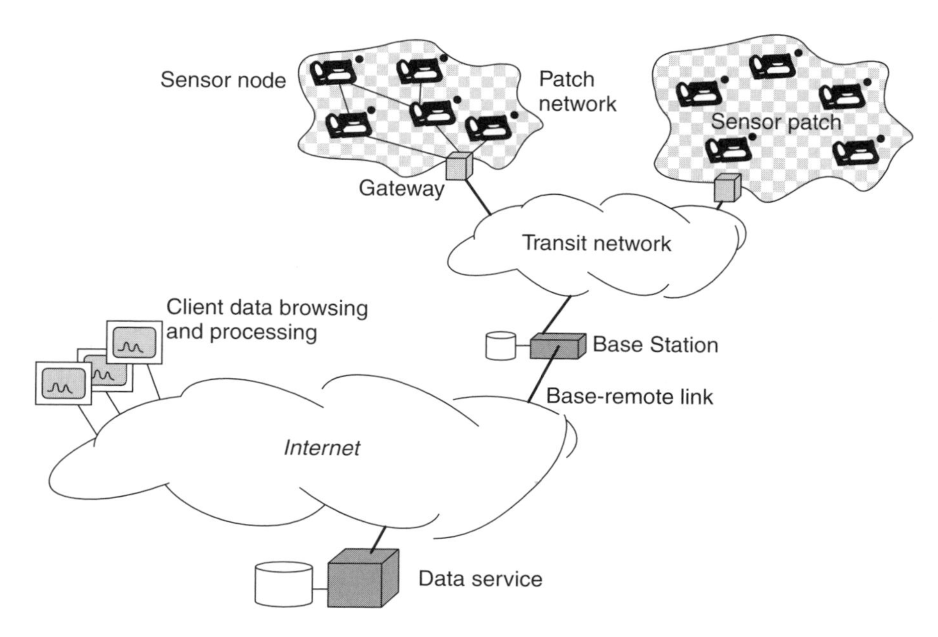

Figure 10.6 System architecture for habitat monitoring.

A tiered architecture is shown in Figure 10.6. The lowest level consists of the sensor nodes that perform general purpose computing and networking in addition to application-specific sensing. The sensor nodes may be deployed in dense patches that are widely separated. The sensor nodes transmit their data through the sensor network to the sensor network gateway. The gateway is responsible for transmitting sensor data from the sensor patch, through a local transit network, to the remote base station that provides WAN (Wide Area Network) connectivity and data logging. The base station connects to database replicas across the Internet. The data is displayed to scientists through a user interface. Mobile devices, referred to as the 'gizmo', may interact with any of the networks whether used in the field or across the world connected to a database replica.

The lowest level of the sensing application is provided by autonomous sensor nodes. These small, battery-powered devices are placed in areas of interest and each collects environmental data, primarily about its immediate surroundings. Because it is placed close to the phenomenon of interest, a sensor can often be built using small and inexpensive individual sensors. High spatial resolution can be achieved by dense deployment of sensor nodes. Compared with traditional approaches, which use a few high quality sensors with sophisticated signal processing, this architecture provides higher robustness against occlusions and component failures.

The computational module is a programmable unit that provides computation, storage, and bidirectional communication with other nodes in the system. The computational module interfaces with the analog and digital sensors on the sensor module, performs basic signal processing (e.g. simple translations based on calibration data or threshold filters), and dispatches the data according to the application's needs. Compared with traditional data logging systems, networked sensors offer two major advantages: they can be retasked in the field and they can easily communicate with the rest of the system. *In-situ* retasking allows the scientists to refocus their observations based on the analysis of the initial results. Suppose that initially we want to collect the absolute temperature readings; however after the initial interpretation of the data we might realize that significant temperature changes exceeding a defined threshold are the most interesting.

Individual sensor nodes communicate and coordinate with one another. The sensors will typically form a multihop network by forwarding each other's messages, which vastly extends connectivity options. If appropriate, the network can perform in-network aggregation (e.g. reporting the average temperature across a region). This flexible communication structure allows us to produce a network that delivers the required data while meeting the energy requirements.

Ultimately, data from each sensor needs to be propagated to the Internet. The propagated data may be raw, filtered, or processed. Bringing direct, wide-area connectivity to each sensor path is not feasible, the equipment is too costly, it requires too much power and the installation of all required equipment is quite intrusive to the habitat. Instead, wide-area connectivity is brought to a base station, adequate power and housing for the equipment is provided. The base station may communicate with the sensor patch using a wireless local area network. Wireless networks are particularly advantageous since often each habitat involves monitoring several particularly interesting areas, each with its own dedicated sensor patch.

Each sensor patch is equipped with a gateway that can communicate with the sensor network and provide connectivity to the transit network. The transit network may consist of a single hop link or a series of networked wireless nodes, perhaps in a path from the gateway to the base station. Each transit-network design has different characteristics with respect to expected robustness, bandwidth, energy efficiency, cost, and manageability.

To provide data to remote end-users, the base station includes WAN connectivity and persistent data storage for the collection of sensor patches. Since many habitats of interest are quite remote, the WAN connection can be wireless (e.g. two-way satellite). The components are reliable, enclosed in environmentally protected housing, and provided with adequate power. In many environments such conditions can be provided relatively easily at a ranger station.

The architecture addresses the possibility of disconnection at every level. Each layer (sensor nodes, gateways, base stations) has some persistent storage which protects against data loss in case of power outage. Each layer also provides data management services. At the sensor level, they take the form of data logging. The base station offers a full-fledged relational database service. The data management at the gateways falls somewhere in between; they offer some database services, but over limited window of data. While many types of communication can be unreliable, when it comes to data collection, long-latency is preferable to data loss. For this kind of communication, a custody transfer model, similar to SMTP (Simple Mail Transfer Protocol) messages or bundles, may be applicable.

Users interact with the sensor network data in two ways. Remote users can access the replica of the base-station database (in the degenerate case they interact with the database directly). This approach allows for easy integration with data analysis and mining tools, while masking the potential wide area disconnections with the base stations. Remote control of the network is also provided through the database interface. Although this control interface is sufficient for remote users, on-site users often require a more direct interaction with the network. A small, PDA-sized device, referred to as a gizmo, enables such interaction. The gizmo can directly communicate with the sensor patch, provide the user with a fresh set of readings about the environment and monitors the network. While the gizmo will typically not take custody of any data, it allows the user interactively to control the network parameters by adjusting the sampling rates, power management parameters, and other network parameters. The connectivity between any sensor node and the gizmo does not have to rely on functioning multi-hop sensor network routing, instead the user often communicates with the mote network directly, relying on single-hop proximity.

10.5.2. Implementation

The motes are used as the sensor nodes. The member of the mote family, called Mica, uses a single channel, 916 MHz radio from RF Monolithics to provide bidirectional communication at 40 kbps, an Atmel Atmega 103 microcontroller running at 4 MHz, and considerable amount of nonvolatile storage (512 kb). A pair of conventional AA batteries and a DC boost converter provide a stable voltage source, though other renewable energy sources can be easily used. The node has a small size and is approximately $2.0 \times 1.5 \times 0.5$ inches.

An environmental monitoring sensor board is used to provide measurements. The Mica Weather Board provides sensors that monitor changing environmental conditions with the same functionality as a traditional weather station. The Mica Weather Board includes temperature, photoresistor, barometric pressure, humidity, and passive infrared (thermopile) sensors.

The barometric pressure module is a digital sensor manufactured by Intersema. The sensor is sensitive to 0.1 mbar of pressure and has an absolute pressure range from 300 to 1100 mbar. The module is calibrated during manufacturing and the calibration coefficients are stored in EEPROM persistent storage. The pressure module includes a calibrated temperature sensor to compensate raw barometric pressure readings.

The humidity sensor is manufactured by General Eastern. It is a polymer capacitive sensor factory calibrated to within 1 picofarad (± 3 % relative humidity). The sensing element consists of an electrode metallization deposited over the humidity sensor polymer. The sensor is modulated by a 555 CMOS timer to sense the charge in the capacitor which is filtered through by RC circuit. The resulting voltage is amplified by an instrumentation amplifier for greater sensitivity over the range of 0 to 100 % relative humidity.

The thermopile is a passive infrared sensor manufactured by Melexis. Heat from black bodies in the sensor's field of view causes a temperature difference between the thermopile's cold junction and the thermopile membrane. The temperature difference is converted to an electric potential by the thermoelectric effect in the thermopile junctions. The sensor does not require any supply voltage. The thermopile includes a thermistor in the silicon mass, and the thermistor may be used to measure the temperature of the cold junction on the thermopile and accurately calculate the temperature of the black body.

The photoresistor is a variable resistor in a voltage divider circuit. The divided voltage is measured by the ADC. The final temperature sensor is a digital calibrated sensor that communicates over the I^2C bus.

The unique combination of sensors can be used for a variety of aggregate operations. The thermopile may be used in conjunction with its thermistor and the photoresistor to detect cloud cover. The thermopile may also be used to detect occupancy, measure the temperature of a nearby object (for example, a bird or a nest), and sense changes in the object's temperature over time. If the initial altitude is known, the barometer module may be used as an altimeter. Strategically placed sensor boards with barometric pressure sensors can detect the wind speed and direction by modeling the wind as a fluid flowing over a series of apertures.

An I^2C analog-to-digital converter is separated from the main Mica processing board to provide greater flexibility in developing components to reduce power consumption. The ADC uses less power than the Atmel processor on the Mica, may be used in parallel with processing or radio transmission on the Mica, and can be operated in various low-power and sleep modes. Additionally, the sensor board includes an I^2C 8×8 power switch that permits individual components on the board to be turned on or off. Each switch can be operated independently of the others, further reducing power consumption.

The Mica Weather Board was designed with interoperability in mind. The Mica includes a 51-pin expansion connector. The connector has the ability to stack sensor boards on top of each other. Instead of allowing each board to compete for pins on the connector, an access protocol is used. The Mica will change the value of a switch on the sensor board using the I^2C bus. Changing the value of the switch triggers the sensor board's hardware logic to access the Mica's resources. When a board has access, it may use the power, interrupt, ADC, and EEPROM lines that are directly connected to the microprocessor and components on the Mica processing board.

Many habitat monitoring applications need to run for 9 months, the length of a single field season. Mica runs on a pair of AA batteries, with a typical capacity of 2.5 amperehours (Ah). A conservative estimate is that the batteries supply 2200 mAh at 3 volts.

The system operates uniformly over the deployment period, and each node has 8.148 mAh per day available for use. The application chooses how to allocate this energy budget between sleep modes, sensing, local calculations and communications. Different nodes in the network have different functions, and they also have very different power requirements. Nodes near the gateway need to forward all messages from a patch, whereas a node in a nest needs to report its own readings. When the power limited nodes exhaust their supplies, the network is disconnected and inoperable.

Minimizing power in sleep mode involves turning off the sensors, the radio, and putting the processor into a deep sleep mode. I/O pins on the microcontroller need to be put in a pull-up state whenever possible, as they can contribute as much as 100 μA of leakage current. Mica architecture uses a DC booster to provide stable voltage from degrading alkaline batteries. With no load, the booster draws between 200 μA and 300 μA, depending on the battery voltage. While this functionality is crucial for predictable sensor readings and communications, it is not needed in the sleep mode. The current draw of the microprocessor is proportional to the supply voltage.

CerfCube is a small, StrongArm-based embedded system that acts as the sensor patch gateway. Each gateway is equipped with a Compact Flash 802.11b adapter. CerfCubes run an embedded version of Linux operating system. Permanent storage is plentiful, and the gateway can use the IBM MicroDrive which provides up to 1 Gb of storage. To satisfy the CerfCube power requirements, a solar panel providing between 60 and 120 watts in full sunlight is connected to a rechargeable battery with capacity between 50 and 100 watt-hours (e.g. sealed lead-acid). The CerfCube with a 12 dbi omni directional 2.4-GHz antenna provides a range of approximately 1000 feet.

In the mote-to-mote solution, a mote is connected to the base station and a mote is in the sensor patch. Both motes are connected to 14 dbi directional 916 MHz Yagi antennae having a range of more than 1200 feet. The differences between the mote and the CerfCube include a different communication frequency and power requirements, and software components. The mote's MAC layer does not require a bidirectional link like 802.11 b. Additionally, the mote sends raw data with a small packet header of four bytes directly over the radio, as opposed to overheads imposed by 802.11b and TCP/IP connections. The mote solution for the gateway is used due to its power efficiency.

The collection of sensor network patches is connected to the Internet through a wide-area link. On GDI, the Internet is connected through a two-way satellite connection. The satellite system is connected to a laptop which coordinates the sensor patches and provides a relational database service. The base station functions as a turn key system, and runs unattended.

The database stores time-stamped readings from the sensors, health status of individual sensors, and metadata (e.g. sensor locations). The GDI database is replicated every 15 minutes over the wide-area satellite link.

In habitat monitoring the ultimate goal is data collection; sampling rates and precision of measurements are often dictated by external specifications. For every sensor there is a cost of taking a single sample. The cost of data processing and compression is traded against the cost of data transmission.

The energy is allocated for sampling the sensors and communicating the results, maintaining the network MAC protocols, health and status, routing tables, and forwarding network messages. These tasks can either be tightly scheduled or run on demand. On one extreme, the system is scheduled at every level, from TDMA access to the channel, through scheduled adaptation of routes and channel quality. Overhead costs are up front and fixed. A TDMA system is expected to perform well if the network is relatively static. On the other extreme, a low-power hailing channel can be used to create on-demand synchronization between a sender and a receiver. The service overhead is proportional to the use of the service. This approach can be more robust to unexpected changes in the network, at the expense of extra cost. Finally, a hybrid approach is possible, where each service runs in an on-demand fashion, but the time period for when the demand can occur is scheduled on a coarse basis.

Power efficient communication paradigms for habitat monitoring include a set of routing algorithms, media-access algorithms, and managed hardware access. The routing algorithms are tailored for efficient network communication while maintaining connectivity when required to source or relay packets.

A simple routing solution for low duty cycle sensor networks is simply to broadcast data to a gateway during scheduled communication periods. This method is the most efficient since data is only communicated in one direction and there is no dependency on surrounding nodes for relaying packets in a multi-hop manner. The routing deployed on GDI is a hierarchical model. The sensor nodes in burrows are transmit only with a low duty cycle and they sample about once per second. The gateway mote is fully powered by solar power, so it is always on and relaying packets to the base station. A multi-hop scheduled protocol is used to collect, aggregate, and communicate data. Methods like GAF (Geographical Adaptive Fidelity) and Span (an energy-efficient coordination algorithm for topology maintenance in dedicated wireless networks) have been used to extend the longevity of the network by selecting representatives to participate in the network, thereby reducing the average per-node power consumption. Although these methods provide factors of 2- to 3-times longer network operation, our application requires a factor of 100-times longer network operation, since the sensor nodes are on for at most 1.4 h per day. GAF and Span do not account for infrequent sampling but rather continuous network connectivity and operation. Scheduled multi-hop routing or low power MAC protocols are augmented with GAF and/or Span to provide additional power savings. GAF and Span are independent of communication frequency, whereas our application requires increased power savings that may be achieved by adjusting the communication frequency.

A power efficient method for scheduling the nodes has to ensure that long multihop paths may be used to relay the data:

- After determining an initial routing tree, set each mote's level from the gateway. Schedule nodes for communication on adjacent levels starting at the leaves. As each level transmits to the next, it returns to a sleep state. The following level is awakened, and packets are relayed for the scheduled time period. The process continues until all levels have completed transmission in their period. The entire network returns to a sleep mode. This process repeats itself at a specified point in the future.
- Instead of a horizontal approach, nodes are woken along paths or subtrees in a vertical approach. Each subtree in turn completes its communication up the tree. This method is more resilient to network contention; however the number of subtrees in the network will likely exceed the number of levels in the network, and subtrees may be disjointed, allowing them to communicate in parallel.

Alternatively, low power MAC protocols can be used. By determining the duty cycle, we can calculate the frequency with which the radio samples for a start symbol. By extending the start symbol when transmitting packets, we can match the length of the start symbol to the sampling frequency. Other low power MAC protocols, such as S-MAC (Sensor-MAC) and Aloha with preamble sampling, employ similar techniques that turn off the radio during idle periods in order to reduce power consumption. The difference between scheduled communication and low power MACs is, instead of having a large power and network overhead to set up a schedule, the overhead is distributed along the lifetime of the node. Both approaches are equivalent in power consumption, the decision as to which to use depends on the end-user interactivity required by the application. A potential trade-off in using a low power MAC is that transmitted packets potentially wake up every node within the cell. Although early rejection can be applied, scheduling prevents unneeded nodes from wasting power processing a packet's headers.

10.6. SUMMARY

The communication-centric design approach in TinyOS is used to build a networking infrastructure for self-organized, deeply embedded collections of devices.

The Tiny active message is used to support dynamic network discovery and multi-hop *ad hoc* routing. Discovery can be initiated from any node, but often it is rooted at gateway nodes that provide connectivity to conventional networks. Each root periodically transmits a message carrying its ID and its distance, which is equal to zero, to its neighborhood. The message handler checks whether the source is the closest node from which it has heard recently (i.e. in the current discovery phase) and, if so, records the source ID as its multi-hop parent, increments the distance, and retransmits the message with its own ID as the source. The discovery component utilizes the buffer swap.

The hardware directly connects the central microcontroller to the radio. This places all of the real-time requirements of the radio onto the microcontroller, which must handle every bit that is transmitted or received in real time. Additionally, it controls the timing of each bit so that any jitter in the control signals that it generates is propagated to the transmitted signal. The TinyOS communication stack handles these constraints while allowing higher level functions to continue in parallel.

The TinyOS approach has proven quite effective in supporting general purpose communication among potentially many devices that are highly constrained in terms of processing, storage, bandwidth, and energy with primitive hardware support for I/O. The event-driven model facilitates interleaving the processor between multiple flows of data and between multiple layers in the stack for each flow while still meeting the severe real-time requirements of servicing the radio. Since storage is very limited, it is common to process messages incrementally at several levels, rather than buffering entire messages and processing them level-by-level. However, events alone are not sufficient; it is essential that an event be able to hand any substantial processing to a task that will run outside the real-time window. This provides logical concurrency within the stack and is used at every level except the lowest hardware abstraction layer. By adopting a nonblocking, event-driven approach, the traditional threads are not supported, with their associated multiple stacks and complex synchronization.

The component approach yields robust operation despite limited debugging capabilities, and facilitates experimentation. The packet components can be swapped with a simple change to the description graph and temporary components can be interposed between existing components, without changing any of the internal implementations. Moreover, the use of components allows essentially an entire subtree of components to be replaced by hardware and vice versa.

The Tiny active message programming model permits experiments with numerous higher level networking layers and fine-grained distributed algorithms. The nodes can be reprogrammed over the network. A node can obtain code capsules from its neighbors or over multihop routes and assemble a complete execution image in its EEPROM secondary store. The node can then use this to reprogram itself. Other examples include a general purpose data logging and acquisition capability, a facility to query nodes by schema, and to aggregate data from a large number of nodes within the network.

Without the traditional layers of abstraction dictating what kinds of capability are available, it is possible to foresee many novel relationships between the application and the underlying system. The adaptive transmission control scheme is a simple example; rejection of the send request causes the application to adjust its rate of originating data. The application level forwarding of multi hop traffic allows the node to keep track of its changing set of neighbors. Moreover, the radio is itself another sensor, since receive signal strength is provided to the ADC. Thus, each packet can be accompanied by signal strength data for use in estimating physical distance or presence of obstructions. The radio is also an actuator, as its signal strength, and therefore cell size, can be controlled.

The lowest layer components are synchronizing all receivers to the transmitter to within a fraction of a bit. Thus, very fine grain time synchronization information can be provided with every packet for control applications.

WINS nodes support battlefield applications, and a variety of vehicle health management and condition-based maintenance applications on industrial, military, and space platforms. For example, a motor and pump test bed for developing component (e.g. bearing), process (e.g. fluid pumping), and system-level (e.g. an overall collection of motors and pumps in a large-scale process) monitoring and diagnostics was constructed at the Rockwell Science Center. This test bed was instrumented with WINS nodes, which incorporate acceleration, pressure, and temperature sensors and algorithms for machinery and process diagnostics. The signal processing algorithms running on the individual nodes provide for incipient detection of a wide variety of faults. The wireless networked communications provide for simple installation and cooperative diagnostics among groups of motors, pumps, and valves in the system. A web-based browser allows the entire system, and any component within the system, to be remotely monitored.

PROBLEMS

Learning Objectives

After completing this chapter you should be able to:

- demonstrate understanding of application and communication support for wireless sensor networks;
- explain the applications of wireless sensor networks to area monitoring and integrated vehicle health management applications;
- discuss what is meant by a WINS development platform;
- discuss building and managing aggregates in wireless sensor networks;
- demonstrate understanding of co-design and reconfiguration;
- discuss the use of wireless sensor networks in habitat and environmental monitoring.

Practice Problems

Problem 10.1: What is the TinyOS concurrency model?
Problem 10.2: What is the role of events?
Problem 10.3: What is the nonblocking approach in TinyOS?
Problem 10.4: What is an active message?

Problem 10.5: Why is the managing of buffer storage difficult?
Problem 10.6: How can power consumption be reduced?
Problem 10.7: What are the node components in a wireless distributed sensor network?
Problem 10.8: What are the requirements for WINS microsensor network?
Problem 10.9: What are the requirements for sensor node deployment?
Problem 10.10: How are the counting targets obtained?
Problem 10.11: What are the requirements for a protocols in a sensor network with diverse target characteristics?
Problem 10.12: What is the sensor state?
Problem 10.13: What are the design criteria?
Problem 10.14: How can sensor networks be deployed?
Problem 10.15: How do the sensor nodes communicate?
Problem 10.16: How can a combination of sensors be used?

Practice Problem Solutions

Problem 10.1:

The TinyOS concurrency model is a two-level scheduling hierarchy, where events preempt tasks, and the tasks do not preempt other tasks. The vast majority of operation is in the form of nonblocking state transitions. Within a task, commands may be called, a command may call subordinate commands, or it may post tasks to continue working logically in parallel with its invocation. By convention, all commands return a status indicating whether the command was accepted, providing a full handshake. Since all components have bounded storage, a component must be able to refuse commands. A command may initiate an operation, for instance, accessing a sensor or sending a message, leaving the operation to be carried out concurrently with other activities, by either using hardware parallelism or tasks.

Problem 10.2:

Events are initiated at the lowest level by hardware interrupts. Events may signal higher level events, call commands, or post tasks. Commands cannot signal events. Thus, an individual event may propagate through multiple levels of components, triggering collateral activity. Whenever the work cannot be accomplished in a small, bounded amount of time, the component should record continuation information in its frame and post a task to complete the work. By convention, the lowest level hardware abstraction components perform enough interrupt processing to re-enable interrupts before signaling the event. Events (or tasks posted within events) typically complete the split-phase operations initiated by commands, signaling the higher-level component that the operation has completed and perhaps passing it the data.

Problem 10.3:

A nonblocking approach is taken throughout TinyOS. There are no locks, and components never spin on a synchronization variable. A lock-free queue data structure is used by the scheduler. Components perform a phase of an operation and terminate, allowing the completion event to resume their execution. Most components are written essentially as reentrant state machines. TinyOS is written in the C programming language with conventional preprocessor macros to highlight the key concepts. The TinyOS execution model is implemented on a single shared stack with a static frame per component.

Problem 10.4:

Active Messages (AM) is a simple, extensible paradigm for message-based communication by using procedure calls. Each message contains the name of a handler to be invoked on a target node upon arrival, and a data payload. The handler function serves the dual purpose of extracting the message from the network and either integrating the data into the computation or sending a response. The AM communication model is event driven and specifically designed to allow a very lean communication stack to process packets directly off the network, while supporting a wide range of applications.

Problem 10.5:

Managing buffer storage is a difficult problem in a communication stack because the following issues must be addressed:

- encapsulating useful data with transport header and trailer information;
- determining when output message data storage can be reused, and
- providing an input buffer for an incoming message before the message has been inspected, to determine where it goes.

Problem 10.6:

Power consumption can be reduced by using periodic listening. By creating time periods when transmitting is not permitted, the nodes must listen only part time. This approach works well when the time scale of the invalid periods is quite large relative to the message transmission time. The downside of this approach is that it limits the used bandwidth.

In sensor networks, a node may act as a router or data processing point, and may need to use the radio bandwidth fully. Low-power listening keeps the same listener duty cycle concept, but greatly reduces the time scale.

To reduce the average power consumption of the network further, low power listening can be combined with periodic listening. Running both schemes simultaneously results in listening at reduced power for only a fraction of the time. The power reductions are multiplicative. These techniques provide a mechanism for trading bandwidth and transmission cost for a reduction in receive power consumption.

Problem 10.7:

Wireless distributed microsensor networks consist of a collection of communicating nodes, where each node incorporates:

(a) one or more sensors for measuring the environment;
(b) processing capability in order to process sensor data into high value information and to accomplish local control, and
(c) a radio to communicate information to/from neighboring nodes and to external users.

Problem 10.8:

The unique aspects of microsensor networks can be examined with significant numbers of prototype devices explicitly designed for this purpose, as opposed to generic computing platforms. Some of the unique requirements for WINS include:

- small, lightweight form factor;
- robustness to wide temperature ranges and other demanding environmental conditions;
- battery or other stand-alone power sources;
- low power operation and access to internal power control mechanisms;
- a small, low-power radio having sufficient range;
- a real-time execution environment;
- the ability to code in a high level language for rapid algorithm hosting and testing, and
- a reasonable cost.

Problem 10.9:

The requirement for simple node deployment necessitates that the network of nodes be capable of self-discovery and self-configuration. Self-organizing procedures for boot-up and automatic node incorporation into the network allow nodes to be added to an operational network for improved coverage or replenishment. Mechanisms for recovering from node failures allow the network to be self healing. WINS uses a power-efficient, time-division, multiple-access scheme supporting multi-hop communication. Routing algorithms avoid creating power consumption hotspots that result in sensors in a neighborhood dissipating battery energy much more rapidly than the rest of the network, causing partitions when their energy is depleted.

Problem 10.10:

The task here is to determine the number of targets in the field, forming an initial count and recomputing the count when targets move, enter, or leave the field. For each distinct target, a sensor leader corresponding to the target is elected. As targets move, new leaders are elected to reflect network changes. Therefore, we can obtain target count by determining the number of leaders elected.

Problem 10.11:

Sensor networks with diverse target characteristics (velocity, moving patterns, etc.) and limited network resources require a protocol with the following characteristics.

- the protocol is distributed and autonomous for scalability;
- the leader election process converges quickly to allow fast leader re-election for fast moving objects;
- the protocol is designed so that minimal amount of inter-sensor communication is needed while keeping application semantics intact;
- a reasonable level of fault tolerance is supported.

Problem 10.12:

Sensor state is a set of parameters that a sensor keeps during each protocol period in order to process packets from other sensors to elect leaders. There is a field to indicate if the sensor node participates in the leader election process.

Problem 10.13:

The design criteria are fast convergence and minimal amount of inter-sensor communications. They are achieved by dropping those packets that will not become a leader at the earliest possible stage.

Problem 10.14:

Sensor networks represent a significant advance over traditional invasive methods of monitoring. Sensors can be deployed prior to the onset of the breeding season or other sensitive period (in the case of animals) or while plants are dormant or the ground is frozen (in the case of botanical studies). Sensors can be deployed on small islets where it would be unsafe or unwise repeatedly to attempt field studies.

Problem 10.15:

Individual sensor nodes communicate and coordinate with one another. The sensors will typically form a multi-hop network by forwarding each other's messages, which vastly extends connectivity options. If appropriate, the network can perform in-network aggregation (e.g. reporting the average temperature across a region). This flexible communication structure allows us to produce a network that delivers the required data while meeting the energy requirements.

Problem 10.16:

The unique combination of sensors can be used for a variety of aggregate operations. The thermopile may be used in conjunction with its thermistor and the photoresistor to detect cloud cover. The thermopile may also be used to detect occupancy, measure the temperature of a nearby object (for example, a bird or a nest), and sense changes in the object's temperature over time. If the initial altitude is known, the barometer module may be used as an altimeter. Strategically placed sensor boards with barometric pressure sensors can detect the wind speed and direction by modeling the wind as a fluid flowing over a series of apertures.

References

Accetta, M., R. Baron, W. Bolosky, D. Golub, R. Rashid, A. Tevanian, and M. Young (1986). Mach: a new kernel foundation for UNIX development. *Proceedings of the Summer 1986 USENIX Conference*, pp. 93–112.

Advanced Configuration and Power Interface. URL: http://www.teleport.com/~acpi.

Adjie-Winoto, W., E. Schwartz, H. Balakrishnan, and J. Lilley (1999). The design and implementation of an intentional naming system. *Proceedings of the ACM Symposium on Operating Systems Principles*, pp. 186–201.

Agre, J. R., L. P. Clare, G. J. Pottie, and N. P. Romanov (1999). Development platform for self-organizing wireless sensor networks. *Proceedings of the SPIE AeroSense '99 Conference on Digital Wireless Communication*, Orlando, FL.

Akyildiz, I., J. McNair, J. Ho, H. Uzunalioglu, and W. Wang (1998). Mobility management in current and future communications networks. *IEEE Network*, **12** (4), 39–49.

Akyildiz, I., W. Su., Y. Sankarasubramaniam, and E. Cayirci (2002). Wireless sensor networks: a survey. *Computer Networks*, **38**, 393–422.

Albrecht, M., M. Frank, and P. Martini (1999). Bluetooth architecture and services overview: the role of IP and quality of service issues. First Workshop on IP Quality of Service for Wireless and Mobile Networks, Aachen, Germany. URL: http://opensource.nus.edu.sg/projects/bluetooth/others/IQWiM99_reprint.pdf.

Amirtharajah, R., T. Xanthopoulos, and A. P. Chandrakasan (1999). Power scalable processing using distributed arithmetic. *Proceedings of the International IEEE Symposium on Low Power Electronics and Design*, pp. 170–75.

Amis, A. D., R. Prakash, T. H. P. Vuong, and D. T. Huynh (2000). Max-min d-cluster formation in wireless ad hoc networks. *Proceedings of the Annual Joint Conference of the IEEE Computer and Communications Societies*, pp. 32–41.

Wireless Sensor Network Designs A. Hać
 ISBN: 0-470-86736-1

ARM Ltd. ARM710T Data Sheet. URL: http://www.arm.com/documentation/datasheets/PDF/DDI10086B.pdf.

Asada, G., M. Dong, T. Lin, F. Newberg, G. Pottie, H. Marcy, and W. Kaiser (1998). Wireless integrated network sensors: low power systems on a chip. *Proceedings of the 24th European Solid-State Circuits Conference, Den Hague*. Elsevier, Amsterdam, pp. 9–12.

Atmel Corporation (2000). Atmel ATmega 103/103L Datasheet.

Bagrodia, R., R. Meyer, M. Takai, Y. Chen, X. Zeng, J. Martin, B. Park, and H. Song (1998). Parsec: a parallel simulation environment for complex systems. *IEEE Computer*, **31** (10), 77–85.

Baker, D. J., and A. Ephremides (1981). The architectural organization of a mobile radio network via a distributed algorithm. *IEEE Transactions on Communication*, **29**, 1694–1701.

Bellovin, S., and M. Merritt (1992). Encrypted key exchange: password-based protocols secure against dictionary attack. *Proceedings of the 1992 IEEE Computer Society Conference on Research in Security and Privacy*, pp. 72–84.

Bellovin, S., and M. Merrit (1993). Augmented encrypted key exchange: a password-based protocol secure against dictionary attacks and password file compromise. *Proceedings of the First ACM Conference on Computer and Communications Security*, pp. 244–50.

Bellows, P. and B. Hutchings (1998). JHDL – an HDL for reconfigurable systems. *Proceedings of the IEEE Symposium ond FPGAs for Custom Computing Machines*, pp. 175–84.

Benini, L., and G. D. Micheli (1997). *Dynamic Power Management Design Techniques and CAD Tools*. Kluwer Academic Publishers, Norwell, MA.

Bennett, F., D. Clarke, J. B. Evans, A. Hopper, A. Jones, and D. Leask (1997). Piconet: embedded mobile networking. *IEEE Personal Communications*, **4** (5), 8–15.

Bershad, B., S. Savage, P. Pardyak, E. Sirer, M. Fiuczynski, D. Becker, C. Chambers, and S. Eggers (1995). Extensibility, safety, and performance in the SPIN operating system. *Proceedings of the ACM Symposium on Operating Systems Principles*, pp. 267–84.

Bhardwaj, M., T. Garnett, and A. Chandrakasan (2001). Upper bounds on the lifetime of sensor networks. *Proceedings of the IEEE International Conference on Communications*, Vol. 3, pp. 785–790.

Bhardwaj, M., R. Min, and A. Chandrakasan (2001). Quantifying and enhancing power-awareness of VLSI systems. *IEEE Transactions on Very Large Scale Integration Systems*, **9**, 757–72.

Bianchi, G. (2000). Performance analysis of the IEEE 802.11 distributed coordination function. *IEEE Journal on Selected Areas in Communications*, **18**, 535–47.

Billinghurst, M., and T. Starner (1999). Wearable devices: new ways to manage information. *IEEE Computer*, **32** (1), 57–64.

Blaze, M., J. Feigenbaum, and J. Lacy (1996). Decentralized trust management. *Proceedings of the 1996 IEEE Symposium on Security and Privacy*, pp. 164–73.

Bloom, B. H. (1970). Space/time trade-offs in hash coding with allowable errors. *Communications of the ACM*, **13**, 422–26.

Bluetooth specification. URL: http://www.bluetooth.com.

Bonnet, P., J. Gehrke, and P. Seshadri (1987). Towards sensor database systems. *Lecture Notes in Computer Science*, Springer Verlag, pp. 3–14.

Bonnet, P., J. Gehrke, and P. Seshadri (2000). Querying the physical world. *IEEE Personal Communications*, **7** (5), 10–15.

Borriello, G., and R. Want (2000). Embedded computation meets the World Wide Web. *Communications of the ACM*, **43** (5), 59–66.

Boser, B. (1997). Electronics for micromachined inertial sensors. *Proceedings of Transducers '97*, pp. 1169–72.

Bruno, J., J. Brustoloni, E. Gabber, A. Silberschatz, and C. Small (1999). Pebble: a component-based operating system for embedded applications. *Proceedings of the USENIX Workshop on Embedded Systems, Cambridge, MA*. URP: http://www.usenix.org/publications/library/proceedings/es99/bruno.html.

Burd, T., T. Pering, A. Stratakos, and R. Brodersen (2000). A dynamic voltage scaled microprocessor system. *Proceedings of the International IEEE Conference on Solid-State Circuits*, pp. 294–95.

Callaway, E., P. Gorday, L. Hester, J. A. Gutierrez, M. Naeve, B. Heile, and V. Bahl (2002). Home networking with IEEE 802.15.4: a developing standard for low-rate wireless personal area networks. *IEEE Communications Magazine*, **40** (8), 70–77.

Case, J. D., M. Fedor, M. L. Scholstall, and C. Davin (1990). RFC 1157: Simple network management protocol. RFC. ITEF.

Chandrakasan, A. P., R. Amirtharajah, S. Cho, J. Goodman, G. Konduri, J. Kulik, W. Rabiner, and A. Wang (1999). Design considerations for distributed microsensor systems. *Proceedings of the IEEE Custom Integrated Circuits Conference*, San Diego, pp. 279–286.

Chang, J., and L. Tassiulas (2000). Energy conserving routing in wireless ad hoc networks. *Proceedings of the Annual Joint Conference of the IEEE Computer Communications Societies*, pp. 22–31.

Chen, B., K. Jamieson, H. Balakrishnan, and R. Morris (2001). Span: an energy-efficient coordination algorithm for topology maintenance in ad hoc wireless networks. *Proceedings of the ACM/IEEE International Conference on Mobile Computing and Networking*, pp. 85–96.

Chu, C. H., N. R. Lo, E. C. Berg, and K. S. J. Pister (1997). Optical communication using micro corner cube reflectors. *Proceedings of the IEEE MEMS Workshop*, Nagoya, Japan, pp. 350–55.

Clausen, T., P. Jaquet, A. Laouiti, P. Minet, P. Muhlethaler, A. Qayyum, and L. Viennot (2001). Optimized link state routing protocol. Internet draft: draft-ietf-manet-olsr-05.txt (November).

Conway, P. and D. Hefferman. CAN and the new IEEE 1451 Smart Transducer Interface Standard. URL: http://www.ul.ie/~pei/pdf_files/fp15.pdf.

Conway, P., D. Hefferman, B. O'Mara, P. Burton, and T. Miao (2000). IEEE 1451.2: An interpretation and example implementation. *Proceedings of the 17th IEEE Instrumentation and Measurement Technology Conference*, Vol. 2, pp. 535–540.

Culler, D. E., J. Hill, P. Buonadonna, R. Szewczyk, and A. Woo (2001). A network centric approach to embedded software for tiny devices. First International Workshop on Embedded Software, Tahoe City, CA. Proceedings. *Lecture Notes in Computer Science*, Springer Verlag, pp. 144–30.

Cummins, T., E. Byrne, D. Brannick, and D. A. Dempsey (1998). An IEEE 1451 standard transducer interface chip with 12-b ADD, two 12.b ADCs, 10 kb flash EEPROM, and 8-b microcontroller. *IEEE Journal of Solid-State Circuits*, **33**, 2112–20.

Czerwinski, S. E., B. Y. Zhao, T. D. Hodes, A. D. Joseph, and R. H. Katz (1999). An architecture for a secure service discovery service. *Proceedings of the ACM/IEEE International Conference on Mobile Computing and Networking*, Seattle, Washington, pp. 24–35.

Dasgupta, P., R. LeBlanc, M. Ahamad, and U. Ramachandran (1992). The Clouds distributed operating system. *IEEE Computer*, **24** (11).

Da Silva Jr, J. L., J. Shamberger, M. J. Ammer, C. Guo, S. Li, R. Shah, T. Tuan, M. Sheets, J. M. Rabaey, B. Nikolic, A. Sangiovanni-Vincentelli, and P. Wright (2001). Design methodologiew for PicoRadio networks. *Proceedings of the IEEE Design Automation and Test in Europe Conference*, pp. 314–23.

Deb, B., S. Bhatnagar, and B. Nath (2001). A topology discovery algorithm for sensor networks with applications to network management. *DCS Technical Report 441*, Rutgers University.

Dick, R. P., G. Lakshminarayana, A. Raghunathan, and N. K. Jha (2000). Power analysis of embedded operating systems. *Proceedings of the IEEE Conference on Design Automation*, pp. 312–15.

Dick, R. P., G. Lakshminarayana, A. Raghunathan, and N. K. Jha (2003). Analysis of power dissipation in embedded systems using real-time operating systems. *IEEE Transactions on Computer-Aided Design of Integrated Circuits and Systems*, **22**, 615–27.

Dijkstra, E. W. (1968). The structure of the multiprogramming system. *Communications of the ACM*, **11**, 341–46.

Diniz, P. S. R. (1997). *Adaptive Filtering Algorithms and Practical Implementation*. Kluwer Academic Publishing.

Dorward, S., R. Pike, and P. Winterbottom (1997). Programming in Limbo. *Proceedings of IEEE Compcon '97*, pp. 245–50.

Dorward, S., R. Pike, D. Presotto, D. Ritchie, H. Trickey, and P. Winterbottom (1997). Inferno. *Proceedings of the IEEE Compcon '97*, pp. 241–44.

Downey, A. (1999). Using pathchar to estimate Internet link characteristics. *Proceedings of the ADM SIGCOMM Conference*, Cambridge, MA, pp. 241–50.

ECos documentation. URL: http://sourceware.cygnus.com/ecos/docs.html.

Elgamal, T. (1985). A public-key cryptosystem and a signature scheme based on discrete logarithms. *Proceedings of CRYPTO '84*, pp. 10–18.

El-Hoiydi, A. (2002). Aloha with preamble sampling for sporadic traffic in ad hoc wireless sensor networks. *Proceedings of the IEEE International Conference on Communications*, pp. 3418–3423.

Engler, D. (1996). VCODE: a retargetable, extensible, very fast dynamic code generation system. *Proceedings of the Conference on Programming Language Design and Implementation*, pp. 160–70.

Engler, D., M. F. Kaashoek, and J. O'Toole Jr (1995). Exokernel: an operating system architecture for application-level resource management. *Proceedings of the ACM Symposium on Operating Systems Principles*, pp. 251–66.

Estrin, D., L. Girod, G. Pottie, and M. Srivastava (2001). Instrumenting the world with wireless sensor networks. *Proceedings of the International IEEE Conference on Acoustics, Speech, and Signal Processing*, Vol. 4, pp. 2033–36.

ETSI HIPERLAN/2 Standard. URL: http://www.etsi.org/technicalactiv/hiperlan2.htm.

Fang, Q., F. Zhao, and L. Guibas (2002). Counting targets: building and managing aggregates in wireless sensor networks. *Palo Alto Research Center Technical Report P2002-10298*.

Farkas, K. I., J. Flinn, G. Back, D. Grunwald, and J. M. Anderson (2000). Quantifying the energy consumption of a pocket computer and a Java virtual machine. *Proceedings of the ACM International Conference on Measurement and Modeling of Computer Systems*, pp. 252–63.

Fleischman, J. and K. Buchenrieder (1999). Prototyping networked embedded systems. *Computer*, **32** (2), 116–19.

Fleischman, J., K. Buchenrieder and R. Kress (1998). A hardware/software prototyping environment for dynamically reconfigurable embedded systems. *Proceedings of the International IEEE Workshop on Hardware/Software Codesign* (CODES/CASHE '98), pp. 105–9.

Fleischman, J., K. Buchenrieder and R. Kress (1999). Java driven codesign and prototyping of networked embedded systems. *Proceedings of the 36th IEEE Conference on Design Automation*, pp. 794–97.

Fleischman, J., K. Buchenrieder and R. Kress (1999). Codesign of embedded systems based on Java and reconfigurable hardware components. *Proceedings of the IEEE Conference and Exhibition on Design, Automation and Test in Europe*, pp. 768–9.

Flinn, J., and M. Satyanarayanan (1999). Energy-aware adaptation for mobile applications. *Proceedings of the ACM Symposium on Operating Systems Principles*, pp. 48–63.

Fox, A., and S. D. Gribble (1996). Security on the move: indirect authentication using Kerberos. *Proceedings of the ACM/IEEE International Conference on Mobile Computing and Networking*, White Plains, NY, pp. 155–64.

Fu, P., A. D. Hope and G. A. King (1998). An intelligent tool condition monitoring system. *The 52nd Meeting of the Society for Machinery Failure Prevention Technology*, pp. 397–406.

Gabber, E., C. Small, J. Bruno, J. Brustoloni, and A. Silberschatz (1999). The Pebble component-based operating system. *Proceedings of the 1999 USENIX Technical Conference*, Monterey, CA. URP: http//www.usenix.org/publications/library/proceedings/usenix99/full_papers/gabber/gabber.pdf.

Gennaro, R. and P. Rohatgi (1997). How to Sign Digital Streams. In *Advances in Cryptology – CRYPTO'97. Lecture Notes in Computer Science*, Vol. 1294, Springer 1997, pp. 180–197.

Gfeller, F., and W. Hirt (1998). A robust wireless infrared system with channel reciprocity. *IEEE Communications Magazine*, **36** (12), 100–06.

Ghose, A., J. Grossklags, and J. Chuang (2003). Resilient data-centric storage in wireless ad-hoc sensor networks. *Proceedings of the Fourth International Conference on Mobile Data Management*, Melbourne, Australia, pp. 45–62.

Gong, L., and N. Shacham (1995). Multicast security and its extension to a mobile environment. *Wireless Networks*, **1**, 281–95.

Goodman, D. (1997). *Personal Communication Systems*. Addison-Wesley, Reading, MA.

Goodman, J., A. Dancy, and A. Chandrakasan (1998). An energy/security scalable encryption processor using an embedded variable voltage DC/DC converter. *Journal of Solid State Circuits*, **33**, 1799–1809.

Gosling, J., B. Joy, and G. Steele (1996). *The Java Language Specification*. Addison-Wesley, Reading, MA.

Gutierrez, J. A., M. Naeve, E. Callaway, M. Bourgeois, V. Mitter, and B. Heile (2001). IEEE 802.15.4: a developing standard for low-power, low-cost wireless personal area networks. *IEEE Network*, **15** (5), 12–19.

Gutnik, V., and A. P. Chandrakasan (1997). Embedded power supply for low-power DSP. *IEEE Transactions on Very Large Scale Integration Systems*, **5**, 425–35.

Haartsen, J., M. Naghshineh, J. Inouye, O. Joeressen, and W. Allen (1998). Bluetooth: vision, goals, and architecture. *ACM Mobile Computing and Communications Review*, **2** (4), 38–45.

Haas, Z. J. (1997). A new routing protocol for the reconfigurable wireless networks. *Proceedings of the International IEEE Conference on Universal Personal Communications*, pp. 562–66.

Hac, A. (2000). *Multimedia Applications Support for Wireless ATM Networks*. Prentice Hall, New Jersey.

Hac, A. (2003). *Mobile Telecommunications Protocols for Data Networks*. John Wiley & Sons, New York.

Heidemann, J., F. Silva, C. Intanagonwiwat, R. Govindan, D. Estrin, and D. Ganesan (2001). Building efficient wireless sensor networks with low-level naming. *Proceedings of the ACM Symposium on Operating Systems Principles*, Vol. 35, No. 5, pp. 146–59.

Heinzelman, W., A. Chandrakasan, and H. Balakrishnan (2000). Energy-efficient communication protocol for wireless microsensor networks. *Proceedings of the 33rd Hawaii International Conference on System Sciences*, pp. 3005–14.

Heinzelman, W., A. Chandrakasan, and H. Balakrishnan (2002). An application-specific protocol architecture for wireless microsensor networks. *IEEE Transactions on Wireless Networking*, **1**, 660–70.

Heinzelman, W., J. Kulik, and H. Balakrishnan (1999). Adaptive protocols for information dissemination in wireless sensor networks. *Proceedings of the ACM/IEEE International Conference on Mobile Computing and Networking*. Seattle, Washington, pp. 174–85.

Heinzelman, W., A. Sinha, A. Wang, and A. P. Chandrakasan (2000). Energy-scalable algorithms and protocols for wireless microsensor networks. *Proceedings of the International IEEE Conference on Acoustics, Speech, and Signal Processing*, Vol. 6, pp. 3722–25.

Helaihel, R. and K. Olukotun (1997). Java as a specification language for hardware – software systems. Proceedings of the IEEE International Conference on Computer-aided Design, pp. 690–97.

Hewlett Packard. Application of Industrial Ethernet. URL: http://www.hpie.com.

Hill, J., R. Szewczyk, A. Woo, S. Hollar, D. Culler, and K. Pister (2000). System architecture directions for networked sensors. *Proceedings of the Ninth ACM International Conference on Architectural Support for Programming Languages and Operating Systems*, Cambridge, MA, pp. 93–104.

Hoare, C. A. R. (1978). Communicating sequential processes. *Communications of ACM*, **21**, 666–77.

Holmquist, L. E., F. Mattern, B. Schiele, P. Alahuhta, M. Beigl, and H.-W. Gellersen (2001). Smart-its friends: A technique for users to easily establish connections between smart artefacts. *Lecture Notes in Computer Science*, pp. 116–22.

IEEE 802.15 Working Group for Wireless Personal Area Networks. URL: http://www.ieee802.org/15.

IEEE Draft Standard for a Smart Transducer Interface for Sensors and Actuators – Digital Communication and Transducer Electronic Data Sheet: Formats for Distributed Multidrop Systems. *IEEE Draft Standard* P1451.3.

IEEE Draft Standard for a Smart Transducer Interface for Sensors and Actuators – Mixed Mode Communication Protocols and Transducer Electronic Data Sheet Formats. *IEEE Draft Standard* P1451.4.

IEEE Standard for a Smart Transducer Interface for Sensors and Actuators – Network Capable Application Processor Information Model. *IEEE Standard* P14451.1_1999.

IEEE Standard for a Smart Transducer Interface for Sensors and Actuators – Transducer to Microprocessor Communication Protocols and Transducer Electronic Data Sheet Formats. *IEEE Standard* 1451.2-1997.

Intanagonwiwat, C., R. Govindan, and D. Estrin. (2000). A sealable and robust communication paradigm for sensor networks. *Proceedings of the ACM/IEEE International Conference on Mobile Computing and Networks*, Boston, Mass, pp. 56–67.

Intel Strong ARM Processors. URL: http://developer.intel.com/design/strong/sa1100.htm.

Intrinsyc Corporation, Vancouver, Canada. Cerfcube embedded Strong ARM system. URL: http://www.intrinsic.com/products/cerfcube.

Jain, R., A. Puri, and R. Sengupta (2001). Geographical routing using partial information for wireless ad hoc networks. *IEEE Personal Communications*, **8** (1), 48–57.

JavaBeans API Specification. Sun Microsystems. URL: http://java.sun.com/beans.

Johnson, R. (1997). Building plug and play networked smart transducers. *Sensors*, October, pp. 40–46.

Kahn, J. H., R. H. Katz, and K. S. J. Pister (1999). Next century challenges: mobile networking for Smart Dust. *Proceedings of the ACM/IEEE International Conference on Mobile Computing and Networking*, Seattle, WA, pp. 171–78.

Kahn, J. H., R. H. Katz, and K. S. J. Pister (2000). Emerging challenges: mobile networking for Smart Dust. *Journal of Communications and Networks*, **2**, 188–96.

Kalavade, A. and P. Moghe (1998). A tool for performance estimation of networked embedded end-systems. *Proceedings of the IEEE Design Automation Conference*, pp. 257–62.

Kalavade, A. and P. A. Subramanyam (1997). Hardware/software partitioning for multi-function systems. Proceedings of the IEEE International Conference on Computer-Aided Design, pp. 516–21.

Karp, B., and H. T. Kung (2000). GPSR: greedy perimeter stateless routing for wireless networks. *Proceedings of the ACM/IEEE International Conference on Mobile Computing and Networking*, pp. 243–54.

Kasten, O., and M. Langheinrich (2001). First experiences with Bluetooth in the Smart-Its distributed sensor network. Second International Workshop on Ubiquitous Computing and Communications, in conjunction with the International Conference on Parallel Architectures and Compilation Techniques, Barcelona, Spain. URL: http://wwwtec.informtik.unirostock.de/RA/pact2001.

Krco, S. Bluetooth based wireless sensor networks – implementation issues and solutions. URL: http://www.telfor.org.yu/radovi/4019.pdf.

Krishnamachari, B., D. Estrin, and S. Wicker (2002). Modeling data-centric routing in wireless sensor networks. *Proceedings of the Annual Joint Conference of the IEEE Computer and Communications Societies*.

Kuhn, T. and W. Rosenstiel (2000). Java-based object-oriented hardware specification and synthesis. *Proceedings of the IEEE ASP-DAC Asia and South Pacific Design and Automation Conference*, pp. 579–81.

Larson, P.-A. (2002). Data reduction by partial preaggregation. *Proceedings of the International Conference on Data Engineering*, pp. 706–15.

Lauer, G. S. (1995). Packet-radio networks. In *Routing in Communications Networks* (M. Steenstrup, Ed.), Prentice-Hall, Englewoods Cliffs, NJ, Chapter 11.

Law, Y. W., S. Dulman, S. Etalle, and P. Havinga. Assessing security-critical energy-efficient sensor networks. *University of Twente Technical Report TR-CTIT-2-18*. URL: http://www.ub.utwente.nl/webdocs/etit/1/00000087.pdf.

Lee, K. (2000). IEEE 1451: A standard in support of smart transducer networking. *Proceedings of the 17th IEEE Instrumentation and Measurement Technology Conference*, Vol. 2, pp. 523–28.

Lee, S.-J., W. Su, and M. Gerla (2000). On-demand multicast routing protocol for ad hoc networks. Internet draft: draft-ietf-manet-odmrp-02.txt (January).

Lee, S., K. Yun, K. Choi, S. Hong, S. Moon and J. Lee (2000). Java-based programmable networked embedded system architecture with multiple application support. *Proceedings of IFIP Conference on Chip Design Automation*. URL: http://ifip.or.at/con2000/icda2000/icda-14-4.pdf.

Lettieri, P., C. Fragouli, and M. B. Srivastava (1997). Low power error control for wireless links. *Proceedings of the ACM/IEEE International Conference on Mobile Computing and Networking*, pp. 139–50.

Liedtke, J. (1995). On micro-kernel construction. *Proceedings of the ACM Symposium on Operating Systems Principles*, pp. 237–50.

Lin, C. R., and M. Gerla (1997). Adaptive clustering for mobile wireless networks. *IEEE Journal on Selected Areas in Communications*, **15**, 1265–75.

Lindsey, S., C. Raghavendra, and K. M. Sivalingam (2002). Data gathering algorithms in sensor networks using energy metric. *IEEE Transactions on Parallel and Distributed Systems*, **13**, 924–35.

Lipmaa, H., P. Rogaway, and D. Wagner. Counter mode encryption. URL: http://csrc.nist.gov/encryption/modes.

Lorch, J. R., and A. J. Smith (1998). Software strategies for portable computer energy management. *IEEE Personal Communications*, **5** (3), 60–73.

Lu, Y. H., L. Benini, and G. D. Micheli (2000). Operating-system directed power reduction. *Proceedings of the International IEEE Symposium on Low Power Electronics and Design*, pp. 37–42.

MacLellan, J., S. Lam, and X. Lee (1993). Residential indoor RF channel characterization. *Proceedings of the 43rd IEEE VTC*, PP. 210–13.

Madden, S., and M. J. Franklin (2002). Fording the stream: an architecture for queries over streaming sensor data. *Proceedings of the International Conference on Data Engineering*, pp. 555–66.

Madden, S., R. Szewczyk, M. J. Franklin, and D. Culler (2002). Supporting aggregate queries over ad-hoc wireless sensor networks. *Proceedings of the Fourth IEEE Workshop on Mobile Computing and Systems Applications*, pp. 49–58.

Mainwaring, A. M., and D. E. Culler (1999). Design challenges of virtual networks: fast, general purpose communication. *Proceedings of the 1999 ACM SIGPLAN Symposium on Principles and Practise of Parallel Programming*, Vol. 34, No. 8, pp. 119–30.

Mainwaring, A., J. Polastre, R. Szewczyk, D. Culler, and J. Anderson (2002). Wireless sensor networks for habitat monitoring. *Proceedings of the First ACM International Workshop on Wireless Sensor Networks and Applications*, Atlanta, GA, pp. 88–97.

Marcy, H. O., J. R. Agre, C. Chien, L. P. Clare, N. Romanov, and A. Twarowski (1999). Wireless sensor networks for area monitoring and integrated vehicle health management applications. AIAA Guidance, Navigation, and Control Conference and Exhibition, Portland, OR. *Collection of Technical Papers*, Vol. 1, (A99-36576 09-63) p. 11.

Martin, T., and D. Siewiorek (1996). A power metric for mobile systems. *Proceedings of the 1996 International IEEE Symposium on Lower Power Electronics and Design*, pp. 37–42.

Mauve, M., A. Widmer, and H. Hartenstein (2001). A survey on position-based routing in mobile ad-hoc networks. *IEEE Network*, **15** (6), 30–39.

Medina, A., I. Matta, and J. Byers (2000). On the origin of power laws in Internet topologies. *ACM Computer Communications Review*, **30** (2), 18–28.

Menezes, A. J., P. van Oorschot, and S. Vanstone (1997). *Handbook of Applied Cryptography*, CRC Press.

MEMS Technology Applications Center. URL: http://mems.mcnc.org.

Min, R., and A. Chandrakasan (2001). Energy-efficient communication for ad-hoc wireless sensor networks. *Proceedings of the 35th Asilomar Conference on Signals, Systems, and Computers*, Vol. 1, pp. 139–43.

Min, R., M. Bhardwaj, S.-H. Cho, A. Sinha, E. Shih, A. Wang and A. Chandrakasan (2000). An architecture for a power-aware distributed microsensor node. *Proceedings of the IEEE Workshop on Signal Processing and Systems*, pp. 581–990.

Min, R., M. Bhardwaj, S. Cho, E. Shih, A. Sinha, A. Wang, and A. Chandrakasan (2001). Low-power wireless sensor networks. *Proceedings of the 14th International Conference on VLSI Design*, Bangalore, India, pp. 205–10.

Min, R., T. Furrer, and A. P. Chandrakasan (2000). Dynamic voltage scaling techniques for distributed microsensor networks. *Proceedings of the IEEE. Computer Society Annual Workshop on VLSI*, pp. 43–46.

MIT: AMPS Project. URL: http://www.mtl.mit.edu/research/icsystems/uamps.

Modal Shop. TEDS Developer Kit Manual, SW-0028. URL: http://www.modal-shop.com.

Naghshineh, M., and M. Willebeek-LeMair (1997). End-to-end QoS provisioning in multimedia wireless/mobile networks using an adaptive framework. *IEEE Communications Magazine*, **35** (11), 72–81.

National Semiconductor: Napa1000 Adaptive Processor. URL: http://www.national.com/appinfo/milaero/napa1000.

Nawab, S. H. and J. M. Winograd (1997). Approximate signal processing. *Journal of VLSI Signal Processing Systems for Signal, Image, and Video Technology*, **15**, 177–200.

Nicol, C. J., P. Larsson, K. Azadet, and J. H. O'Neill (1997). A low power 128-tap digital adaptive equalizer for broadband modems. *Proceedings of the International IEEE Conference on Solid-State Circuits*, pp. 94–95.

Ogier, R. G., F. L. Templin, B. Bellur, and M. G. Lewis (2001). Topology broadcast based on reverse-path forwarding. Internet draft: draft-ietf-manet-tbrpf-03.txt (November).

Ousterhout, J. K. (1994). *Tcl and the Tk Toolkit*. Addison-Wesley.

Pados, D., K. W. Halford, D. Kazakos, and P. Papantoni-Kazakos (1995). Distributed binary hypothesis testing with feedback. *IEEE Transactions on Systems, Man and Cybernetics*, **25** (1), 21–42.

Park, V. D., M. S. Corson (1997). A highly adaptive distributed routing algorithm for mobile wireless networks. *Proceedings of the Annual Joint Conference of the IEEE Computer and Communications Societies*, Vol. 3, pp. 1405–13.

Passerone, C., R. Passerone, C. Sansoe, J. Martin, A. Sangiovanni-Vincentelli, and R. McGreer (1998). Proceedings of the International IEEE Workshop on Hardware/Software Design Codesign (CODES/CASHE '98), pp. 15–19.

Pering, T., T. Burd, and R. Brodersen (1998). The simulation and evaluation of dynamic voltage scaling algorithms. *Proceedings of the IEEE International Symposium on Low Power Electronics and Design*, pp. 76–81.

Perrig, A., R. Canetti, J. D. Tygar, and D. Song (2000). Efficient authentication and signing of multicast streams over lossy channels. *Proceedings of the IEEE Symposium on Security and Privacy*, pp. 56–73.

Perrig, A., R. Canetti, D. Song, and J. D. Tygar (2001a). Efficient and secure source authentication for multicast. *Proceedings of the Network and Distributed System Security Symposium*, pp. 35–46.

Perrig, A., R. Szewczyk, V. Wen, D. Culler, and J. D. Tygar (2001b). SPINS: Security protocols for sensor networks. *Proceedings of the ACM/IEEE International Conference on Mobile Computing and Networking*, Rome, Italy, pp. 189–99.

Pfleeger, C. P. (1997). *Security in Computing*. Prentice Hall, New York.

Pottie, G. J. (1998). Wireless sensor networks. *Proceedings of the IEEE Information Theory Workshop*, Killarney, Ireland, pp. 139–40.

Pottie, G. J., and W. J. Kaiser (2000). Wireless integrated network sensors. *Communications of the ACM*, **43** (5), 51–58.

Probert, D., J. Bruno, and M. Karaorman (1991). SPACE: a new approach to operating system abstractions. *Proceedings of the International Workshop on Object Orientation in Operating Systems*, pp. 133–37.

Pu, C., T. Autrey, A. Black, C. Consel, C. Cowan, J. Inouye, L. Kethana, J. Walpole, and K. Zhang (1995). Optimistic incremental specialization: streamlining a commercial operating system. *Proceedings of the ACM Symposium on Operating Systems Principles*, pp. 314–24.

Rabaey, J. M., M. J. Ammer, J. L. da Silva, D. Patel, and S. Roundy (2000). PicoRadio supports ad hoc ultra-low power wireless networking. *IEEE Computer*, **1** (7), 42–48.

Raghunathan, V., C. Schurgers, S. Park, and M. B. Srivastava (2002). Energy-aware wireless microsensor networks. *IEEE Signal Processing Magazine*, **19** (2), 40–50.

Ratnasamy, S., P. Francis, M. Handley, R. Karp, and S. Shenker (2001). A scalable content-addressable network. *Proceedings of the ACM SIGCOMM Conference*, Vol. 31, No. 4, pp. 161–72.

Ratnasamy, S., B. Karp, L. Yin, F. Yu, D. Estrin, R. Govindan, and S. Shenker (2002). GHT: a geographic hash table for data-centric storage. *Proceedings of the First ACM International Workshop on Wireless Sensor Networks and Applications*, pp. 78–87.

RF Monolithics, Inc. Tr1000 916.50 MHz Hybrid Transceiver. URL: http://www.rfm.com/products/data/tr1000.pdf.

Rivest, R. L. (1992). The MD5 message-digest algorithm. Internet request for comments, April, RFC 1321.

Rivest, R. L. (1995). The RC5 encryption algorithm. *Proceedings of the First Workshop on Fast Software Encryption*, pp. 86–96.

Rivest, R. L., A. Shamir, and L. M. Adleman (1978). A method for obtaining digital signatures and public-key crypto systems. *Communications of the ACM*, **21**, 120–26.

Rodoplu, V., and T. H. Meng (1999). Minimum energy mobile wireless networks. *IEEE Journal on Selected Areas in Communications*, **17**, 1333–44.

Rohatgi, P. (1999). A compact and fast hybrid signature scheme for multicast packet authentication. Proc. Sixth ACM Conference on Computer and Communications Security (CCS'99), pp. 93–100.

Rose, B. (2001). Networks: a standard perspective. *IEEE Communications Magazine*, **39** (12), 78–85.

Royer, E., and C. K. Toh (1999). A review of current routing protocols for ad hoc mobile wireless networks. *IEEE Personal Communications*, **6** (2), 46–55.

Rozier, M., V. Abrossimov, F. Armand, I. Boule, M. Gien, M. Guillemont, F. Herrmann, C. Kaiser, S. Langlois, P. Leonard, and W. Neuhauser (1988). Chorus distributed operating system. *Computing Systems*, **1**, 305–70.

Salonidis, T., P. Bhagwat, and L. Tassiulas (2000). Proximity awareness and fast connection establishment in Bluetooth. *Proceedings of the ACM International Symposium on Mobile Ad Hoc Networking and Computing*, Boston, Mass., pp. 141–42.

Savarese, C., J. M. Rabaey, and J. Beutel (2001). Location in distributed ad-hoc wireless sensor networks. *Proceedings of the International IEEE Conference on Acoustics, Speech and Signal Processing*, Vol. 4, Salt Lake City, Utah, pp. 2037–40.

Schmidt, A., K. A. Aidoo, A. Takaluoma, U. Tuomela, K. Van Laerhoven, and W. Van de Velde (1999). Advanced interaction in context. *Proceedings of the First International Symposium on Handheld and Ubiquitous Computing*, Karlsruhe, Germany, pp. 89–101.

Schneier, B. (1996). *Applied Cryptography*, John Wiley & Sons, New York.

Secure Microcontrollers for SmartCards. URL: http://www.atmel.com/atmel/acrobat/1065s.pdf.

Shah, R. C., and J. M. Rabaey (2002). Energy aware routing for low energy ad hoc sensor networks. *Proceedings of the IEEE Wireless Communications and Networking Conference*, Vol. 1, Orlando, FL, pp. 350–55.

Sharony, J. (1996). An architecture for mobile radio networks with dynamically changing topology using virtual subnets. *Mobile Networks and Applications*, **1**, 75–86.

Shen, C.-C., C. Srisathapornphat, and C. Jaikaeo (2001). Sensor information networking architecture and applications. *IEEE Personal Communications*, **8** (4), 52–59.

Siep, T. M., I. C. Gifford, R. C. Braley, and R. F. Heile (2000). Paving the way for personal area network standards: an overview of the IEEE P802.15 Working Group for Wireless Personal Area Networks. *IEEE Personal Communications*, **7** (1), 37–43.

Singh, S., M. Woo, and C. S. Raghavendra (1998). Power aware routing in mobile ad hoc networks. *Proceedings of the ACM/IEEE International Conference on Mobile Computing and Networking*, pp. 181–90.

Sinha, A. and A. Chandrakasan (2000). Energy aware software. *Proceedings of the Thirteenth IEEE International Conference on VLSI Design*, pp. 50–55.

Sinha, A. and A. Chandrakasan (2001). Operating system and algorithmic techniques for energy scalable wireless sensor networks. *Proceedings of the Second International Conference on Mobile Data Management*, Hong-Kong, pp. 199–209.

Sinha, A. and A. Chandrakasan (2001). Dynamic power management in wireless sensor networks. *IEEE Design and Test of Computers*, **18** (2), 62–74.

Sinha, A., A. Wang, and A. Chandrakasan (2000). Algorithmic transforms for efficient energy scalable computation. *Proceedings of the International IEEE Symposium on Low Power Electronics and Design*, pp. 31–36.

Slijepcevic, S., M. Potkonjak, V. Tsiatsis, S. Zimbek, and M. B. Srivastava (2000). On communication security in wireless ad-hoc sensor networks. *Proceedings of the Eleventh IEEE International Workshop on Enabling Technologies: Infrastructure for Collaborative Enterprises*, pp. 139–44.

Smart-Its Project. URL: http://www.smart-its.org.

Sohrabi, K., and G. J. Pottie (1999). Performance of a novel self-organization protocol for wireless ad-hoc sensor networks. *Proceedings of the IEEE Vehicular Technology Conference*, pp. 1222–26.

Sohrabi, K., J. Gao, V. Ailawadhi, and G. J. Pottie (2000). Protocols for self-organization of a wireless sensor network. *IEEE Personal Communications*, **7** (5), 16–27.

Spike Homepage. URL: http://www.spike-wireless.com.

Stinson, D. (1996). *Cryptography: Theory and Practice*. CRC Press.

Stoica, I., R. Morris, D. Karger, F. Kaashoek, and H. Balakrishnan (2001). Chord: a scalable peer-to-peer lookup service for Internet applications. *Proceedings of the ACM SIGCOMM Conference*, Vol. 31, pp. 149–60.

Sukhatme, G. S., and M. J. Mataric (2000). Embedding robots into the Internet. *Communications of the ACM*, **43** (5), 67–73.

Sun Microsystems, Inc. Embedded Java Application Environment. URL: http://java.sun.com/products/embeddedjava.

Sun Microsystems, Inc. Java 2 SDK documentation. URL: http://java.sun.com/products/jdk/download-pdf-ps.html.

Swaszek, P. F., and P. Willett (1995). Parley as an approach to distributed detection. *IEEE Transactions on Aerospace and Electronic Systems*, **31**, 447–57.

Tan, T. K., A. Raghunathan, and N. K. Jha (2002a). Embedded operating system energy analysis and macro-modeling. *Proceedings of the IEEE Conference on Computer Design: VLSI in Computers and Processors*, pp. 515–22.

Tan, T. K., A. Raghunathan, G. Lakshminarayana, and N. K. Jha (2002b). High-level energy macromodeling of embedded software. *IEEE Transactions on Computer-Aided Design of Integrated Circuits and Systems*, **21**, 1037–50.

Telenor. Telenor's H.263 Software. URL: http://www.nta.no/brukere/DVC/h263_software.

Tennenhouse, D. L. (2000). Proactive computing. *Communications of the ACM*, **43** (5), 43–50.

Tennenhouse, D. L., J. M. Smith, W. D. Sincoskie, D. J. Wetherall, and G. J. Minden (1997). A survey of active network research. *IEEE Communications Magazine*, **35** (1), 80–86.

Toh, C. K. (2001). Maximum battery life routing to support ubiquitous mobile computing in wireless ad hoc networks. *IEEE Communications Magazine* (June), 138–47.

Transvirtual Technologies, Kaffe Open VM. URL: http://www.transvirtual.com/kaffe.html.

Tseng, Y.-C., S.-Y. Ni, and E.-Y. Shi (2001). Adaptive approaches to relieving broadcast storms in a wireless multihop mobile ad hoc network. *Proceedings of the IEEE International Conference on Distributed Computing Systems*, pp. 481–88.

Unidirectional Link Routing Protocol Working Group home page. URL: http://www-sop.inria.fr/rodeo/udlr.

US National Institute of Standards and Technology (1999). Data Encryption Standard. Draft Federal Information Processing Standards Publication 46-3, January.

US National Institute of Standards and Technology. Advanced encryption standard development effort. URL: http://csrc.nist.gov/encryption/aes.

Vahdat, A., A. Lebeck, and C. S. Ellis (2000). Every joule is precious: the case for revisiting operating system design for energy efficiency. Proceedings of ACM SIGOPS European Workshop.

Van Dyck, R. E., and L. E. Miller (2001). Distributed sensor processing over an ad hoc wireless network: simulation framework and performance criteria. *Proceedings of the IEEE Military Communications Conference*, Washington DC, pp. 894–98.

Vasilko, M. Dynamically reconfigurable hardware. WWW Library, Bournemouth University. URL: http://dec.Bournemouth.ac.uk/drhw_lib.

Viswanathan, R., and P. K. Varshney (1997). Distributed detection with multiple sensors I. Fundamentals. *Proceedings of the IEEE*, **85**, 54–63.

Von Eicken, T., D. E. Culler, S. C. Goldstein, and K. E. Schauser (1992). Active messages: a mechanism for integrated communication and computation. *Proceedings of the 19th Annual International Symposium on Computer Architecture*, Queensland, Australia, pp. 256–66.

Wang, A., S.-H. Cho, C. Sodini, and A. Chandrakasan (2001). Energy efficient modulation and MAC for asymmetric RF microsensor systems. *Proceedings of the International IEEE Symposium on Low Power Electronics and Design*, pp. 106–11.

Wang, A., W. Heinzelman, and A. Chandrakasan (1999). Energy-scalable protocols for battery-operated microsensor networks. *Proceedings of the IEEE Workshop on Signal Processing and Systems*, pp. 483–92.

Warneke, B., B. Atwood, and K. S. J. Pister (2001). Smart Dust mote forerunners. *Proceedings of the IEEE International Conference on MEMS*, pp. 357–60.

Warneke, B., M. Last, B. Liebowitz, and K. S. J. Pister (2001). Smart dust: Communicating with a cubic-millimeter computer. *IEEE Computer*, **34**(1), 44–51.

Warrior, J. (1996) IEEE P1451 Network Capable Application Processor Information Model. *Proceedings Sensors Expo*, Anaheim. Helmers Publishing, pp. 15–21.

Wei, G. and M. Horowitz (1996). A low power switching supply for self-clocked systems. *Proceedings of the International IEEE Symposium on Low Power Electronics and Design*, pp. 313–17.

Weiser, M., B. Welch, A. Demers, and S. Shenker (1994). Scheduling for reduced CPU energy. *Proceedings of the first USENIX Symposium on Operating Systems Design and Implementation*, pp. 13–23.

Weiss, K., T. Steckstor, and W. Rosenstiel (1999). Performance analysis of a RTOS by emulation of an embedded system. *Proceedings of the International Workshop on Rapid System Prototyping*, pp. 146–51.

Weste, N. and K. Eshraghian (1993). *Principles of CMOS VLSI Design: A Systems Perspective*, Addison-Wesley, Reading, Mass.

Willig, A., J. Rabaey, R. Shah, and A. Wolisz (2002). Altruists in the PicoRadio sensor network. *Proceedings of the IEEE International Workshop on Factory Communication Systems*, Vasteras, Sweden. URL: http://www-tkn.ee.tuberlin.de/publications/papers/andreas_willig.pdf.

Wright, B. and M. Dillon. Application of IEEE P1451 Smart Transducer Interface Standard in condition based maintenance. URL: http://www.arofe.army.mil/conferences/CBM_Abstract/papers/Ryback_Oceana.pdf.

Xilinx, Inc. Low power documentation. URL: http://www.eilinx.com/products/xaw/pwr/pwr_doc.htm.

Xilinx, Inc. Virtex 2.5 V Field Programmable Gata Arrays Datasheet v1.9. URL: http://www.xilinx.com/partinfo/ds003.pdf.

Xu, Y., J. Heidemann, and D. Estrin (2001). Geography-informed energy conservation for ad hoc routing. *Proceedings of the ACM/IEEE International Conference on Mobile Computing and Networking*, Rome, Italy, pp. 70–84.

Yao, K., R. E. Hudson, C. W. Reed, D. Chen, and F. Lorenzelli (1998). Blind beamforming on a randomly distributed sensor array system. *IEEE Journal on Selected Areas in Communications*, **16**, 1555–67.

Ye, W., J. Heidemann, and D. Estrin (2002). An energy-efficient MAC protocol for wireless sensor networks. *Proceedings of the Annual Joint Conference of the IEEE Computer and Communications Societies*, Vol. 3, pp. 1567–76.

Yi, Y., M. Gerla, and T. J. Kwon (2002). Efficient flooding in ad hoc networks using on demand (passive) cluster formation. ACM International Symposium on Mobile Ad Hoc Networking and Computing, Lausanne, Switzerland. URL: http://www.cs.ucla.edu/NRL/wireless/uploads/mobihoc-yiyi.pdf.

Yi, Y., T. J. Kwon, and M. Gerla (2001). Passive clustering in ad hoc networks. Internet draft: draft-ietf-yi-manet-pac-00.txt (November).

Yip, K., and F. Zhao (1996). Spatial aggregation: theory and applications. *Journal of Artificial Intelligence Research*, **5**, 1–26.

Young, J. S., J. MacDonald, M. Shilman, A. Tabbara, P. Hilfinger, and A. R. Newton (1998). Design and specification of embedded systems in Java using successive, formal refinement. Proceedings of the IEEE Design Automation Conference, pp. 70–75.

Zhao, F., J. Shin, and J. Reich (2002). Information-driven dynamic sensor collaboration. *IEEE Signal Processing Magazine*, **19** (2), 61–72.

Zhong, L. C., J. Rabaey, C. Guo, and R. Shah (2001). Data link layer design for wireless sensor networks. *Proceedings of the IEEE Military Communications Conference. Communications for Network-Centric Operations: Creating the Information Force*, Vol. 1, Washington, DC, pp. 352–56.

Zyuban, V., and P. Kogge (1997). The energy complexity of register files. *Proceedings of the International IEEE Symposium on Low Power Electronics and Design*, pp. 305–10.

Index

access point (AP), 152
Active Message (AM), 327, 364
actuator, 34, 58, 110, 112
Ad Hoc On Demand Distance Vector Routing (AODV), 103, 105, 190
address space identifier (ASID), 254, 255
Advanced Configuration and Power Interface (ACPI), 78
AES-128 (advanced encryption standard 128-bit cryptographic keys), 302
aggregate queries, 120, 121
altruist, 109, 114, 115
analog to digital converter (ADC), 33, 35, 52, 57, 66, 323, 334, 340, 356, 357, 361
application programming interface (API), 11, 15, 16, 24, 47, 54, 122, 197, 198, 230, 291, 196, 318
application specific integrated circuit (ASIC), 23, 28, 93
Asynchronous Connectionless (ACL) link, 276, 282–284, 315, 317
attribute-based addressing, 101
authenticated broadcast, 219, 222
authentication, 241
Automatic Repeat Request (ARQ), 276, 281, 283

base station (BS), 215, 225, 337
base station transceiver (BTS), 151, 156–163
Baseband (BB) protocol, 278
Berkeley Software Distribution (BSD), 258, 259
Binary Phase Shift Keying (BPSK), 305
bit error rate (BER)
Bluetooth, 110, 142, 144–153, 162, 275, 277–288, 290, 293–298, 300, 314–319
broadcasting, 84, 222

C/OS, 265, 271, 341
Carrier Sense Multiple Access (CSMA), 117, 119, 120, 122, 199, 302, 332
carrier sense multiple access / collision avoidance (CSMA/CA), 104, 110, 159, 302
Cellular IP, 287, 288
central processor unit (CPU), 19, 79, 80, 144, 243, 251
certifying authority (CA), 241
channel interface module (CIM), 46
cipher-block chaining (CBC), 217
class-based addressing, 104, 105

Wireless Sensor Network Designs A. Hać
© 2003 John Wiley & Sons, Ltd ISBN: 0-470-86736-1

cluster head, 71, 72, 181–184, 198, 203–205, 207, 209
cluster, 171, 178, 185, 191, 192, 203, 208, 209
clustering, 81, 181, 198, 199, 202, 319
code generation, 20
codesign and reconfiguration, 2, 9
Complementary Metal-Oxide Semiconductor (CMOS), 67, 68, 97, 98, 334, 343, 356
condition based maintenance, 24, 58, 325
connectivity map, 177
continuous variable slope delta modulation (CVSD), 276, 281, 284
controller area network (CAN), 46, 47, 54, 55, 56, 59
Corner Cube Retroreflector (CCR), 156, 157, 160
cosynthesis method and prototyping platform, 2, 4
counter mode (CTR), 217, 218
cue, 189, 343
Cyclic Redundancy Check (CRC), 276, 281, 283, 301, 317

data aggregation, 67, 70, 193, 194, 293, 319, 349
Data Encryption Standard (DES), 242
Data Encryption Standard – Cipher Block Chaining (DES-CBC), 218, 242
data fusion, 72
data link layer (DLL), 298, 299, 320
data-centric storage (DCS), 276, 306, 308–310, 314, 322
DES cipher-block chaining (DES-CBC), 242
DES electronic code book (DES-ECB), 242
design integration, 4
Destination Sequenced Distance Vector Routing protocol (DSDV), 105
Digital Cordless Telephone (DCT), 338, 339, 341
Digital Signal Processing (DSP), 64, 75, 93, 95, 192, 196
direct sequence spread spectrum (DSSS), 152, 302, 305, 321
directed diffusion, 81
Distance Vector Multicast Routing Protocol (DVMRP), 161
distributed aggregate management (DAM), 346
distributed hash-table (DHT), 308, 309
Distributed Multidrop System (DMS), 38
distributed sensor networks, 141
dynamic power management (DMP), 76, 79
Dynamic Source Routing (DSR), 190
dynamic voltage scaling (DVS), 64, 68, 69, 73–75, 79, 95, 97–99
dynamically reconfigurable field-programmable gate array (DPGA) board, 4–7, 10, 11, 14–19, 29

Electronically Erasable Programmable Read Only Memory (EEPROM), 54, 122, 333, 356, 357, 361
embedded application, 235, 258, 275
embedded Cygnus operating system (eCOS), 67, 74, 271
embedded device, 12
embedded operating system, 26, 264, 265
embedded system, 1–8, 12, 64, 261, 357
embedded systems platform, 17
encryption algorithm, 2
energy aware routing (EAR), 101–103, 106, 107, 109–113, 115–117, 119, 120, 136
energy-efficient communication, 81
energy-quality (E-Q), 64, 65, 69, 70, 76, 208
epoch, 224
Ethernet, 47, 56, 111
External Storage (ES), 308

Fast Fourier Transform (FFT), 33, 194
Field Programmable Gate Array (FPGA), 5, 9, 10, 14, 16, 17, 20, 22–30, 93
Finite Impulse Response (FIR), 17, 69
FIR filter, 17, 69, 72, 75, 196

First Node Dies (FND), 181, 184
flooding, 84, 112, 198, 199, 200, 211
Forward Error Correction (FEC), 195, 196, 276, 281, 283, 317
FPGA architecture, 25
Frame Check Sequence (FCS), 301
Frequency Hopping Spread Spectrum (FHSS), 153, 154, 275, 283
friendly neighbor, 109, 114

garbage collection, 260
gateway, 147, 198, 204, 205, 290, 291, 293, 295–297, 318, 319, 336, 359
General Purpose Interface Bus (GPIB), 35
Geographical Adaptive Fidelity (GAF), 359
global positioning system (GPS), 186, 199
Global Standard for Mobile (GSM), 188
Great Duck Island (GDI), 351, 352, 358, 359
Greedy Perimeter Stateless Routing (GPSR), 306, 308, 309, 315, 322
grouping, 133
guaranteed time slots (GTS), 301

Half of the Nodes Alive (HNA), 181, 184
hardware abstraction layer (HAL), 340, 341
hardware and software codesign, 3
Heating, Ventilation, and Air Conditioning (HVAC), 298
high-level synthesis (HLS), 20, 30
HiperLAN/2, 152
Host Controller Interface (HCI), 144, 146

IEEE 1451 Standards for Smart Transducer Interface for Sensors and Actuators, 32–49, 52, 54–61
IEEE 802.11, 82, 114, 115, 150, 152, 206, 298, 299, 358
IEEE 802.15, 152, 276, 298–306, 314–316, 320, 321
IETF Unidirectional Link Routing Working Group, 161
implicit entry-exit pair (IEEP), 267, 268
IMT2000 (International Mobile Telecommunication), 23, 24
in-network aggregation, 125, 133, 135
Industry Scientific Medical (ISM), 67, 143, 152, 153, 280, 281, 283, 294, 302, 316, 321, 340
Inferno operating system, 236, 239–241, 261, 274
Information Society Technologies Advisory Group (ISTAG), 276
infrared data association (IrDA), 279, 286
infrared object exchange (IrOBEX), 279
Integrated Circuit (IC), 302
Integrated Device Technology (IDT), 261
Integrated Electronics, PiezoElectric (IEPE), 40, 41
Inter Integrated Circuit, 143, 356, 357
International Telecommunication Union – Telecommunication Standardization Sector (ITU-T), 301
Internet Engineering Task Force (IETF), 161, 287
Internet Protocol (IP), 42, 203, 204, 286–288, 306, 307, 322
Internet, 4, 12, 46, 283, 317
interprocess communication (IPC), 243, 265, 266, 269, 270
interprotection domain call, 262, 264
interrupt handler, 263
interrupt latency, 263, 264
interrupt service routine (ISR), 265
interrupt, 251, 252, 254

JaCoP (Java driven codesign and prototyping environment), 2, 12, 13, 16–18, 29
Java Beans specification, 13, 17
Java Native Interface (JNI), 11, 15, 24, 27
Java programming language, 12, 39, 259, 260
Java virtual machine (JVM), 10, 15, 19, 22, 27, 29
Joint Test Action Group (JTAG), 339

Large Scale Office Scenario (LSOSC), 118, 119, 120
Laser Mirror Scanner (LMS), 70
Last Node Dies (LND), 181, 184
Light Emitting Diode (LED), 145
line-of-bearing (LOB), 194
line-of-sight, 161, 164
Link Manager (LM), 278
Link State Routing (LSR), 105
Linked Cluster algorithm (LCA), 191, 192
Linux, 11, 14, 18, 146, 149, 265, 268, 271, 357
local area network (LAN), 46, 150, 286, 287
Local Storage (LS), 308
logical link control (LLC), 278, 299, 300, 320, 321
Logical Link Control and Adaptation Protocol (L2CAP), 146
Low Energy Adaptive Clustering Hierarchy (LEACH), 81, 181, 182, 184, 193
Low Power Oscillator (LPO), 281, 282, 284
low-rate wireless personal area network (LR-WPAN), 298, 301, 306, 314

Management Information Base (MIB), 168
master, 146, 153, 163, 281–285, 294–296, 317, 319
maximum transmission unit (MTU), 279, 286, 317
Media Access Control (MAC), 81, 82, 83, 96, 103, 104, 109–111, 113, 114, 116, 117, 120, 137, 159, 186, 187, 292, 298–304, 314, 320, 321, 324, 332, 358, 359, 360
 MAC common part sublayer (MCPS-SAP), 300, 321
 MAC footer (MFR), 300
 MAC header (MHR), 300
 MAC layer management entity (MLME-SAP), 300, 321
 MAC protocol data unit (MPDU), 300
 MAC service data unit (MSDU), 300
message authentication code (MAC), 216, 218–225, 232
Message Digest 4 (MD4), 241
Message Digest 5 (MD5), 221, 241, 242
Micro Controller Unit (MCU), 145
Micro Electro Mechanical Systems (MEMS), 65, 67, 75, 95, 141, 151, 154, 157, 161, 165, 192, 288, 292, 314, 334, 345
micro-Adaptive Multi-domain Power-aware Sensors (μAMPS), 71, 72, 73, 84, 95
micro-TESLA, 213–215, 217, 220–222, 224, 230, 231, 233
microcontroller, 35
microprocessor, 31, 58
microsensor, 35, 64, 65, 72, 77, 81, 192, 338
Million Instructions Per Second (MIPS), 92, 144, 236, 243, 245, 253, 254, 261
minimum shift keying (MSK), 305
mobile ad hoc network (MANET), 185, 187, 190, 198–201, 203, 208
Mobile IP, 287
mote, 121, 152, 156–158, 359
Motion Pictures Experts Group (MPEG), 237, 260
MPR Node (MPRN), 201
multifunction systems, 2
multimode systems, 2
Multipoint Relay (MPR), 201–103

Network Capable Application Processor (NCAP), 36–38, 43, 45–49, 52–56, 59, 61
networked embedded system, 2, 9, 11–30
nucleus, 251, 253, 255

object-oriented design, 3, 13, 88
Offset Quadrature Phase Shift Keying (O-QPSK), 305
Open Shortest Path First (OSPF), 161

open systems interconnection (OSI) reference model, 299
operating system (OS), 235, 236, 242–245, 264, 266–270, 273, 341
operation, administration, and maintenance (OA&M), 237
output feedback mode (OFB), 217

passive clustering, 198–200, 203–207, 210
PC Interface (PCI), 5, 11, 15, 18, 23, 24
Pebble operating system, 235, 242–245, 252, 253, 255, 257–259, 261, 271, 273
Perimeter Refresh Protocol, 309
Personal Area Network (PAN), 152, 301, 302
Personal Computer (PC), 4, 5, 14, 18, 19, 125, 277, 285, 291, 297–299, 320, 341
personal digital assistant (PDA), 23, 24, 153, 188, 259, 275, 277, 286, 291, 298, 352, 355
Phase Lock Loop (PLL), 68, 69, 73
physical (PHY) layer, 298, 300, 314, 321
physical layer protocol data unit (PPDU), 304
physical layer service data unit (PSDU), 304
piconet, 146, 147, 153, 282, 284, 294, 296, 316, 317, 319
Plan 9 operating system, 238, 244
plug-and-play, 58
Point Coordination Function (PCF), 114
portal manager, 253, 274
portal traversal, 248
portal, 244, 246, 247, 249, 250, 252, 255–258, 261, 271, 272, 273, 336
power management (PM), 76, 78
power-aware design, 65
power-aware wireless sensor networks, 63
Printed Circuit Board (PCB), 73
profiling, 3
programming language C++, 242, 260, 327
programming language C, 145, 238–240, 242, 258, 260, 271, 340, 341, 364
programming language Limbo, 238, 239, 240, 259
programming language Pascal, 238
protection domain (PD), 246–249, 272
Pseudo-Noise (PN), 305

Quality of Service (QoS), 2, 93, 152, 283, 305, 315
Quantum Effect Design (QED), 261

radio frequency (RF), 75, 95, 154, 162, 192, 215, 275, 278, 288, 290, 314, 318, 323, 330, 340
random access memory (RAM), 7, 18, 52, 54, 57, 58, 67, 133
reachability map, 177
Read Only Memory (ROM), 54, 67
Received Signal Strength Indicator (RSSI), 84, 281
Reduced Instruction Set Computer (RISC), 18, 93, 144, 153, 335, 339
request to send / clear to send (RTS/CTS), 159
Resilient Data-Centric Storage (R-DCS), 276, 307, 310, 312
Resource Reservation Protocol (RSVP), 288
reuse library, 3
RF Module (RFM), 122
RFCOMM, 279
RFM (RF Monolithics), 330, 331, 355
Route Reply (RREP), 106
Route Request (RREQ), 105, 106
Routing Information Protocol (RIP), 161
routing, 123, 190
RS232, 143, 144, 339, 341
run-time system (RTS), 20, 22, 23
runtime management, 15

S-MAC (sensor-MAC), 360
scatternet, 153, 294, 295, 316, 319
scratch-pad memory, 7
Secure Hash Algorithm (SHA), 241, 242

Secure Network Encryption Protocol (SNEP), 213, 217–219, 221, 223, 225, 230
Security Protocols for Sensor Networks (SPINS), 213, 216, 217, 223, 230, 231
self-configuring wireless sensor network, 109
self-organizing wireless network, 276
semaphore, 247, 250, 262
sensor fusion, 338
Serial Peripheral Interface (SPI) protocol, 52, 339
service access point (SAP), 299, 300
Service Discovery Protocol (SDP), 296
service-specific convergence sublayer (SSCS), 299, 300, 321
signal to interference ratio (SIR)
signal to noise ratio (SNR), 66, 96
Simple Mail Transfer Protocol (SMTP), 355
simulated annealing, 8
sink, 112, 118, 176
slave, 146, 147, 153, 163, 281–285, 294, 295, 317, 319
Small to Medium Enterprise (SME), 56
Smart Dust, 151, 152, 154, 158, 159, 164
smart sensor, 31, 34, 58, 121
Smart Transducer Interface Module (STIM), 37–41, 43, 45–49, 51–56, 58, 61
software synthesis, 4
source, 112, 138
special function register (SFR), 57, 58
specification, 3
static RAM (SRAM), 144, 339, 340
station-to-station (STS), 241
steam-based function (SBF), 25
StrongARM, 67–69, 72, 76, 78, 84, 339, 357
Structured Query Language (SQL), 102, 121, 124, 126, 134, 135, 347
Structured Replication in DCS (SR-DCS) scheme, 309
Styx protocol, 238
Surface Mount Device (SMD), 151
symmetric block cipher (RC6), 227
symmetric key stream cipher (RC4), 242
Synchronous Connection Oriented (SCO) link, 276, 282–284, 315, 317
synthesis, 9
system call entry-exit pair (SCEEP), 267, 268
System Developers Toolkit (SDT), 341
system programming interface (SPI), 53, 145
Systems Performance Evaluation Consortium (SPEC92), 92

target tracking, 214
thread, 6, 190, 248, 251, 262–264
Time Division Duplex (TDD), 283, 294
time division multiple access (TDMA), 160, 337, 341, 358
Timed, Efficient, Streaming, Loss-tolerant Authentication Protocol (TESLA), 215, 220
TinyOS, 152, 323, 324, 326–328, 330–333, 360, 362, 364
topology discovery, 166, 169, 171, 177, 209
Transaction Control Protocol (TCP), 288
Transaction Control Protocol/Internet Protocol (TCP/IP), 36, 46, 279, 286, 317, 358
Transducer Bus Interface Module (TBIM), 38, 43, 47, 48, 60
Transducer Electronic Data Sheet (TEDS), 32, 33, 38, 39, 42–44, 47–49, 51, 54, 57–59, 60
Transducer Independent Interface (TII), 37, 49, 51, 54, 57
translation lookaside buffer (TLB), 243, 253–255
transmit power control (TPC), 152

ubiquitous computing, 325
Ultra Wide Band (UWB), 288, 290, 314, 318

universal asynchronous receiver
transmitter (UART), 144, 145, 326, 330
Universal Serial Bus (USB), 144
Unix, 236, 240, 256, 257, 260, 262
untrusted location, 215
User Datagram Protocol (UDP), 286

Very Large Scale Integration (VLSI), 64,
85, 92, 95
VHDL (VHSIC Hardware Description
Language), 4, 10, 14, 22, 30, 325
VHSIC (Very High Scale Integrated
Circuit), 325
VHSIC Hardware Description Language
(VHDL), 4, 10, 14, 22, 30, 325
vibration sensors, 32–34
Video Cassette Recorder (VCR), 298
virtual memory (VM), 11, 15, 243, 248,
249, 253
VLSI systems, 64, 85, 97
voltage controlled oscillator (VCO), 68

weak freshness, 219, 232
Web-based applications, 3
wide area network (WAN), 353, 354
wireless application protocol (WAP),
279, 286
Wireless Integrated Network Sensors
(WINS), 162, 324, 334–338, 340, 341,
343–345, 362, 363, 365, 366
Wireless Local Area Network (WLAN),
150, 152
Wireless World Research Forum
(WWRF), 276
World Wide Web (WWW), 2

zone routing protocol (ZRP), 190, 191